SELF-ASSEMBLING PEPTIDE SYSTEMS IN BIOLOGY, MEDICINE AND ENGINEERING

edited by

AMALIA AGGELI

NEVILLE BODEN

University of Leeds, United Kingdom

and

SHUGUANG ZHANG

Massachusetts Institute of Technology, Cambridge, MA, U.S.A.

SPRINGER-SCIENCE+BUSINESS MEDIA, B.V.

A C.I.P. Catalogue record for this book is available from the Library of Congress.

ISBN 978-90-481-5740-2 ISBN 978-0-306-46890-2 (eBook)
DOI 10.1007/978-0-306-46890-2

Printed on acid-free paper

SELF-ASSEMBLING PEPTIDE SYSTEMS IN BIOLOGY, MEDICINE AND ENGINEERING

TABLE OF CONTENTS

Self-assembling peptide systems in biology medicine and engineering

FOREWORD

One of the major drivers in biological research is the establishment of structures and functions of the 50,000 or so proteins in our bodies. Each has a characteristic 3-dimensional structure, highly "ordered" yet "disordered"! This structure is essential for a protein's function and, significantly, it must be sustained in the competitive and complex environment of the living cell.

It is now being recognised that when a cell loses control, proteins can self-assemble into more complex supermolecular structures such as the amyloid fibres and plaques associated with the pathogenesis of prion (CJD) or age-related (Alzheimer's) diseases. This is a pointer to the wider significance of the self-assembling properties of polypeptides. It has been long known that, in silk, polypeptides are assembled into ß-sheet structures which impart on the material its highly exploitable properties of flexibility combined with high tensile strength. But only now emerging is the recognition that peptides can self-assemble into a wide variety of non-protein-like structures, including fibrils, fibres, tubules, sheets and monolayers. These are exciting observations and, more so, the potential for materials and medical exploitations is so wide ranging that over 80 scientists from Europe, USA, Japan and Israel, met 1- 6 July 1999 in Crete, to discuss the wide-ranging implications of these novel developments.

There was a spirit of excitement about the workshop indicative of an important new endeavor. The emerging perception is that of a new class of materials set to become commercially viable early in the 21st century. This stems from the opportunities for processing by the self-assembly route combined with the fact, just as in the case of proteins, that functionality can be designed into the self-assembled structures. They can be made responsive to pH change, mechanical forces, temperature, pressure, electro-chemical potential, electrical and magnetic fields, and light. They can function as sensors and actuators and can act as molecular motors capable of interconverting energies (*vis-à-vis* metabolism). They can even be programmed for biodegradation! Extraordinary and widely exploitable properties. Particularly in view of the exceptional thermal stability of peptides (up to 350°C). Nor are production costs insurmountable. Large scale production by the fermentation or transgenic plant or animal routes, at the ton level if needed, are already being developed. Projected costs are as low as a few pounds per kilogram. Applications in tissue engineering, biomedical devices, industrial fluids and personal care products are all under development. Could these new materials become the wonder polymers of the 21st century!

The workshop was charged with a vibrant atmosphere as may be expected of this newly developing interdisciplinary area. As Francis Crick best put it "In Nature hybrid species are usually sterile, but in science the reverse is often true. Hybrid subjects are often astonishingly fertile, whereas if a scientific discipline remains too pure it usually wilts".

In view of the exciting prospects for this new area of endeavor, it was felt that it would be useful to record the proceedings of the workshop for those unable to attend.

The articles summarising lectures are presented in alphabetical (author) order. Arranging them under subject headings is inappropriate as most of them focus on generic issues.

DR AMALIA AGGELI, *University of Leeds, UK*
PROFESSOR NEVILLE BODEN, *University of Leeds, UK*
& DR SHUGUANG ZHANG, *Massachusetts Institute of Technology, USA*

July, 2000

EXPLOITING PEPTIDE SELF-ASSEMBLY TO ENGINEER NOVEL BIOPOLYMERS: TAPES, RIBBONS, FIBRILS AND FIBRES

A.AGGELI, I.A.NYRKOVA, M.BELL, L.CARRICK, T.C.B.MCLEISH, A.N.SEMENOV & N.BODEN
Centre for Self-Organising Molecular Systems, University of Leeds LS2 9JT, UK

1. Abstract

A generic model is presented for the self-assembly of chiral units into intrinsically twisted tapes, ribbons, fibrils and fibres. Rationally designed self-assembling β-strand-forming peptides are shown to behave as chiral rod-like objects, exhibiting the entire hierarchy of these structures, which can form nematic fluids or gels. These observations provide new insight into the generic self-assembling properties of β-sheets, and the factors governing the structures and extraordinary stability of pathological amyloid fibrils *in vivo*. More generally, they provide a prescription of routes to novel tape and fibril-like macromolecules based on a wide variety of self-assembling chiral units.

2. Introduction

Prospects for the large-scale production of low-cost peptides by genetic engineering [1] open up new opportunities for exploiting protein-like self-assembly as a route to novel biomolecular materials [2],[3],[4],[5]. In this context, the small-oligopeptide route has distinct processing advantages over the use of longer polypeptides. Previously we have demonstrated that oligopeptides can be designed to self-assemble into micrometer-long β-sheet tapes [6],[7]. Using rationally designed peptides, comprising 11 L-amino-acid residues, we now wish to show that, as a consequence of the amino acid chirality, an entire hierarchy of twisted self-assembling macromolecular structures is accessible, with tapes as the most primitive form. The tapes are seen to associate into ribbons (double tapes), then into fibrils comprised of twisted stacks of ribbons, and finally into fibres formed by the entwining of these fibrils. These polymers are shown to give rise to nematic fluids and gels at concentrations determined by the characteristic flexibility and length of each type of polymer. A theoretical model is presented which allows to predict and control the morphology and properties of these self-assembling structures from the molecular parameters of the peptide monomers. Protocols are described to extract the peptide-

A. Aggeli et al. (eds.), Self-assembling Peptide Systems in Biology, Medicine and Engineering, 1–17.

peptide interaction energies which govern the stabilities of the self-assembled structures, their average sizes and the concentration ranges over which they are observable. The finite width and helicity of the fibrils is shown to stem from a competition between the free energy gain from attraction between ribbons and the penalty due to elastic distortion of the intrinsically twisted ribbons on incorporation into a growing fibril. Fibres are stabilised in a similar way.

3. Materials and Methods

3.1 THEORY

(a) The tape (Fig. 1c′) has a helical pitch h_{tape} and a radius r_{tape}

$$h_{tape} = b_2\left(\frac{2\pi}{\gamma_0}\right)\left(1+\left(\frac{\gamma_v}{\gamma_\theta}\right)^2\right)^{-1}, r_{tape} = b_2\left(\frac{\gamma_v}{\gamma_\theta^{\,2}}\right)\left(1+\left(\frac{\gamma_v}{\gamma_\theta}\right)^2\right)^{-1}, \quad (1),(2)$$

where γ_v and γ_θ are respectively the tape bend and twist angles (in radians) per monomer rod, along the tape and b_2 is the distance between adjacent rods in the tape.

(b) The fibril width is determined by a balance between the gain in attraction energy (coming from $\varepsilon_{fibril}^{attr}$) and the elastic cost associated with fibril formation. Assuming that the ribbon contour length is fixed and the deformations are weak, from symmetry arguments we find that this cost is

$$\varepsilon_{elast} = \frac{1}{2}k_{bend}(v - v_0)^2 + \frac{1}{2}k_{twist}(\theta - \theta_0)^2 \quad (3)$$

(per unit length of each ribbon in the fibril), where v and θ are the local curvature and the local twist strength of the ribbon within fibril, $\theta_0 = 2\pi/h_{ribbon}$ is the twist strength of an isolated ribbon, while its bend strength is zero ($v_0 = 0$), and k_{bend}, k_{twist} are the ribbon elastic constants. For a ribbon in a fibril: $v=\gamma^2\rho/(1+\gamma^2\rho^2)$, $\theta=\gamma/(1+\gamma^2\rho^2)$, where $\gamma=2\pi/h_{fibril}$, h_{fibril} is the fibril's helical pitch and ρ is the distance of the ribbon from the central axis of the fibril. The thicker the fibril is, the larger the typical ρ are, and hence the higher the cost ε_{elast}. The net energy gain ε_{fibril} per peptide in a fibril,

$$\varepsilon_{fibril} = \frac{p-1}{2p}\varepsilon_{fibril}^{attr} - \varepsilon_{fibril}^{elast}, \quad (4)$$

has a maximum at some p (p is the number of ribbons in the fibril). Hence, well-defined width of fibrils, corresponding to this optimal p, see diagram Fig.2.

(c) The concentration and character of inter-structure transformations are determined by the microscopic energies ε_j. E.g., given ε_{trans} *is high enough* ($\varepsilon_{trans} > 4$, all energies here are measured in k_BT units) and ε_{ribbon} is small (≤1), the single tapes emerge abruptly at

$$c_{cr}^{tape} \cong \nu_{tape}^{-1} \exp\left(-\varepsilon_{tape} + \varepsilon_{trans}\right) \tag{5}$$

and their typical aggregation number is

$$\langle m_{tape} \rangle \cong \left[\left(c / c_{cr}^{tape}\right) - 1\right]^{1/2} \exp\left(\varepsilon_{trans} / 2\right), \tag{6}$$

if $c_{cr}^{tape} < c < c_{cr}^{ribbon}$ (c is the total peptide concentration and ν_{tape} is the 'freedom' volume of the bonds forming the tape). Next, given the tape bend and twist are not very high, i.e. ε_{elast} (cp.eq.(3)) is small enough, the net ribbon energy

$$\varepsilon_{ribbon} = \frac{1}{2}\varepsilon_{ribbon}^{attr} - \varepsilon_{ribbon}^{elast} \tag{7}$$

is positive. Hence, at concentration

$$c_{cr}^{ribbon} \cong c_{cr}^{tape} + c_{tape}^{\max}, \qquad c_{tape}^{\max} \cong \nu_{tape}^{-1} \varepsilon_{ribbon}^{-2} \exp\left(-\varepsilon_{tape}\right) \tag{8}$$

the ribbons emerge; above c_{cr}^{ribbon}, the population of peptide in single tapes saturates at c_{tape}^{max} and all extra peptide goes into ribbons, simultaneously the average aggregation number of ribbons grows as

$$\langle m_{ribbon} \rangle \approx \left[\left(c \big/ c_{cr}^{ribbon} \right) - 1\right]^{1/2} \varepsilon_{ribbon}^{2} \exp\left(\left(\varepsilon_{trans} + \varepsilon_{tape}\right)/2\right) \tag{9}$$

whereas the length of tapes saturates at

$$\langle m_{tape} \rangle \cong \varepsilon_{ribbon}^{-1} \tag{10}$$

The formulae in eqs.(6,8,9,10) mentioned here are asymptotic. The detailed theory generalizing the classic isodesmic model [8] can be found in references [9] and [10]. This theory allows for predictions for any combination of the energetic parameters, and it was used for the fits in Figs.3c,d, 4e, and 5c,d.

3.2. PEPTIDE PRODUCTION AND DATA COLLECTION

Preparation of P_{11}-I and P_{11}-II (molecular weights 1,498Da and 1,594Da respectively, purity >95%) as well as TEM and far-UV CD techniques have been described elsewhere [11]. The solubility of P_{11}-I and P_{11}-II polymers are pH and ionic strength dependent, due to charges on Arg^+ (pK ~ 12.5) and Glu^- (pK ~ 4) residues. These peptides in distilled water are positively charged due to acidity of the solution arising from residual trifluoroacetic acid associated with the peptide. They precipitate in buffers with pH in the range $4.3 < pH < 12.5$, as a result of strong attractions between aggregates due to the ampholitic nature of the peptide. This attraction is much weaker outside this pH range: stable fibril dispersions occur in the range of $pH \approx 2\text{-}3$, even in buffers. Primary structure of P_{11}-I: CH_3CO-Gln-Gln-Arg-Gln-Gln-Gln-Gln-Gln-Glu-Gln-Gln-NH_2. Primary structure of P_{11}-II: CH_3CO-Gln-Gln-Arg-Phe-Gln-Trp-Gln-Phe-Glu-Gln-Gln-NH_2. β-AP (1-40) peptide [12] (purity >98%) was purchased from ICN Chemicals and used as received. For P_{11}-II samples, the solvent was double distilled with conductivity $1.5 \cdot 10^{-6}$ S · cm^{-1}.

4. Results and Discussion

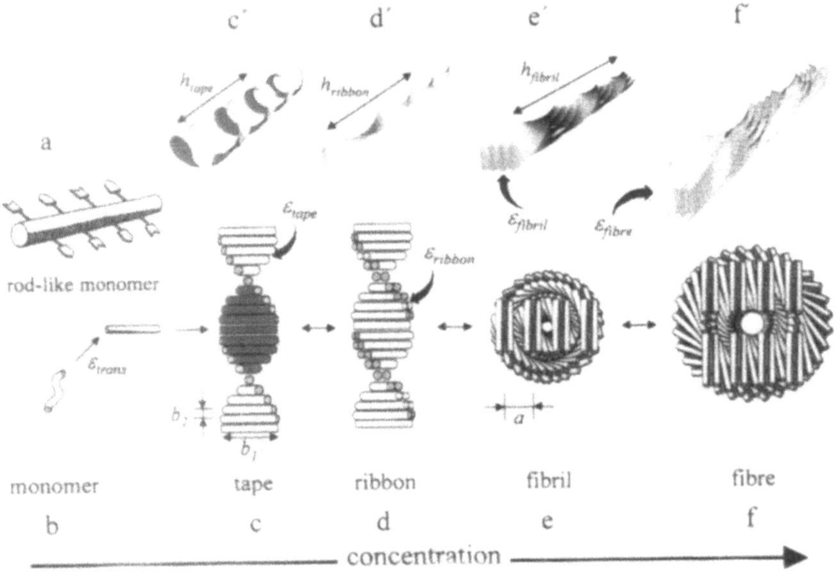

Figure 1. **Model of hierarchical self-assembly of chiral rod-like units. A : Local arrangeents (c-f) and the corresponding global equilibrium conformations (c'-f') for the hierarchical self-assembling structures formed in solutions of chiral molecules (a), which have complementary donor and acceptor groups, shown by arrows, via which they interact and align to form tapes (c). The black and the white surfaces of the rod (a) are reflected in the sides of the helical tape (c) which is chosen to curl towards the black side (c'). The outer sides of the twisted ribbon (d), of the fibril (e) and of the fibre (f) are all white. One of the fibrils in the fibre (f') is drawn with darker shade for clarity. (e) & (f) show the front views of the edges of fibrils and fibres, respectively. Geometrical sizes (the numbers in parentheses show the values of the corresponding geometric sizes for P_{11}-I and P_{11}-II peptides, based on X-ray diffraction data and molecular modeling) : inter-rod separation in a tape b_2 (b_2=0.47nm); tape width, equal to the length of a rod, b_1 (b_1=4nm); inter-ribbon distance in the fibril, a (a=1.6-2nm for P_{11}-I, and a=2-2.4nm for P_{11}-II).**

We start by considering a peptide in a β-strand conformation as a hypothetical chiral rod-like unit, with complementary donor and acceptor groups aligned on opposing sides, and having chemically different upper and lower surfaces (Fig. 1a). This unit is able to undergo one-dimensional self-assembly in solution and to form the hierarchical set of structures depicted in Fig.1. Generally, an isolated monomer in solution will be in a different conformation (b), with lower free energy than in the rod-like state: the corresponding conformational free energy change is ε_{trans}. The rod-like "monomers" self-assemble *via* recognition between complementary donor and acceptor groups, to form long twisted tapes (c): the association free energy change is ε_{tape} per inter-monomer bond. The tape twist stems from the chirality of the monomers (Fig.1a, eg right-handed peptide β-strands, due to the L-chirality of naturally-occurring amino acids) which give rise to a left-handed twist around the long axis of the tape (c). The tape has distinct faces (white and black in Fig.1c), whenever the upper and lower surfaces of the monomer are chemically different (assuming that the white sides of the monomers have higher affinity for each other than for the black sides). The differences in the chemical structures of the two faces of the tape and in their affinity to the solvent give rise to a cylindrical curvature, causing the tape to *curl* into a helical configuration (c'). One face of the tape (colored black in Fig.1c) is expected to be less soluble than the other (i.e. black is more hydrophobic if the solvent is water). This results in inter-tape attraction and hence in double tape (*ribbon*, (d)) formation (energy $\varepsilon^{attr}_{ribbon}$ per peptide). Both faces of the ribbon are identical (white in Fig. 1d), and are characterised by a saddle curvature. Hence the ribbon does not bend and its axis is straight at equilibrium (d'). The white sides of the ribbons are, in turn, mutually attractive (energy $\varepsilon^{attr}_{fibril}$ per pair of interacting peptides) leading to stacking of ribbons into *fibrils* (e). Furthermore, the ends of the rods decorating the edges of the fibrils can also be mutually attractive, causing fibrils to entwine into *fibres* (f), stabilised by attraction energy $\varepsilon^{attr}_{fibre}$.

All of the self-assembling structures in Fig.1 are left-handed twisted due to *chirality* of the rod-like monomer. If the ribbons were not twisted, an unlimited growth of fibril and fibre widths would be expected. Instead, when twisted ribbons aggregate into stacks, *fibrils* with well-defined widths are formed. Fibres are formed in a similar way from twisted fibrils. Indeed, in order to aggregate, twisted objects must bend and adjust their twist in response to the packing constraints imposed by its twisted neighbours. Hence, there is an elastic energy cost ε_{elast}, which must be compensated for, by the gain in attraction energy (coming from $\varepsilon^{attr}_{ribbon}$, $\varepsilon^{attr}_{fibril}$ and $\varepsilon^{attr}_{fibre}$) upon stacking. The distortion energy ε_{elast} is higher for thicker stacks. This serves to stabilise the widths of fibrils and fibres.

The "state diagram" of possible aggregate structures calculated using this model is shown in Fig.2. Fibrils with finite diameter are seen to be stable for a wide range of values of $\varepsilon^{attr}_{fibril}$ provided that the intrinsic pitch h_{ribbon} of the lone ribbon strongly exceeds the inter-ribbon gap a in the fibril. For low $\varepsilon^{attr}_{fibril}$ the ribbons do not stack into fibrils, whilst for high $\varepsilon^{attr}_{fibril}$ the ribbons form infinite aggregates (sheet-like crystallites) in which the ribbons are completely untwisted. The optimum number p of stacked ribbons per fibril, and hence the fibril diameter, increases with h_{ribbon} and $\varepsilon^{attr}_{fibril}$. This is usually accompanied by an increase in the fibril helix pitch h_{fibril}.

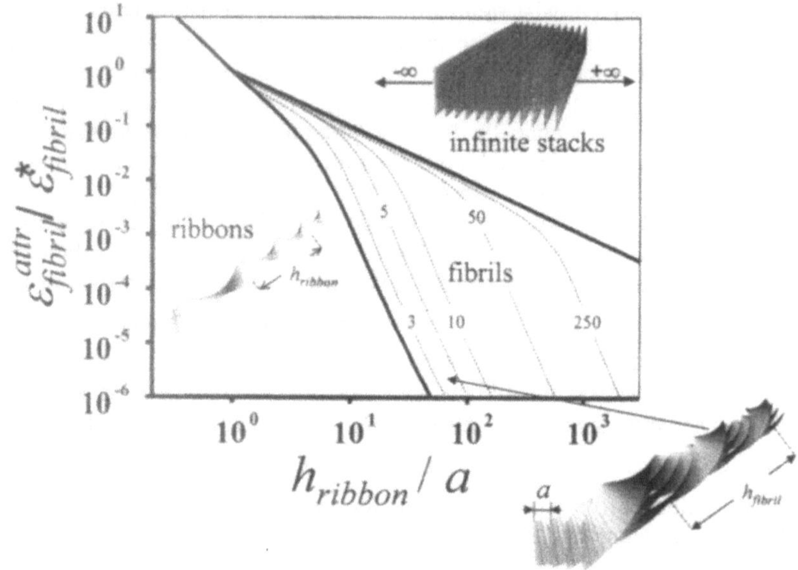

Figure 2 : Phase diagram of a solution of twisted ribbons which form fibrils. The scaled variables are: relative helix pitch of isolated ribbons h_{ribbon}/a, and relative side-by-side attraction energy between ribbons : $\varepsilon^{attr}_{fibril} / \varepsilon^{*}_{fibril}$ ($\varepsilon^{*}_{fibril} = (2\pi^2 b_2 / a^2)\, k_{twist}$, see the text and Fig. 1d,e' for notations). The areas divided by the thick lines reveal the conditions where ribbons, fibrils and infinite stacks of completely untwisted ribbons are stable. The dotted lines are lines of stability for fibrils containing p ribbons (p are written on the lines); $k_{bend} / k_{twist} = 0.1$.

The concentration ranges over which the various self-assembled structures are observable, their contour lengths, and abruptness of interstructure transformations with concentration, are determined by the energy parameters ε_j. In order to realise sequentially the entire hierarchy of structures depicted in Fig.1, with increasing monomer concentration, it is essential that $\varepsilon_{tape} >> k_BT >> \varepsilon_{ribbon} >> \varepsilon_{fibril} >> \varepsilon_{fibre}$ (ε_{ribbon}, ε_{fibril} and ε_{fibre} are the net energies gained per one peptide inside the corresponding structures as compared to a peptide inside the structure of the previous level), otherwise some structures may not appear.

To illustrate the predictions of our theoretical model, and to demonstrate how the ε_j can be measured for real self-assembling systems, we shall consider the behavior of two rationally designed peptides P_{11}-I and P_{11}-II (Table 1). The primary structure of P_{11}-I is based on a sequence of glutamine (Gln) residues, whose side-chains are believed to interact strongly in water [13], presumably via hydrophobic and complementary hydrogen bonding interactions. Arginine (Arg) and glutamate (Glu) residues have been placed in positions 3 and 9, to provide molecular recognition between adjacent antiparallel β-strand peptides in tape-like aggregates, in order to prevent random peptide association. These favorable intermolecular side-chain interactions coupled with the co-operative intermolecular hydrogen bonding between peptide backbones will result in high scission energy ε_{tape}, thus promoting β-sheet tape formation (Fig.1c). Furthermore, one side ("black") of the tape will be lined by the $CONH_2$ groups of the Gln residues, whilst its other side ("white") will be lined by the $CONH_2$, the guanido and the COOH groups of the Gln, Arg and Glu residues, respectively. At low pH in particular, there will also be a net positive charge per peptide. The high hydrophilicity of both surfaces of the tape combined with the electrostatic repulsion between positively charged surfaces, will result in very small $\varepsilon_{ribbon}^{attr}$ and $\varepsilon_{fibril}^{attr}$ energies compared to k_BT, thus promoting predominantly single tape formation for low enough peptide concentration in acidic solutions.

At very low concentrations, P_{11}-I is seen to be predominantly in the monomeric random coil conformation (Figs.1b & 3a), whereas at higher concentrations $c \geq 0.01$mM, it forms semi-flexible tapes (Fig.3b) with a width $W \approx 4$nm, equal to the expected length of an 11-residue peptide in a β-strand conformation, and persistence length $\tilde{l} < 0.3\mu$m. The different chemical nature of the two sides of the tape seems to cause it to bend and twist simultaneously, resulting in curly tapes with a left-handed twist, a helical pitch $h_{tape} \approx (30\pm15)$nm, and a radius $r_{tape} \approx 5$nm. At $c \geq 1$mM, loose ribbons are also observed, with $\tilde{l} \sim 0.3$-1μm, and $h_{ribbon} \approx (50\pm20)$nm. These values in conjunction with the theoretical model were used to derive the magnitudes of the bend γ_v and twist γ_θ angles for the single tapes and the ribbons (Table 1).

Aqueous solutions of P_{11}-I tapes produce FTIR spectra with absorption maxima in amide I′ at 1630 and 1690cm^{-1}, demonstrative of a predominantly antiparallel β-sheet structure. They also exhibit characteristic β-sheet CD spectra [14] with minimum and maximum ellipticities at 218nm and 195nm respectively (Fig.3a). The fraction of the peptide in β-sheet tapes starts to grow abruptly at a critical concentration $c_{cr}^{tape} \approx 0.008$ mM (Fig. 3c). The two-state transition from random-coil to β–sheet with increasing concentration, has an isodichroic point at 211nm. We treated ε_{trans} and ε_{tape} as fitting parameters, and were able to describe well the growth of the β-sheet CD band with concentration (solid line in Fig. 3c). The best-fit energy values obtained are: $\varepsilon_{trans} = (6.5\pm1.5)k_BT$, and $\varepsilon_{tape} = (31.0\pm1.5)k_BT$. The ε_{trans} energy results in the nucleated growth of tapes, manifested by a "sudden" onset of β-sheet tapes' formation at c_{cr}^{tape}. Using these values of energetic parameters, this single tape model predicts a mean tape contour length for a given peptide concentration, which agrees well with the observed range of contour lengths in the TEM images for the same concentration. At $c_{cr}^{ribbon} \approx 1$mM loose ribbons start appearing, implying a weak attraction between tapes. This may be mediated by multiple, co-operative, complementary hydrogen bonding between the $-CONH_2$ groups of glutamine side-chains, which line completely one of the two polar sides of the tapes. Van der Waals forces are likely to be involved, too. From the value of c_{cr}^{ribbon}, we estimate that the ribbons are stabilised by ε_{ribbon}=(0.0035±0.0015)k_BT. Fibrils (Fig.1e′) are not observed up to c=25mM, hence: $\varepsilon_{fibril} < 10^{-3}k_BT$, and $\varepsilon_{fibril}^{attr} \leq 0.1k_BT$.

TABLE 1. Comparative molecular parameters and macroscopic properties of the aqueous solutions of the two *de novo* designed self-assembling β-tape forming peptides P_{11}-I and P_{11}-II, shown schematically in a β-strand conformation. White and black circles represent polar and apolar amino-acid residues respectively. The upper side of the β-strand is identical for both peptides. c_{cr} is the threshold concentration above which the corresponding type of self-assembling structure emerges in substantial quantities. For explanation of the various energetic parameters ε_j, see the text and Fig.1. All values of ε_j are given in k_BT. The errors in the values of ε_j arise from the errors in the value of v_{tape}, and from the uncertainty of measurements of the real lengths of the polymers in solution at a given concentration. The values of ε_{trans} and ε_{tape} for both peptides, and ε_{ribbon} for P_{11}-II are received from the fits of the self-assembly curves (Figs. 3c & 5c), and the observed tape/ribbon lengths, with v_{tape}~$5\cdot10^{-6}$ nm^3 (this value is estimated from the observed persistence lengths of the ribbons). ε_{fibril} is estimated from c_{cr}^{fibril} and ε_{ribbon} for P_{11}-I from the corresponding c_{cr}^{ribbon}. $\tilde{l}$, h, γ_v, and γ_θ are either directly measured from electron micrographs or calculated with the help of the model similar to Fig.2. Elastic constants are estimated from the observed persistence lengths of ribbons and fibrils. The contour lengths L agree with both the data and the predictions of the theory. Directly measured constants are indicated with asterisk.

	Peptide P_{11}-I 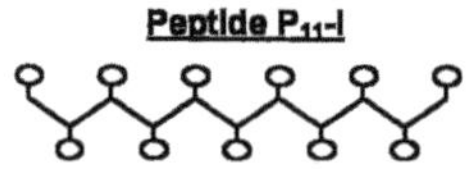			Peptide P_{11}-II 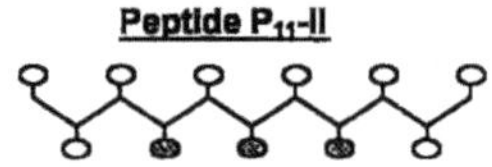		
	β-Tapes	Ribbons	Fibrils	β-Tapes	Ribbons	Fibrils
c_{cr} / μM	8*	1,000*	c > 25,000*	< 90*	90*	700 ± 200*
ε_{trans}		6.5 ± 1.5			3 ± 1	
ε_{tape}	31.0 ± 1.5			24.5 ± 1.0		
ε_{ribbon}		(3.5±1.5)10^{-3}			0.6 ± 0.3	
ε_{fibril}			< 10^{-3}			(2.0±0.3)10^{-4}
Pitch h / nm	30 ± 15*	50 ± 20*			160 ± 40	160 ± 40*
Twist angle γ_θ/°	3*	3*		1	1	1*
Bend angle γ_v/°	3*	0*			0*	0*
$\tilde{l}$ / μm	< 0.3*	0.6 ± 0.2*			1.0 ± 0.3*	20-70*
L / μm (c=6mM)	10^{-1}	10^{-1} - 10^{0}		10^{-3}	10^{0}	10^{17}
Properties of aq. solution	Isotropic fluid (c <13mM) Nematic fluid / gel (c>13mM)			Isotropic fluid (c < 0.9mM) Nematic fluid (c = 0.9 – 6mM) Nematic gel (c> 6mM)		

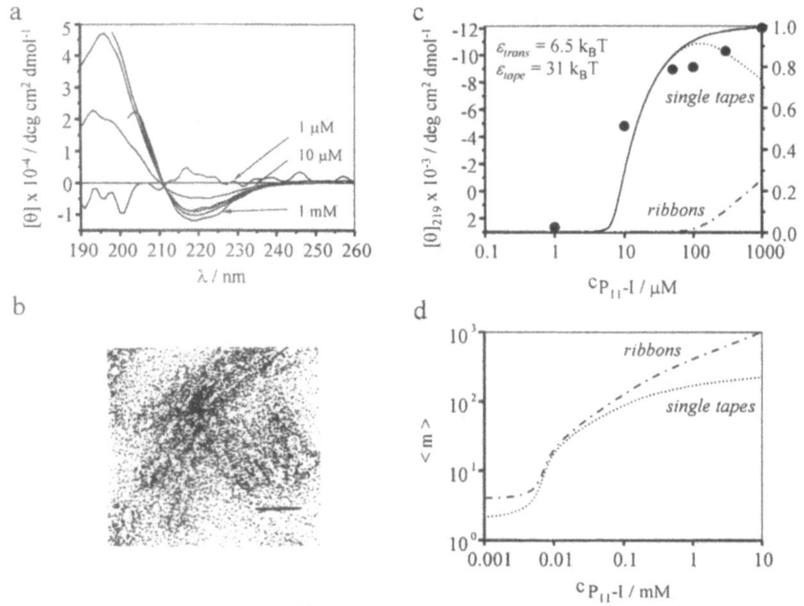

Figure 3. **Self-assembly of P_{11}-I. (a) Far-UV CD spectra as a function of peptide concentration. The solutions were prepared by mixing the dry peptide with the required volume of water adjusted to pH=2 with phosphoric acid. Data were collected with one-month old solutions stored at 20°C. For interpretation of the CD spectra, see the text and legend of Fig.5(a) & (b). (b) Negatively-stained TEM image of single "curly" tapes, reminiscent of Fig.1c′ (the scale bar corresponds to 50nm). (c) Plot of the β-sheet fraction in solution (black circles) as a function of total peptide concentration, based on the CD data (the mean residue ellipticity [θ] at 219nm is taken as a linear function of the β-sheet fraction in solution). The solid line is the fit of the data with the single tape theory. The best-fit values of the energetic parameters ε_{trans} and ε_{tape}, which were chosen to comply with the concentration dependence of the CD data and with the observed lengths of tapes at c = 5 mM, are shown in the panel. (d) Theoretical concentration dependence of the average number ⟨m⟩ of peptides per single tape (dotted line) and in ribbons (dash-dot line), based on the energetic parameters derived from the fit (c). Minimum number of peptides in tapes is 2, and in ribbons is 4. The predicted lengths of tapes and ribbons are in agreement with the observed lengths in the TEM pictures for the same c.**

To increase the tendency of the peptide to associate into ribbons, we need to increase the magnitude of ε_{ribbon} either by decreasing $\varepsilon_{ribbon}^{elast}$ or by increasing $\varepsilon_{ribbon}^{attr}$. The latter can be achieved by addition of salts or of appropriate cosolvents, but more elegantly by replacing the glutamines at positions 4, 6 and 8 by phenylalanine, tryptophan and phenylalanine respectively. This new peptide, P_{11}-II (Table 1) will form β-tapes with a hydrophobic "adhesive" stripe running along one side of the tape and this will promote their association into ribbons in water. At $c \geq 0.1$ mM in water, P_{11}-II is indeed found to form long, stable semi-flexible β-sheet ribbons with a width of 2-4nm, which fits with the expected cross section of *ca* 2x4nm^2 of these ribbons, and a persistence length $\tilde{l}$ ~1μm, as shown using both rotary shadowed (Fig. 4a) and negatively stained TEM specimens. At $c \geq 0.6$ mM, a second transition from ribbons to fairly rigid fibrils is observed (Fig. 4b,c). The fibrils have a well-defined screw-like structure with typical minimum and maximum widths $W_1 \approx$ 4nm and $W_2 \approx$ 8nm respectively. At even higher concentrations still, a third structural transition takes place and fibres are detected, typically comprised of two entwined fibrils (Fig. 4d). The sequence of these structural transitions is also supported by distinctive far- and near-UV CD spectra, corresponding to P_{11}-II monomers, ribbons and fibrils (Fig. 5a,b).

Focussing on the behavior at low concentrations, we see that P_{11}-II is predominantly in the monomeric random coil conformation (Fig.1b), whereas the fraction of peptide in β-sheet structures starts to grow abruptly at $c \approx 0.07$mM (Fig. 5a,c). We treated ε_{trans} and ε_{tape} as fitting parameters, and were able to describe well the growth of the β-sheet CD band with concentration. However, this single tape model yields a mean tape length of about 20 nm at $c = 0.2$ mM, (Fig.5d), much shorter than the observed length ≥ 500 nm (Fig.4a). It is possible, however, to describe the CD data (solid line in Fig. 5c) and simultaneously to predict the occurrence of these long aggregates (Fig.4a) by inclusion of a third energetic parameter ε_{ribbon} associated with ribbon (double tape) formation (Fig. 1d). These long aggregates then turn out to be double tapes rather than single ones (Fig. 5c,d). The CD spectra as a function of c have no isodichroic point (Fig.5a), further supporting that more than two states, *ie* peptide monomers, β-tapes and ribbons, are involved in the conformational transition. The best-fit energy values obtained are: $\varepsilon_{trans} = (3\pm1)k_BT$, $\varepsilon_{tape} = (24.5\pm1.0)k_BT$ and $\varepsilon_{ribbon} = (0.6\pm0.3)k_BT$.

The estimated ε_{trans} is higher for P_{11}-I than for P_{11}-II (Table 1). Although both peptides have the same length, they may have different propensity to form a random coil in the monomeric state which could account for this difference in ε_{trans}. The magnitude of ε_{tape} is also higher for P_{11}-I than for P_{11}-II, which indicates that the intermolecular glutamine side-chain interactions between P_{11}-I peptides are more efficient at promoting self-assembly compared to intermolecular aromatic side-chain interactions between P_{11}-II peptides. ε_{ribbon} is at least two orders of magnitude lower for P_{11}-I compared to P_{11}-II, as predicted by peptide design. This difference explains the shorter (by one order of magnitude) length of P_{11}-I ribbons compared to P_{11}-II ones (Table 1). It also accounts for the one order of magnitude difference in critical concentrations for ribbon formation between the two peptides. This results in stabilisation of single, curly β-tapes in a wide range of P_{11}-I concentrations. In contrast, P_{11}-II tapes are not observed because they convert to ribbons as soon as they are 3-4 peptides long, at very low concentration.

The formation of fibrils (Fig.4b) at higher concentrations of P_{11}-II, implies the presence of a weaker attraction between the polar sides of P_{11}-II ribbons ($\varepsilon^{attr}_{fibril}$, Fig.1e′). From the concentration at which they appear, we calculate $\varepsilon_{fibril} = (2.0\pm0.3)10^{-4}$ k_BT. Despite this attraction, the fibril dispersions are stable and the fibril diameter is finite (rather than growing indefinitely). Furthermore, the fibril width W_1 corresponds to the expected length of an 11-residue β-strand, whilst W_2 corresponds to roughly 4 ribbons (i.e. 8 single tapes, each tape with a thickness of *ca* 1nm) per fibril, and is concentration independent (at least from 0.6 to 7 mM). The energy required to break such a fibril, scission energy ε_{sc}, is $\varepsilon_{sc} = 8\varepsilon_{tape} \sim 200k_BT$, (comparable to covalent bond energies!), and is much higher than that of a single ribbon $\varepsilon_{sc} = 2\varepsilon_{tape} \sim 50k_BT$. This results in fibrils of extraordinary predicted equilibrium average length : L_{fibril} ~10^8 km!, compared to L_{ribbon}~1 μm, for c=6mM.

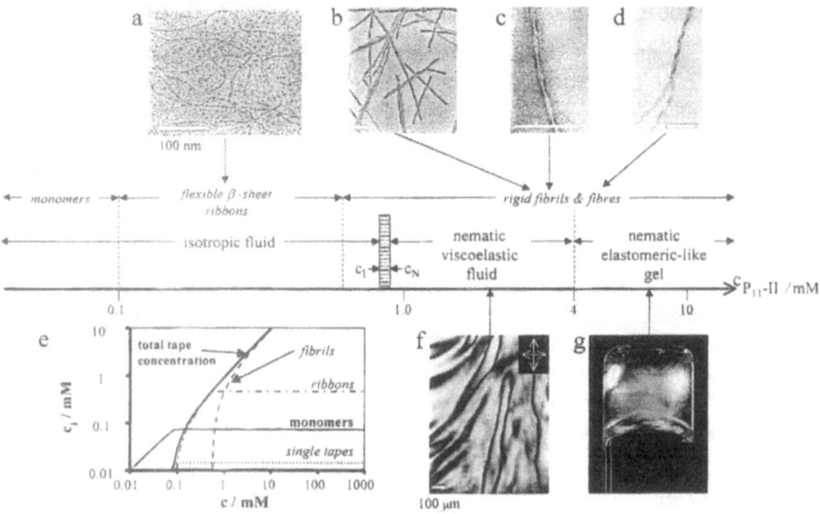

Figure 4. Aggregate structures and liquid crystalline phase behavior observed in solutions of P_{11}-II in water with increasing peptide concentration c (log scale). The electron micrographs (**a**) of ribbons (c = 0.2 mM), and (**b**) of fibrils (c = 6.2 mM) were obtained with a four month old solution after platinum rotary shadowing. The observed micrometer-long contour length may be limited by multiple ruptures of the fibrils during preparation of the samples for TEM imaging. Higher resolution TEM images of ribbons were also obtained using negatively stained samples (data not shown). Micrographs (**c**) (c = 6.2 mM) and (**d**) (c = 6.2 mM) were obtained with a one-month-old solution after uranyl acetate negative staining. CD and FTIR have confirmed that the fibrils are made of β-sheet structures. X-ray diffraction data have also shown arcs corresponding to 0.47nm periodicity, consistent with the expected interstrand distance in a β-sheet (unpublished data). The TEM micrographs show the principle aggregate structures whose populations $c_j = f_j c$ (f_j is the fraction of peptides incorporated in the j-th structure) change with peptide concentration, as depicted in (**e**). The curves in (e) were calculated with the generalised model described in the text (see also Fig. 5d). The aggregation behavior of the peptide, probed using time resolved fluorescence anisotropy and CD of filtered solutions, is fully consistent with the expectations of the model (unpublished data). The polarising optical micrograph (**f**) shows the thick thread-like texture observed for a solution with c = 3.7 mM in a 0.2 mm pathlength microslide. (**g**) shows a self-supporting birefringent gel (c = 6.2 mM) in an inverted 10 mm o.d. glass tube, viewed between crossed polarisers. The scale bars in a, b, c and d correspond to 100 nm, in f to 100 μm.

Fibril formation is readily explicable by our model of stabilisation by twist (Fig.1b). Indeed, β-sheet ribbons have an intrinsically left-handed twist, due to the L-chirality of peptides [16]. The fibrils also exhibit a left-handed twist with a helix pitch h_{fibril} of *ca.* 120–200nm (Fig. 4c). From the observed geometrical characteristics of P_{11}-II ribbons and of fibrils, our theory estimates h_{ribbon} ~ 120–200nm, elastic constants k_{bend} and k_{twist}, and twist angle γ_θ =1° for isolated P_{11}-II ribbons and $\varepsilon^{attr}_{fibril}$ ~ 0.015k_BT for fibrils (Table 1). The magnitude of $\varepsilon^{attr}_{fibril}$ is expected to be similar both for P_{11}-I and P_{11}-II, because of the identity of their "white"polar sides. However P_{11}-I ribbons are three times more twisted than P_{11}-II ones (compare twist angles γ_θ in Table 1). The higher elastic penalty $\varepsilon^{elast}_{fibril}$ associated with untwisting P_{11}-I ribbons compared to P_{11}-II ones, seems to result in lower overall magnitude of ε_{fibril} for P_{11}-I compared to P_{11}-II, and thus prevent the stacking of P_{11}-I ribbons into fibrils. This explains why P_{11}-I ribbons do not combine into fibrils up to *c*=25mM, whilst P_{11}-II ribbons form fibrils at *c*<1mM. We have also increased ε_{fibril} even further by designing a third peptide with glutamate side-chains on its polar side (as opposed to glutamines for P_{11}-II.) This change resulted in stabilisation of wider fibrils each composed of more than 8 ribbons, as opposed to typically 4 ribbons per P_{11}-II fibril.

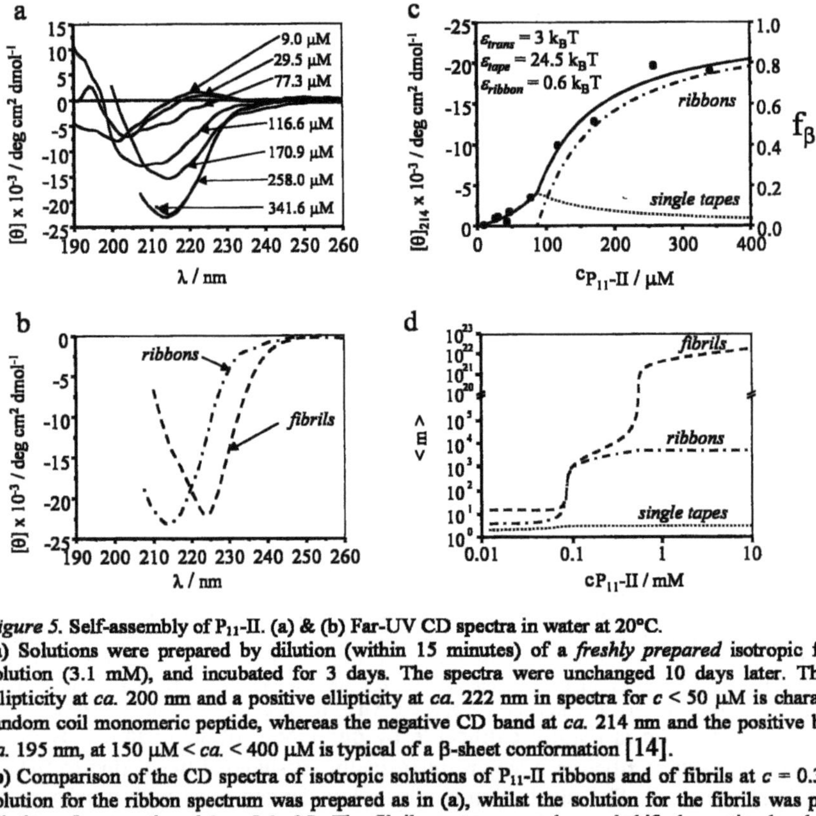

Figure 5. Self-assembly of P_{11}-II. (a) & (b) Far-UV CD spectra in water at 20°C.

(a) Solutions were prepared by dilution (within 15 minutes) of a *freshly prepared* isotropic fluid, stock solution (3.1 mM), and incubated for 3 days. The spectra were unchanged 10 days later. The negative ellipticity at *ca.* 200 nm and a positive ellipticity at *ca.* 222 nm in spectra for *c* < 50 μM is characteristic of random coil monomeric peptide, whereas the negative CD band at *ca.* 214 nm and the positive band below *ca.* 195 nm, at 150 μM < *ca.* < 400 μM is typical of a β-sheet conformation [14].

(b) Comparison of the CD spectra of isotropic solutions of P_{11}-II ribbons and of fibrils at *c* = 0.3 mM. The solution for the ribbon spectrum was prepared as in (a), whilst the solution for the fibrils was prepared by dilution of a nematic gel (*c* = 3.1mM). The fibril spectrum reveals a red-shifted negative band (centred at 224nm, compared to 214nm for ribbons). We believe this red shift accompanying formation of fibrils, arises partly from the superposition of a strong aromatic CD band on the classical β-sheet CD spectrum. The appearance of the aromatic CD band suggests a change in the packing of tryptophan side chains[15], which are in a more constrained chiral environment in fibrils than in ribbons. Near-UV CD spectra were also collected to monitor the environment of the aromatic residues in the different polymers. The ribbons and the fibrils produced distinct near-UV CD spectra, which further supports that a change of packing of aromatic side-chains accompanies fibril formation.

(c) Fit of the theoretical model (solid line) for the self-assembly of peptides into single and double β-sheet tapes, to the measured concentration dependence of the mean residue ellipticity [θ] of the negative CD band at λ = 214 nm. $[\theta]_{214}$ is taken to be a linear function of the fraction f_β of peptides in β-sheet tapes. The fractions of peptides involved in single and double tapes are represented with dotted and dash-dot lines, respectively.
(d) Theoretical concentration dependencies of the average number $\langle m \rangle$ of peptides in single tapes (dotted line) and in ribbons (dash-dot line) and in fibrils made of p = 4 ribbons (dashed line). The molecular parameters were chosen to comply with the fit (c) and with the observed lengths of ribbons at c = 0.2 mM and with the transition concentration from ribbon to fibrils at c = 0.6 mM. Note that the minimum number of peptides in tapes is 2, in ribbons is 4 and in fibrils is 8.

The value of ε_{fibril} for P_{11}-II is such that the corresponding energy of attraction between ribbons per their persistence length, $\varepsilon_{fibril} \cdot \tilde{l}_{ribbon} / b_2 \sim 20 - 30\ k_BT$, -where b_2 is the distance between adjacent β-strands along the tape (see Fig.1c)-, is sufficiently high to stabilise the fibrils against splitting, once the ribbons become longer than a few hundreds of peptide β-strands. It is apparent that in such a hierarchical system of structures, the growth of the more primitive self-assembled structures up to a certain critical size, enables new (weaker) interactions to come into play, and this leads to the formation of the next hierarchical structure. Once the growth of this new structure is established, the concentration and the size of the previous structure becomes fixed. Indeed, Fig. 4e shows that the concentrations of peptide in monomeric and single tape states saturate at $c \sim 0.09$ mM, at which point ribbons emerge. The single tape length is also seen to saturate at 3-4 peptides per tape (Fig. 5d). Similarly, the population of peptides in ribbons (Fig. 4e) and the ribbon lengths (Fig. 5d) become constant when fibrils are first formed. The transitions to higher order aggregates are thus reminiscent of (but not identical to) the formation of micelles in surfactant solutions. We can also compare the stabilisation of fibrils and fibres by twist, to the stabilisation of micelles by amphiphilic interactions.

The fibrils have a persistence length $\tilde{l}_{fibril} \approx$ 20-70μm, as opposed to $\tilde{l}_{ribbon} \approx$ 0.7-1.5μm, which makes the fibrils considerably more rigid than the ribbons. Indeed a plywood-like stack made of p layers is expected to be between p and p^3 times more rigid than the primary layers (p is for a loose stack, and p^3 for a stack with well-glued layers). Hence the expected persistence length of the P_{11}-II fibrils made of p=4 ribbons, is up to 64 times higher than the ribbon persistence length, *ie* in good agreement with our observations (Table 1).

The rigidity of the fibrils/fibres gives rise to the formation of a nematic phase at $c \geq 0.9$ mM (0.001 v/v). The texture in the optical micrograph (Fig. 4f) and its dependence on flow, is characteristic of viscoelastic nematic fluids of semi-rigid polymers [17]. The isotropic-to-nematic phase separation gap is narrow: 0.8 mM < c_I < c_N < 0.9mM (relative gap width $w \equiv c_N/c_I - 1 < 0.13$), and is insensitive to temperature variations up to at least 60°C. Polydisperse *rigid-rod* solutions have much wider phase separation gaps ($w \sim 2$) [18]. The fibrils behave more like typical semi-rigid (worm-like) chains with hard-core excluded volume interactions, for which ($w \sim 0.09$) [18]. The isotropic-to-nematic transition of such chains with rectangular cross-section $W_1 X$

W_2 is predicted [18] to occur at volume fractions $\Phi_{IN} \approx 5.5W / \tilde{l}_{fibril}$ (where $W \approx 2W_1W_2/(W_1+W_2)$, provided that $L \geq \tilde{l}_{fibril}$); this yields for P_{11}-II, $\Phi_{IN} \approx 0.0004 - 0.0015$v/v (corresponding to $c_{IN} \approx 0.4$-1.5mM), in agreement with our observations. $\tilde{l}_{ribbon}$ for P_{11}-I is one to two orders of magnitude shorter than $\tilde{l}_{fibril}$ of P_{11}-II. The isotropic-to-nematic transition of solutions of such semi-flexible ribbons of P_{11}-I is predicted to occur at $\Phi_{IN} \approx 0.015$-0.05v/v (corresponding to 15-50mM). Indeed we find that P_{11}-I forms nematic phase at c ~13mM (Table 1).

At $c \geq 4$ mM, the birefringent solution of P_{11}-II becomes a self-supporting birefringent gel (Fig. 4g). Dynamic and steady-state rheological measurements show that fibril-based gels are brittle, and do not relax even after days, behavior reminiscent of permanent gels of semi-rigid polymers (A.Aggeli, N.Boden, T.C.B.McLeish & P.Mawer, unpublished data). In contrast, tape-based gels are more extendable [6] and relax slowly with time, behavior indicative of transient gels of semi-flexible polymers. The different brittleness between the two types of gels can be ascribed to the roughly two orders of magnitude difference in flexibility between tapes and fibrils. The different relaxation behaviours of the gels stem from the distinctive natures of the crosslinks which make the two kinds of gels. In tape-based gels, the crosslinks are of topological origin, and thus have a limited life-time, allowing relaxation of the gel network at long times. In fibril-based gels, the considerably more rigid fibrils entwine around each other to form fibre-like junctions (Fig.1f'). This crosslinking mechanism is far more stable and permanent compared to the topological entanglements, and thus makes the gel behave like a permanently-crosslinked network. We conclude that the type of polymer (tape, ribbon or fibril, each associated with its own characteristic flexibility, contour length and crosslinking mechanism) determines the liquid crystalline and gelation properties of its solution.

Fibril and fibre formation is characterised by slow kinetics. Fibril formation takes up to several weeks to complete, depending on concentration, as monitored extensively by CD and TEM. The formation of nematic phases of P_{11}-II, following fibril formation, takes several hours for $c \sim 4$ mM, and several weeks for $c \leq 1$ mM. Gelation occurs even more slowly, e.g. for $c = 6$ mM it takes a week. Transitions from one self-assembling structure to another take even longer than their initial formation: when an aged (two months old) P_{11}-II nematic gel ($c = 6$ mM) was diluted to a concentration where ribbons are intrinsically stable ($c = 0.2$ mM), CD and TEM revealed that, even after standing for four months at the lower concentration, numerous fibrils coexist still with ribbons. The extremely slow kinetics originates from the multiplicity of molecular interactions in fibrils. The high tape scission energy $\varepsilon_{tape} \approx 24.5\ k_BT$ of P_{11}-II peptide, ensures that peptide dissociation from free ends of tapes, ribbons or fibrils is a rare event: the effective dissociation rate $\tau_0^{-1} \exp(-\varepsilon_{tape}) \sim 10^{10} \tau_0^{-1}$ corresponds to a dissociation time τ of the order of one second for $\tau_0 \sim 100$ps (τ_0 is a typical diffusion time of a peptide 'rod' over a distance of 0.1 nm). The energy required to break the ribbon, $\varepsilon_{sc} = 2\varepsilon_{tape} \sim 50k_BT$, gives rise to a scission (breaking) time of a ribbon, of: $\tau_0 \exp(2\varepsilon_{tape}) \sim 10^4$ years. Nucleation of short ribbons in solutions of monomer peptide, in the absence of fibril seeds, is expected to be faster than nucleation of fibrils. To form fibrils, ribbons need to wind around each other, which is likely to be highly hindered in a dense system of entangled ribbons. Alternatively, they can dissociate into smaller fragments followed by re-assembly; this process will be slow due to the high ε_{sc}. We therefore expect very slow fibril formation with defective 'fresh' fibrils. Aged fibrils will be far more *perfect* and hence even more stable: to disintegrate they need to lose monomers at their ends. This is a very slow process as the fibrils are very long (Fig. 5d and Table 1).

There are a variety of biological and designed proteins which similarly to our model peptides, self-assemble into twisted amyloid β-sheet polymers typically 5-10nm in diameter [19], [20], [21], [22], eg characterising Alzheimer's and prion diseases, and of the *de novo* β-sandwich protein betabellin 15D [23]. The scheme (Figs.1 & 2) provides an explanation of the factors governing the structures, and stabilities of these fibrous aggregates. In the case of Alzheimer's disease, the pathological, amyloid-forming peptide β-AP (1-39 / 42) is cleaved from the much longer precursor protein APP, and subsequently self-assembles into amyloid fibrils (protofilaments) comprised of stacks of usually 2 to 4 twisted β-sheet tapes (Fig. 1*e'*). Three or more of theses fibrils are wound around each other in the fibres (Fig. 1*f'*). A central hollow core is observed in some amyloid fibres[19] and is particularly enhanced in the fibres formed *in vitro* by the SH3 domain of P13 kinase[22]. The size of the hole will depend on whether the fibrils in the fibre interact with their β-strands parallel to each other as in Fig. 1*f* & *f'*, or at an angle determined by the interactions between segments of the proteins bordering the edges of the fibrils. Whatever the details of each specific system, there is little doubt that it is the intrinsic twist of the unifying, underlying β-sheet scaffold of the constituent β-tapes which governs the way tapes stack and ultimately the distinctive morphology of amyloid fibrils and fibres as elaborated in Figs.1 & 2. It has been pointed out that the β-sheets in amyloid fibrils tend to be flatter than the β-sheets in globular proteins[22]. This is in accordance with our model (Fig. 2): the β-sheets in the tapes have to untwist in order to stack together in the fibrils. The energies involved in fibril formation, together with the presence of the twist in β-sheets as shown above for P_{11}-II peptide, also explain the remarkable stability and accumulation of amyloids *in vivo*, as well as the slow kinetics, and the nucleation-growth mechanism of amyloid self-assembly found by *in vitro* studies[12],[24]. To complete the list of similarities, we have found that a 1.2 mM solution of fibrils of Alzheimer's β-AP (1-40) peptide, when sucked into an optical capillary, produces a banded nematic texture, similar to that observed for P_{11}-II. We propose that systematic analysis of the self-assembly of amyloid peptides using the scheme summarised in Figs.1 & 2, can provide unique quantitative information on the various energies that stabilise amyloid fibrils, and thus enhance our understanding of the associated life-threatening diseases.

We are thus led to the belief that a wide variety of peptides could be designed to behave according to the scheme in Figs.1 & 2, opening up an exciting new class of chiral tape-like and fibrillar macromolecular structures which can form liquid crystals and gels. Moreover, by appropriate peptide design, a combination of desirable properties, such as biological functionality, high stability of fibrils or fibres and responsiveness to external triggers such as solvent polarity, temperature, pH or ionic strength can be incorporated, making these new materials much more versatile than existing peptidic biomaterials such as collagen [25]. We envisage this unique combination of properties makes these materials ideally suited for use in biomedical and other applications, such as for the production of bioscaffolds to control the shape and alignment of cells for tissue engineering, new suture materials, templates for helical crystallisation of macromolecules, and matrices for separation of chiral molecules.

We close by emphasizing that the scheme (Figs. 1 & 2) is not restricted to β-sheet forming peptides and proteins. It is applicable to any chiral molecule able to undergo one-dimensional self-assembly, such as the hemoglobin S protein which forms 14-stranded twisted filaments in sickle cell anaemia [26], and a variety of synthetic organic molecules [27], as well as non-chiral molecules which interact with chiral counterions [28]. As such it contributes to our understanding of the way in which chirality is expressed in the form of self-assembled structures. Moreover, it provides a prescription for measurement of the interaction energies governing the self-assembly process.

Acknowledgements

We wish to thank the UK Engineering and Physical Sciences Research Council and Schlumberger Cambridge Research for financial support, and the Royal Society for the award of a Dorothy Hodgkin Fellowship to A.A. We also thank Dr P.McPhie for assistance with the TEM, and Dr R.Harding for collection of the optical micrograph.

References

1. Artsaenko, O., Kettig, B., Fiedler, K., Conrad, K. & During, K. (1998) Potato tubers as a biofactory for recombinant antibodies., *Molecular Breeding* **4**, 313-319.
2. Urry, D. (1999) Elastic molecular machines in metabolism and soft-tissue restoration., *Trends in Biotechnology* **17**, 249-257.
3. Panitch, A., Yamaoka, T., Fournier, M. J., Mason, T. L. & Tirrell, D. A. (1999) Design and biosynthesis of elastin-like artificial extracellular matrix proteins containing periodically spaced fibronectin CS5 domains., *Macromolecules* **32**, 1701-1703.
4. Krejchi, M. T., Atkins, E. D. T., Waddon, A. J., Fournier, J., Mason, T. L. & Tirrell, D. A. (1994) Chemical sequence control of beta-sheet assembly in macromolecular crystals of periodic polypeptides., *Science* **265**, 1427-1432.
5. Zhang, S. G. & Altman, M. (1999) Peptide self-assembly in functional polymer science and engineering, *Reactive & functional polymers* **41**, 91-102.
6. Aggeli, A., Bell, M., Boden, N., Keen, J. N., Knowles, P. F., McLeish, T. C. B., Pitkeathly, M. & Radford, S. E. (1997) Responsive gels formed by the spontaneous self-assembly of peptides into polymeric beta-sheet tapes, *Nature* **386**, 259-262.
7. Aggeli, A., Bell, M., Boden, N., Harding, R., McLeish, T. C. B., Nyrkova, I., Radford, S. E. & Semenov, A. (2000) Exploiting protein folding and misfolding to engineer novel nanostructured materials, *The Biochemist* **22**, 10-14.
8. Israelachvili, J. N. (1985) *Intermolecular and Surface Forces* (Academic Press, London).
9. Nyrkova, I., Semenov, A. N., Aggeli, A., Bell, M., Boden, N. & McLeish, T. C. B. (2000) Self-assembly and structure transformations in living polymers forming fibrils, *Eur. Phys.J.B* in press.

10. Nyrkova, I., Semenov, A., Aggeli, A. & Boden, N. Fibril stability in twisted tape systems : a new kind of micelle found in solutions of self-assembled peptide ß-sheet tapes, *Eur.Phys.J.B* **in press.**

11. Aggeli, A., Bell, M., Boden, N., Keen, J. N., McLeish, T. C. B., Nyrkova, I., Radford, S. E. & Semenov, A. (1997) Engineering of peptide beta-sheet nanotapes, *Journal of Materials Chemistry* **7**, 1135-1145.

12. Terzi, E., Holzemann, G. & Seelig, J. (1995) Self association of beta-amyloid peptide (1-40) in solution and binding to lipid membranes., *Journal of Molecular Biology* **252**, 633-642.

13. Perutz, M. F., Johnson, T., Suzuki, M. & Finch, J. T. (1994) Glutamine repeats as polar zippers: Their possible role in inherited neurodegenerative diseases., *Proc. Natl. Acad. Sci USA* **91**, 5355-5358.

14. Manning, M. C., Illangasekare, M. & Woody, R. W. (1988) Circular dichroism studies of distorted alpha-helices, twisted beta-sheets and beta-turns., *Biophysical Chemistry* **31**, 77-86.

15. Grishina, I. B. & Woody, R. W. (1994) Contributions of Tryptophan Side Chains to the Circular Dichroism of Globular Proteins: Exciton Couplets and Coupled Oscillators, *Faraday Discussions* **99**, 245-262.

16. Chothia, C. (1973) Conformation of twisted beta-pleated sheets in proteins, *Journal of Molecular Biology* **75**, 295-302.

17. Dobb, M. G., Johnson, D. J. & Saville, B. P. (1977) Supramolecular structure of a high-modulus polyaromatic fiber (Kevlar 49), *Journal of Polymer Science* **15**, 2201-2211.

18. Semenov, A. N. & Khokhlov, A. R. (1988) Statistical physics of liquid-crystalline polymers., *Sov. Phys. Usp.* **31**.

19. Sunde, M. & Blake, C. F. (1998) From the globular to the fibrous state: protein structure and structural conversion in amyloid formation, *Quarterly Review of Biophysics* **31**, 1-39.

20. Kirschner, D. A., Elliott-Bryant, R., Szumowski, K. E., Gonnerman, W. A., Kindy, M. S., Sipe, J. D. & Cathcart, E. S. (1998) In vitro amyloid fibril formation by synthetic peptides corresponding to the amino terminus of apoSAA isoforms from amyloid-susceptible and amyloid-resistant mice., *Journal of Structural Biology* **124**, 88-98.

21. Seilheimer, B., Bohrmann, B., Bondolfi, L., Muller, F., Stuber, D. & Dobeli, H. (1997) The Toxicity of the Alzheimer's Beta-Amyloid Peptide Correlates with a Distinct Fibre Morphology, *Journal of Structural Biology* **119**, 59-71.

22. Jimenez, J., Guijarro, J., Orlova, E., Zurdo, J., Dobson, C., Sunde, M. & Saibil, H. R. (1999) Cryo-electron microscopy structure of an SH3 amyloid fibril and model of the molecular packing, *The EMBO Journal* **18**, 815-821.

23. Lim, A., Saderolm, M. J., Makhov, A. M., Kroll, M., Yan, Y., Perera, L., Griffith, J. D. & Erickson, B. W. (1998) Engineering of betabellin-15D: a 64 residue beta sheet protein that forms long narrow multimeric fibrils, *Protein Science* **7**, 1545-1554.

24. Harper, J. & Lansbury, P. (1997) Models of amyloid seeding in Alzheimer's disease and scrapie: Mechanistic Truths and Physiological Consequences of the Time-dependence Solubility of Amyloid Proteins., *Annual Review of Biochemistry* **66**, 385-407.

25. Byrom, D. (1991) *Biomaterials: Novel materials from biological sources* (Stockton Press).

26. Eaton, W. A. & Hofrichter, J. (1990) in *Advances in Protein Chemistry* (Academic Press, Inc., San Diego), Vol. 40.

27. Terech, P. & Weiss, R. G. (1997) Low molecular mass gelators of organic liquids and the properties of their gels, *Chemical Reviews* **97**, 3133-3159.

28. Oda, R., Huc, I., Candau, S. & MacKintosh, F. C. (1999) Tuning bilayer twist using chiral counterions, *Nature* **399**, 566-569.

RIBBON-LIKE LAMELLAR STRUCTURES FROM CHAIN-FOLDED POLYPEPTIDES

E. D. T. ATKINS
H. H. Wills Physics Laboratory, University of Bristol,
Tyndall Avenue BRISTOL BS8 1TL, UK

1. Introduction and Background

For long, flexible chain molecules, crystallising in semi-quiescent conditions, the entropic component, within the free-energy equation, for the generation of fully extended-chain crystals, is usually too large to overcome. As a consequence, long flexible chain molecules often settle into a free-energy (sub) minimum that expresses itself in the form of regular chain-folded lamellar-like crystals. These nanometer-thin, wafer-like crystals, with their large surfaces decorated with two-dimensional, repetitive folds, was first visualised for the chemically unpretentious polyethylene molecule, when crystallised from solution [1]. Electron microscopy and electron diffraction from individual lamella, coupled with X-ray diffraction from aggregated lamellae (mats), oriented by uniaxial compression, have been the basic methods used for establishing the structures and morphologies of these chain-folded lamellar crystals [1-5]. If amide units are now implanted into the alkane chain, as is the case for the nylon family, the lamellar thickness reduces, as a consequence of the stronger straight-stem interactions: less are needed to compensate for the adjacent re-entry fold. There is an approximate linear correlation between the lamellar thickness and frequency of implanted amide units: starting at ≈10 nm for polyethylene, with only van der Waals interactions in the straight-stems, and reducing in value through the higher nylons down to ≈ 6 nm for nylon 6, and to a projected lamellar thickness of 3-4 nm for nylon 2 (polyglycine). Indeed, the protein silk egg-stalk of the green lace-wing fly *Chrysopa* forms ribbon-like, chain-folded lamellae 3 nm in thickness [6].

In this contribution, a review is presented of recent results concerned with the controlled generation of regular chain-folded, antiparallel (ap) β–sheet protein crystalline structures from repetitive amino sequences that bear a relationship to silk-like β–proteins [7-11]. Keith, Lotz and coworkers [12-15] have already shown, that a number of repetitive-sequence polypeptides in the β–conformation could be crystallised in the form of chain-folded lamellae with estimated thicknesses ≈ 5 nm. It is also well-known, from crystallography of globular proteins, that the polypeptide chains can make sharp folds (turns), with either one or two amide units (two or three amino acid C_α atoms) in the turn; known as β– or γ–turns, respectively.

Some of the perceived advantages creating such chain-folded lamellae are outlined as follows.

(1) A repetitively chain-folding structure would contain two (β–strand and chain turn) of the three prominent and recognisable protein conformations (the third being the α–helix). Thus, the results would be pertinent to protein structure and folding in general.

(2) As already explained, most flexible polymers crystallise by repetitive, adjacent re-entry chain-folding, and there is a great deal known about the kinetics and thermodynamics of the folding process. It would be useful to know how subservient proteins are to these general rules of behaviour.

(3) The apβ–sheet has established X-ray diffraction and spectroscopic fingerprints, so it should

A. Aggeli et al. (eds.), Self-assembling Peptide Systems in Biology, Medicine and Engineering, 19–33.

be possible to test for its presence during and after crystallisation.

(4) The surfaces could be decorated with useful amino acid to form particular spatial arrangements on the surface; upper and lower surfaces could be decorated differently.

(5) Chain-folded lamellae are membrane-like and therefore have many potential biological uses.

In the first instance, the chain-folded crystal structures from a group of four monodisperse, genetically-engineered polypeptides, represented by the repetitive amino acid sequence: -[$(AG)_x$EG]-, for integer x = 3, 4, 5 and 6, are discussed. The effect of the increasing length of the alanylglycyl (AG) segments is examined. In a separate set of experiments the changes in crystal structure, when the glutamic acid unit (E) is replaced by other amino acids, are investigated. Finally, for poly$(AG)_3$EG, the consequences on the crystal structure when every other $(AG)_3$ segment is reversed will be considered.

2. Choice of Poly$(AG)_X$EG Sequence

The alanylglycyl (AG) dyads were selected because they are common in many β–silks and form extended β–strands, which usually assemble into apβ–sheets and stack to form stable crystals [16,17]. Adjacent chains in an apβ–sheet, as the notation implies, run in opposite directions, a fundamental ingredient in the crystal design. Thus, it was hoped that chain-folded lamellae could be prepared by crystallising from solution and that the $(AG)_x$ segments would create the straight-stems and the additional EG dyad would lie in the fold. The range of 3 to 6 alanylglycyl dyads (8- to 14-mer repeating sequences) was chosen after consideration of the anticipated thickness of the chain-folded lamellae. Glutamic acid (E) was incorporated into the repetitive sequences because of its bulk, polar nature, and chemical reactivity. The large size of E relative to A and G should also deter its inclusion into the crystalline lamellar interior on steric grounds: the A-to-A intersheet spacing for polyAG [18] is 0.498 nm, whereas the smallest intersheet distance reported [12] for the β–form of poly(E) is 0.782 nm, nearly 60% larger. During crystal growth, interactions with the solvent should also act to segregate the glutamic acid units to the lamellar surface.

The detailed crystallisation procedures and preparation of oriented mats suitable for X-ray diffraction analysis is reported elsewhere [7].

In order to aid the understanding of these chain-folded polypeptide structures the salient features of the apβ–sheet silk structures are first summarised.

3. Antiparallel β–Sheet Structures found in Silks

These structures consist of sheets composed of relatively extended protein chains with a 2-fold helical conformation (β-strand or β-conformation). Adjacent, antiparallel chains zip together via linear C=O.....H-N hydrogen bonds to form the so-called apβ–sheet structure. Figure 1 is a computer-drawn model showing a view orthogonal to an apβ–sheet with a repetitive alanylglycyl sequence.

The repeating distances along the chain axis (*c*) and in the hydrogen bond direction (*a*) are essentially invariant and therefore act as characteristic and diagnostic values for apβ–sheet structures. Reported *a* and *c* values [17] for *Bombyx mori* silk are 0.948 nm and 0.695 nm, respectively. There is a noticeable out of plane puckering of the sheets, as shown in Figure 2.

For repetitive alanylglycyl sequence there are two types of apβ–sheet. The first arrangement is when the side group decoration patterns on upper and lower surfaces differ, i.e. the glycyl hydrogen atoms on one surface and the alanyl methyl groups on the other, as shown in Figures 1 & 2a. This may be described as a polar apβ–sheet. The second is when the side group decoration patterns on the upper and lower surfaces of the apβ–sheet are identical, i.e. an equal mixture of glycyl hydrogen atoms and alanyl methyl groups on both surfaces, as illustrated in Figure 2b.

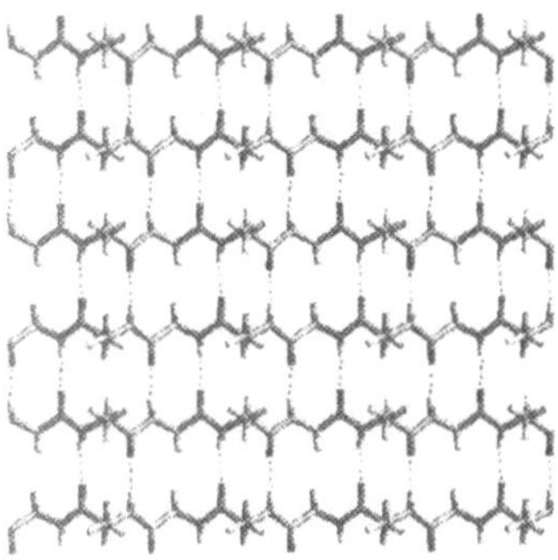

Figure 1. View of a single polar apβ-sheet, with a repetitive alanylglycyl sequence, orthogonal to the *ac*-plane.

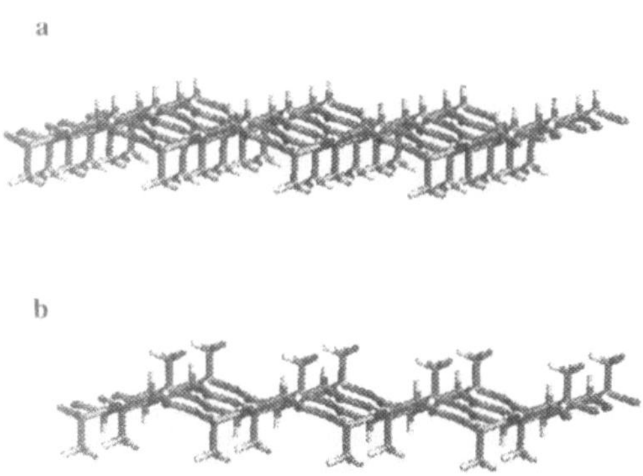

Figure 2. Oblique view of apβ-sheets with repetitive alanylglycyl β-strands; (a) polar apβ-sheet and (b) apolar apβ-sheet.

This can be described as an apolar apβ–sheet. In the full three-dimensional structure the sheets stack with in-phase puckering but usually with alternating or random shears of ca. ± *a*/4 in the *ac* plane. The sheet stacking distance, which is related to the crystallographic *b*-axis, is a function of the size and distribution of the amino acid side groups that decorate the upper and lower apβ–sheet surfaces. In the orthorhombic protein apβ–sheet structures, where ***a***, ***b*** and ***c*** are mutually orthogonal, there are two apβ–sheet stacking categories that need to be considered for the benefit of the structural analysis of poly(AG)$_3$EG. Apolar apβ–sheets would be expected to stack as shown in Figure 3a, i.e. with equal (*b*/2) spacing. In this structure, the 0*k*0 diffraction signals for *k* odd would be systematically absent. In particular, the 010 would not be observed. On the other hand, polar apβ–sheets stacked with like surfaces together as shown in Figure 3b, and as reported [18,17] for polyAG, would have the centre sheet in the unit cell displaced from the *b*/2 position and a 010 diffraction signal would be observed; furthermore, its relative intensity would be a convenient indicator of the deviation from equally-spaced apβ–sheet stacking in the *b* direction. Thus, these β–silk-like structures are held together with

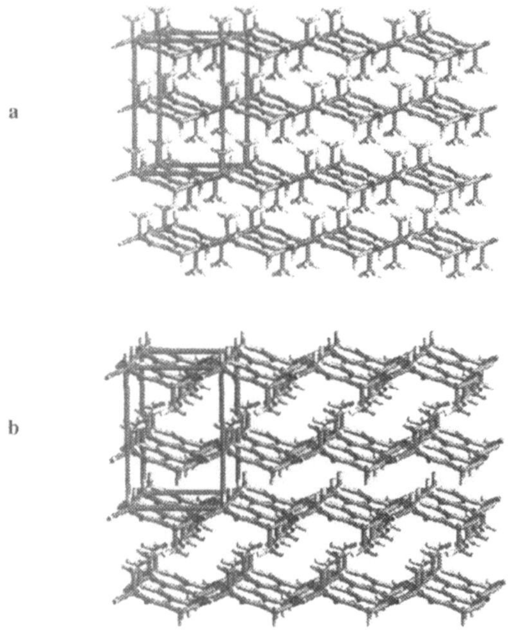

Figure 3. (a) View of stacked apolar apß-sheets; the orthorhombic unit cell is drawn as a box. The 010 diffraction signal would be absent in this structure. (b) View of polar apβ-sheets, stacked face to face and illustrating how the hydrophobic methyl groups form layers and displace the centre sheet from the *b*/2 position. Thus in this case, the 010 diffraction signal would be expected to be present.

covalent bonds in the chain (*c*-axis) direction, hydrogen bonds at right-angles (*a*-axis) and van der Waals forces in the third mutually orthogonal *b*-direction.

4. X-ray Diffraction Results

4.1. POLY(AG)$_3$EG

The wide-angle X-ray diffraction pattern of poly(AG)$_3$EG, taken with the x-ray beam directed parallel to the plane of a oriented mat and with the mat normal horizontal, exhibits discrete

Bragg diffraction signals consistent with an oriented crystalline polymer and all the observed diffraction signals index on an orthorhombic unit cell with dimensions a = 0.948nm, b = 1.060nm, and c = 0.695nm. A schematic diagram showing the indexing of the diffraction signals is shown in Figure 4. A noticeable feature is that the orientation direction is along a. Thus, the directionally coincident a and a^* axes lie along the meridian line and therefore the

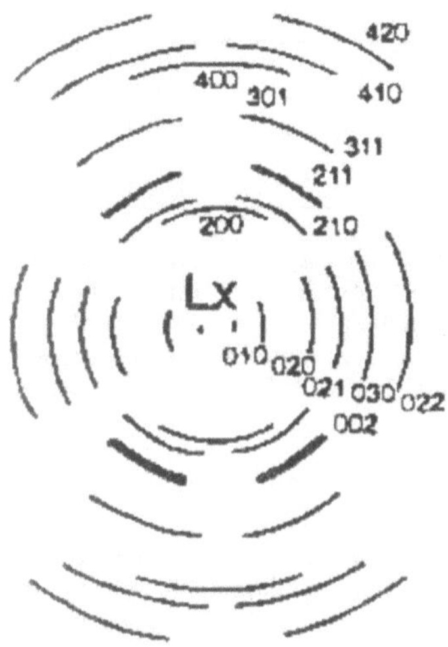

Figure 4. Schematic illustration of the X-ray diffraction pattern from poly(AG)$_3$EG with the beam directed parallel to the plane of an oriented mat: mat normal horizontal. The diffraction signals are indexed and the low-angle (Lx) diffraction signal appears on the equator at 3.6 nm.

h00 diffraction signals, the easily seen 200 and 400 in particular, appear as arcs, centred on that meridian. Furthermore the families of 0k0 and 00l diffraction signals appear on the equator. Differences in the line broadening of the various diffraction signals, which relate in a consistent way to their assigned Miller indices, are observed. The 200 and 400 diffraction arcs are particularly sharp. This demands long-range correlation along the a axis. Diffraction signals with indices of the general hkl type are considerably broader than those with indices hk0. In particular the 211 diffraction signal is estimated to be tenfold broader than its 210 near-neighbour. Comparison of the line-widths, leads to the conclusion that the coherent scattering length in the c direction is small, with an estimated length of $\approx$ 4 nm, when allowance is made for instrumental line broadening. The relative breadth of the prominent 010 and 020 diffraction signals suggests that the coherent scattering length in the b direction is intermediate between those in the a and c directions. A low-angle X-ray diffraction peak, at a spacing of 3.6 $\pm$ 0.1 nm, is found on the equator.

4.2. POLY(AG)$_X$EG FOR x = 4, 5 & 6

The overall features of the wide-angle X-ray patterns are similar to those of poly(AG)$_3$EG and the diffraction signals all index on orthorhombic unit cells with the same a and c parameters.

There is a continual reduction in the value of *b* within the family as shown in Table 1.

Table 1. Unit cell *b*-values, in nanometers, for the poly(AG)$_X$EG series, with the value [9] for poly(AG) for comparison.

Polypeptide sequence	*b*-value in nanometers
Poly(AG)$_3$EG	1.060
Poly(AG)$_4$EG	1.028
Poly(AG)$_5$EG	0.970
Poly(AG)$_6$EG	0.962
Poly(AG)	0.922

Also, diffraction signals with indices of the type *hkl* sharpen as the length of the repetitive alanylglycyl segment increases. Specifically, the diffraction signal indexed as 211 is approximately twice as sharp in poly(AG)$_4$EG as in poly(AG)$_3$EG, and the relative sharpness increases further through the series, i.e. as x increases. Low-angle arcs, from interlamellar stacking, also occur on the equator but with increasing in spacing (up to 6.2 nm) with increasing x value.

5. Structure of Poly(AG)$_3$EG Crystals

The line broadening of diffraction signals with Miller index $l \neq 0$, the 211 in particular, indicates that there are interruptions to the crystallographic lattice in the *c*-direction, that is, the chain direction, at distances $\leq$ 4 nm. This is ca. 25 times less than the length of the protein chain. The low-angle signal at a spacing of 3.6 nm, resembles an inter-particle interference function. This diffraction peak can be made to disappear with swelling agents without affecting the underlying crystal structure. This behaviour is reminiscent of inter-lamellar stacking and is similar to that reported for chain-folded lamellar crystals of polyethylene [19,20], nylons [21,22] and *Chrysopa* insect silk [6]. Adjacent chains in the structure are antiparallel and the straightforward explanation for these results is that the chains in poly(AG)$_3$EG protein are folding regularly in a distance less than the measured inter-lamellar periodicity of 3.6 nm. The nature of the changes in the X-ray diffraction patterns from the other members of the poly(AG)$_X$EG family add considerable support to the chain-folded model. As the number of alanylglycyl dyads is increased, the line-width of the 211 diffraction signal systematically decreases; a change that is to be expected as the straight-stems in the chain-folded crystals increase in length, assuming the geometry of the folds remains intact. In addition, the measured spacing of the low-angle diffraction peak increases to accommodate the thicker crystal.

5.1. THE β–TURN MODEL

A model of a chain-folded apβ–sheet that is commensurate with the dimensions deduced from the X-ray diffraction data, and which folds in phase with the repetitive octapeptide, is illustrated in Figure 5.

The lamellar thickness is approximately 3 nm. There are three pairs of alanylglycyl units in each straight stem and the remaining glutamylglycyl dyad forms a β–turn. Thus, all the glutamic acid residues occur at the fold surfaces. At first sight, the arrangement appears to be a perfect solution for the structure; however, it is an apolar sheet with a similar distribution of hydrogen atoms and methyl side groups on the upper and lower surfaces. The apolar nature of the apβ–sheet is a direct consequence of a 2-amino acid, β–turn. Thus, regular stacking of the chain-folded apβ–sheets in the *b*-direction with *b*/2 spacing would be anticipated. This would result in the systematic absence of the 010 diffraction signal, as discussed in section 3. A strong 010 diffraction signal is observed experimentally; thus, this β–turn model is fundamentally flawed.

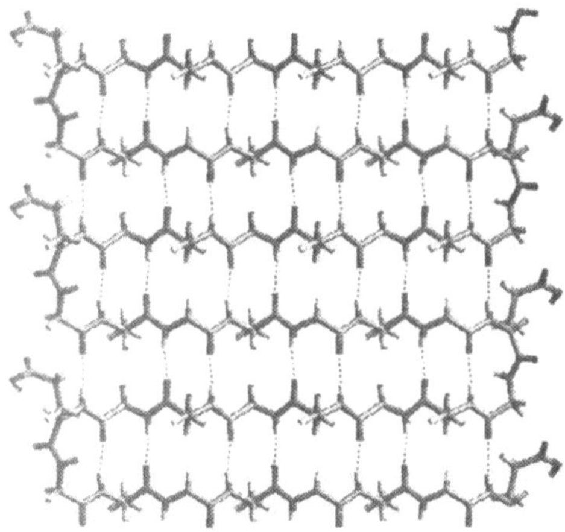

Figure 5. A single apolar apβ-sheet of poly(AG)$_3$EG folding in-phase with the octapeptide periodicity and forming β-turns via the glutamic acid-glycine diads. The lamellar thickness (horizontal width) is commensurate with the interlamellar stacking periodicity of 3.6 nm.

5.2. AN ALTERNATIVE SOLUTION

If the chain-folded apβ–sheets were polar in character, with one surface decorated entirely with methyl side groups and the other with hydrogen atoms, and like surfaces were in contact, then the sheets would automatically stack with a displacement of sheets off the *b*/2 periodicity in an alternating manner, as illustrated in Figure 3b. This type of stacking would generate the desired 010 diffraction signal. Indeed, this is precisely the model proposed [18] for poly(L-alanylglycine) over thirty years ago, although in that case no chain-folding was invoked. In such an arrangement, the sheets are stacked in pairs with the hydrophobic alanyl methyl side groups sandwiched together.

An apolar apβ–sheet, with an AG repetitive sequence can be converted into a polar apβ–sheet if successive adjacent (antiparallel) intersheet chains are translated by *c*/2 (one amino acid unit) as illustrated schematically in Figure 6a. The consequence of this manoeuvre results in an extra amino acid occurring in the hairpin-like fold, i.e. a 3-amino acid turn as shown in lower part of Figure 6b. (A single-amino acid turn is stereochemically not possible). In practice, if successive chains in the sheet need to be simultaneously translated by *c*/2 and rotated around the *c*-direction by π, owing to the 2-fold helical nature of the β–conformation of the chain, as illustrated in Figure 2.

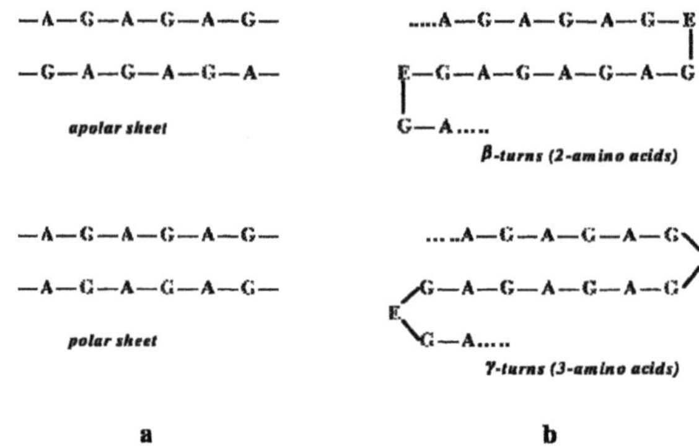

Figure 6. (a) Relative chain translation and reversal of sequence; (b) Consequences on fold geometry.

5.3. THE γ-TURN MODEL

Figure 7a illustrates such a chain-folded apβ–sheet structure for poly(AG)$_x$EG. Each fold now consists of three amino acids, rather than two. These tripeptide folds have been referred to as γ-turns [23] and are common in globular proteins [24]. Thus, changing from a β–turn to a γ-turn has the effect of converting an apolar apβ–sheet into a polar apβ–sheet. Clearly, the nature of the reverse turn is coupled to the type of apβ–sheet in these sequential polypeptides: the apolar apβ–sheet with β–turns, and the polar apβ–sheet with γ-turns. If the glutamic acid group remains in the fold there are three possible arrangements for a straight-stem plus fold, namely: (AG)$_2$A GEG, (GA)$_2$G AGE and (GA)$_2$G EGA. The resolution of the X-ray data is not able to delineate between the three possibilities. The GEG apβ–sheet shown in Figure 7a, with the glutamic acid unit at the apex of the fold and with the smaller glycine units on either side (see Figure 7b), was chosen as the most likely folded γ-turn structure.

These apolar, γ-turn apβ–sheets will stack face to face to form three-dimensional crystal as shown in Figure 7c. The methyl side groups from the alanyl units form hydrophobic layers interspersed with layers of hydrogen atoms from the glycyl units. In this structure a strong 010 diffraction signal is generated, consistent with the X-ray diffraction data [8].

6. Phase Relationship between Sequence and Folding

All of the experimental evidence supports a regular chain-folded apβ-sheet architecture for the crystalline forms of the poly(AG)$_x$EG series. The X-ray diffraction analyses argue strongly for stacking of polar apβ–sheets with like surfaces in contact. This requirement demands an odd number of amino acids in each fold, irrespective of whether the glutamic acid units remain on the fold surfaces. It is clear that the γ-turn is the only serious contender. The *b*-value of poly(AG)$_3$EG of 1.060 nm is 19.5% larger than that reported [18] for polyAG (and 14% larger than *Bombyx mori* silk fibroin [25]) and so this may be used to raise the question of whether glutamic acid residues enter the crystalline core of the lamellae; i.e., whether the folding periodicity is out of phase with the octapeptide sequence. This is an important issue,

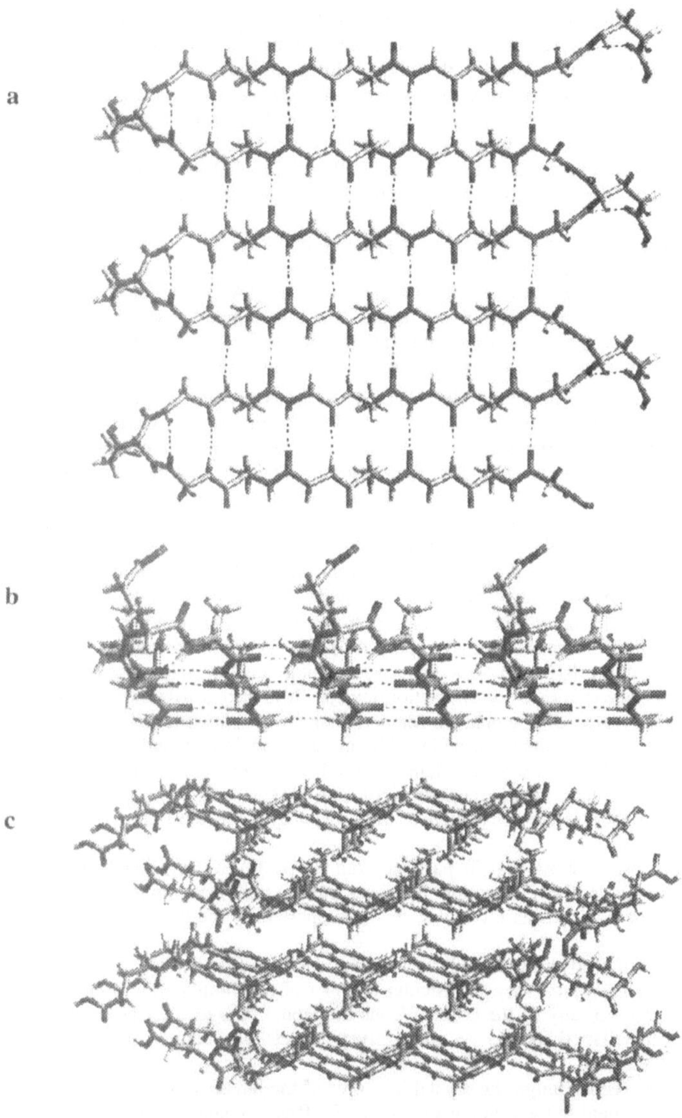

Figure 7. (a) A single polar apβ-sheet of poly(AG)$_3$EG folding in-phase with the octapeptide periodicity and forming γ-turns with GEG sequences. (b) An oblique view of the γ-turns in a polar apβ-sheet. (c) View of refined chain-folded lamellar crystal of the proposed structure for poly(AG)$_3$EG with polar apβ-sheets and γ-turns [8].

since the choice of the amino acid sequence was part of the strategy for creating crystalline chain-folded lamellae with useful groups (in this case glutamic acid residues) wholly decorating the lamellar surfaces. Clearly the chain-folding periodicity is confined within certain limits, as imposed by the relative line-width of the 211 diffraction signals, by the occurrence of low-angle diffraction peaks and by the experimental evidence in general. On the other hand, if a mismatch of two amino acids (equivalent to a *c*-direction translation of 0.695 nm) were permitted, it would be enough to allow the bulkier $-CH_2$-CH_2-COOH glutamic acid side groups to impregnate the crystalline lattice in the form of spatially distributed localised defects.

There are a number of pieces of evidence against the decoupling of folding periodicity from the repetitive amino acid sequence, that is, in favour of keeping the glutamic acid units at the fold surfaces. Although the *b*-parameter value is 19.5% greater for poly$(AG)_3$EG, than for polyAG, this difference reduces, as more alanylglycyl dyads are incorporated, to only 8.5% for poly$(AG)_6$EG (see Table 1). Thus, with the folds taking three-eighths of the available amino acids in poly(AG)3EG, it is perhaps not surprising that the rather short straight-stems (five amino acids) cannot develop sufficient van der Waals interactions to pull the apβ-sheets closer together against the expected overcrowding near the fold surfaces. As the straight-stems get progressively longer throughout the series, the *b*-value approaches that of polyAG [18]; a pattern of crystalline behaviour consistent with the evidence. There is a counter-argument, of course, that the relative dilution of glutamic acid units, in passing from poly$(AG)_3$EG to poly$(AG)_6$EG, results in a corresponding reduction in the number of localised defects, allowing closer packing. This counter-argument is weakened by the following consideration. If the glutamic acid units enter the lattice they occur selectively only in the inter-alanyl decorated layers (since E effectively replaces an A unit in an otherwise wholly repetitive AG sequence). Thus, the middle apβ-sheet in the unit cell (see Figure 3b) would move proportionally closer to the *b*/2 position in passing from poly$(AG)_3$EG to poly$(AG)_6$EG. This would change the ratio of the relative intensities of the 010 and 020 diffraction signals, and to a lesser extent other diffraction signals, contrary to the X-ray diffraction results. It is also worthwhile to consider what the lowest possible value of *b* would be for a structure with glutamic acid units in the straight-stems. If all the alanyl units were replaced by glutamic acid units, and the folds were left off. A computer-modelling exercise calculates the minimum value to be 1.26 nm, which is 18.9% and 30.7% greater than those measured for poly$(AG)_3$EG and for poly$(AG)_6$EG, respectively. Figure 8 illustrates the consequences of allowing glutamic acid units to penetrate the lamellar core.

The X-ray diffraction results show that in passing from poly$(AG)_3$EG to poly$(AG)_6$EG, the coherent scattering length in the *c*-direction increases substantially, as highlighted by the relative sharpening of the 211 diffraction signal; this is supported by the increased spacing of the low-angle signal.

If glutamic acid units can feed through and be accommodated in the crystalline core of the lamellae, there would be no reason why a particular member of this family should have a fold length different from those of any of the others. Thus the model fails to offer an explanation for the increase in lamellar thickness as alanylglycyl dyads are added in going from poly$(AG)_3$EG to poly$(AG)_6$EG.

If glutamic acid residues feed through the crystalline cores of the lamellae we would expect to see a measurable improvement in the relative sharpness of diffraction signals containing Miller index *k*, the 010 and 020 in particular, on passing from poly$(AG)_3$EG to poly$(AG)_6$EG, as the proportion of glutamic acid units is reduced. No such improvement is noticed.

7. Effect of Amino Acid Side-chain Volume on Sheet Stacking

Comparison of the *b*-values (see Table 1) of the poly$(AG)_x$EG series showed that as the AG straight-stem lengths increased the stacking periodicity decreased. This implies that the large steric size of the glutamic acid units on the fold surface influences the sheet stacking

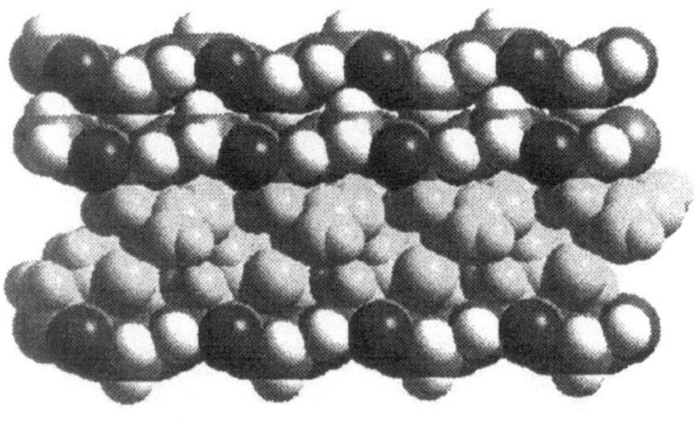

Figure 8. Projection of a space-filling model, parallel to the *a*-direction, of stacked apβ-sheets with the glutamic acid side groups shaded in light gray. If the glutamic acid units were to penetrate the crystal core they can only do so in the inter-alanyl layer.

periodicity within the crystals. One way to examine this feature in more detail is to keep the AG straight-stem length invariant and change the size (volume) of the amino acid side-chain [26] at the fold surface. Thus, a series of monodisperse polypeptides, based on $poly(AG)_3EG$, but with the glutamic acid unit replaced with other amino acids were synthesised, crystallised and their crystalline structures analysed along the lines described previously. All the diffraction signals indexed on the orthorhombic unit cell similar to $poly(AG)_3XG$ with the exception of the sheet stacking parameter *b*. Table 2 lists the values for the various structures.

Table 2. Unit cell *b*-values, in nanometers, for the various amino acids at position X in $poly(AG)_3XG$.

Amino acid unit at X	b-value in nanometers
Alanine [9]	0.922
Serine	0.926
Asparagine	0.974
Valine	1.016
Glutamic acid	1.060
Glutamic acid & Lysine (alternating)	1.111
Phenylalanine	1.188
Tyrosine	1.198

The replacement amino acids chosen, in order of increasing volume, are as follows: alanine, serine, asparagine, valine, alternating lysine and glutamic acid, phenylalanine and tryosine. Computer-generated space filling models of the γ-folds, illustrating the variation in size of the side-chains, is shown in Figure 9. Figure 10 is a plot of the average intersheet distance (*b*/2) in against the volume for eight different polypeptides based on $poly(AG)_3XG$ for varying X. A linear correlation exists between these two parameters. Thus, fine-tuning of molecular architecture can be achieved by appropriate choice of the substituent amino acids in these chain-folded periodic polypeptide lamellar crystals.

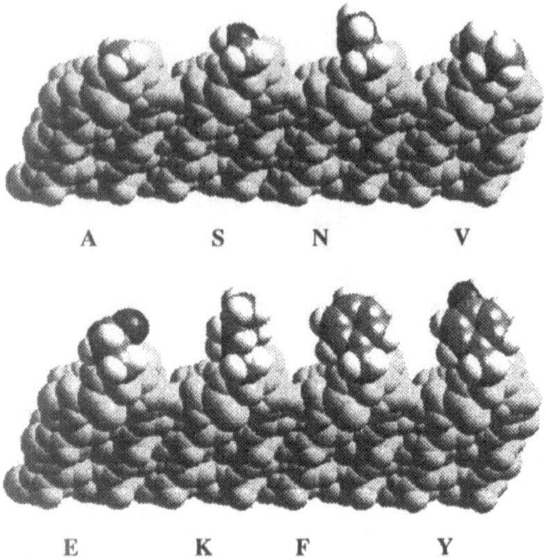

Figure 9. Perspective views of the three amino acid, GXG γ-turn for the eight different amino acid side groups substituted at position X. The other atoms in the hairpin fold are shaded gray.

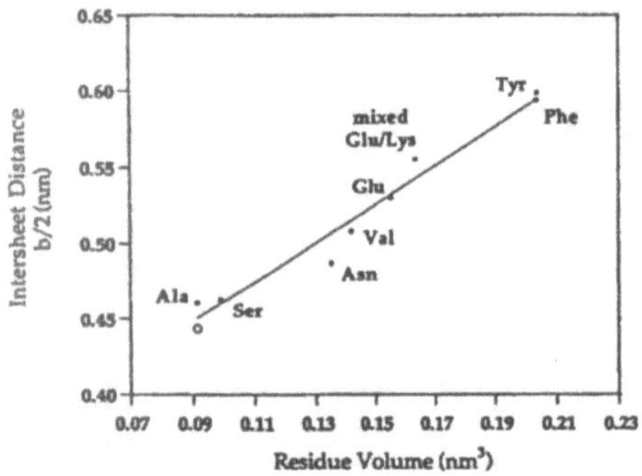

Figure 10. Linear correlation between the average intersheet distance and the volume [26] of the substituted amino acid.

8. Why γ-folds in these Chain-folded Structures?

The question that naturally arises from the structural studies on poly(AG)$_3$EG is why it adopts γ-turns? There are two possibilities to consider. (1) Is it because the tighter, 2-amino acid β-turn in the apolar apβ-sheets is not possible, because it is strained stereochemically, in these structures? Or (2), are intersheet interactions between alanyl units the dominant forces? In this latter case polar apβ-sheets would be desired since they have one surface completely decorated with hydrophobic methyl side groups from the alanyl units.

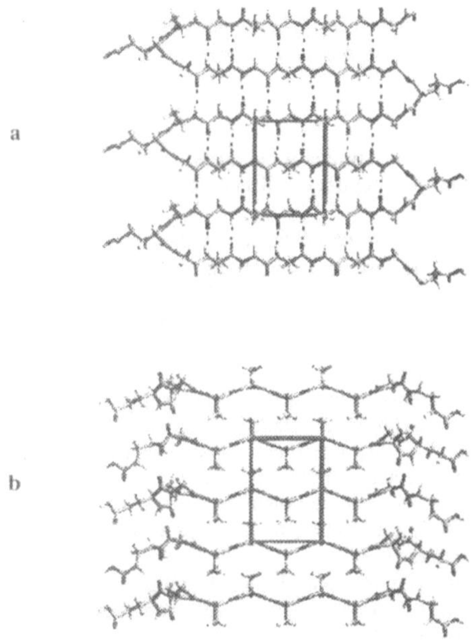

Figure 11. Projections of the proposed apolar apβ-sheet structure for poly(±AG)$_3$EG [11] with γ-turns in-phase with the octapeptide sequence. (a) View orthogonal to *ac*-plane. (b) View orthogonal to *bc*-plane of apolar apβ-sheets stacked at 0.526 nm apart.

The sequence poly[(AG)$_3$EG(GA)$_3$EG], which can be simplify in terms of notation to poly(±AG)$_3$EG, was designed in an attempt to resolve this dilemma. In this sequence, alternate (AG)$_3$ segments in poly(±AG)$_3$EG are reversed. Figure 6a illustrates the effect of (AG)$_3$ sequence reversal on apβ-sheets: polar sheets become apolar or visa versa. (AG)$_3$ sequence reversal mimics relative chain translation by *c*/2 and rotation through π. Thus, repetitive chain folding with the glutamic acid residues in the folds leads to either of two possibilities. (1) β-turns and polar apβ-sheets, or (2) γ-turns and apolar apβ-sheets. The first of these two alternatives should be favoured if intersheet interactions between alanyl methyl groups are dominant, while observation of apolar sheets would suggest that turn geometry is the most important factor in the crystalline architecture.
These two structures can be distinguished by a straightforward X-ray diffraction experiment. The major difference between the two stacking modes, shown in Figures 3b and 3c, is the

absence or presence of a 010 diffraction signal, respectively. Chain-folded crystalline lamellae of poly(±AG)$_3$EG were prepared and the X-ray diffraction pattern compared with that of poly(AG)$_3$EG. The pattern obtained from poly(±AG)$_3$EG has no 010 diffraction signal consistent with a structure with apolar apß-sheets and γ-turns.

Computer-generated projections of the apolar chain-folded apβ-sheet of poly(±AG)$_3$EG, folding in-phase with the pseudo-octapeptide repeat, and with glutamic acid units at the apices of the folds, are shown in Figure 11a. The X-ray diffraction results [11] support the apβ-sheet stacking arrangement illustrated in Figures 11b. The apolar apβ-sheets stack regularly at an intersheet distance of 0.526 nm but with random ≈ ±a/4 shears [11].

9. Conclusions

The X-ray diffraction experiments and structural analysis shows that these genetically-engineered monodisperse, sequential polypeptides will crystallise in the form of chain-folded lamellae.

In all cases the preferred fold is a three amino acid γ-turn rather than the tighter, two amino acid β-turn. In the case of poly(AG)$_3$EG the apβ-sheets are polar with one surface decorated with hydrophobic alanyl units and the other surface with glycyl units. The amino acid sequence is found to be in-phase with the repetitive folding and the glutamic acid units are confined to the fold surface. These chain-folded polar apβ-sheets stack together in pairs with the hydrophobic alanyl surfaces together.

Increasing the length of the AG segments increases the lamellar thickness proportionally and also the sheets stack closer together. This latter effect suggests that the bulky glutamic acid side chain can influence the crystalline core intersheet stacking distance. This is indeed the case. When the glutamic acid unit is replaced by other amino acid units a linear correlation is discovered between the average intersheet stacking distance and the volume of the amino acid side chain at the fold surface. Thus, fine tuning of the crystal architecture is possible by appropriate choice of amino acid sequence.

X-ray diffraction analysis of poly(±AG)$_3$EG chain-folded lamellar crystals showed the apβ-sheets were apolar, with both surfaces decorated equally with alanyl and glycyl units, so that again the γ-turn was the preferred fold conformation.

The results show that the study of the three-dimensional architectures from sequence-designed polypeptides is proving to be a successful methodology for testing and discovering the rules that govern the general folding behaviour of proteins. The relative ease by which amino acid sequence changes can be performed makes it now possible to test ideas and concepts regarding protein folding.

Acknowledgements

This work emanates from a collaborative research venture between the Physics Department, University of Bristol and the Department of Polymer Science & Engineering at the University of Massachusetts at Amherst. The structural investigation described in this review was a team effort and relies heavily on the contributions of my colleagues Dr Sharon Cooper & Mr Pawel Sikorski at Bristol, and Prof. David Tirrell, Drs Eric Cantor, Mark Krejchi, Alyssa Panitch & Ajay Parkhe at Amherst as described in [7-11]. Professors M. J. Fournier and T. L. Mason at the Biochemistry and Molecular Biology Department, Amherst were responsible for the polypeptide biosynthesis using genetic engineering. The work at Bristol was supported by the Engineering & Physical Sciences Research Council.

References

1. A. Keller, A. *Kolliod-Z u Z Polymere* **231** (1969) 386.
2. Keller, A. *Kolliod-Z u Z Polymere* **197** (1964) 98.
3. Bassett, D. C. *Phil. Mag.* **12** (1965) 119.
4. Atkins, E. D. T.; Keller, A.; Sadler, D. M. *J. Polym. Sci. Pt. A-2* **10** (1972) 863.
5. Organ, S. J.; Keller, A. *J. Mater. Sci.* **20** (1985) 1571.
6. Geddes, A. J.; Parker, K. D.; Atkins, E. D. T.; Beighton, E. *J. Molec. Biol.* ***1968***, *32*, 343.
7. Krejchi, M. T.; Atkins, E. D. T.; Waddon, A. J.; Fournier, M. J.; Mason, T. L.; Tirrell, D. A. *Science*. **1994**, *265*, 1427.
8. Krejchi, M. T.; Cooper, S. J.; Deguchi, Y.; Atkins, E. D. T.; Fournier, M. J.; Mason, T. L.; Tirrell, D. A. *Macromolecules* **1997**, *30*, 5012.
9. Panitch, A.; Matsuki, K.; Cantor, E. J.; Cooper, S.J.; Atkins, E. D. T.; Fournier, M. J.; Mason, T. L.; Tirrell, D. A. *Macromolecules* **1997**, *30*, 42.
10. Cantor, E. J.; Atkins, E. D. T.; Cooper, S. J.; Fournier, M. J.; Mason, T. L.; Tirrell, D. A. *J. Biochem.* **1997**, *122*, 217.
11. Parkhe, A. D; Cooper, S. J.; Atkins, E. D. T.; Fournier, M. J.; Mason, T. L.; Tirrell, D. A. *Int. J. Biol. Macromolecules* **1998**, *23*, 251.
12. Keith, H. D.; Paddon, F.J.; Giannoni, G. *J. Molec. Biol.* **1969**, *43*, 423.
13. Keith, H. D.; Giannoni, G.; Paddon, F.J. *J. Biopolymers* **1969**, *7*, 775.
14. Lotz, B.; Keith, H. D. *J. Molec. Biol.* **1971**, *61*, 195.
15. Lotz, B.; Keith, H. D. *J. Molec. Biol.* **1971**, *61*, 201.
16. Fraser, R. D. B.; MacRae, T. P.; Stewart, F. H. C. *J. Molec. Biol.* **1966**, *19*, 508.
17. Fraser, R. D. B.; MacRae, T. P. *Conformation in Fibrous Proteins*, Academic Press, New York, **1973**.
18. Fraser, R. D. B.; MacRae, T. P.; Stewart, F. H. C.; Suzuki, E. *J. Molec. Biol.* **1965**, *191*, 706.
19. Geil, P. H. *Polymer Single Crystals*, Interscience, New York, **1963**.
20. Keller, A. *Reports Prog. Phys.* **1968**, *31*, 624.
21. Atkins, E. D. T.; Hill, M. J.; Hong, S. K.; Keller, A.; Organ, S. J. *Macromolecules* **1992**, *25*, 917.
22. Bellinger, M. A.; Waddon, A. J.; Atkins, E. D. T.; Macknight, W. J. *Macromolecules* **1994**, *27*, 2130.
23. Nemethy, G.; Printz, M. P. *Macromolecules* **1972**, *5*, 755.
24. Blundell, T.L. private communication.
25. Marsh, R. E. Corey, R. E.; Pauling, L. *Biochim.Biophys. Acta* **1955**, *16*, 1.
26. Richards, F. M. *Annu. Rev. Biophys. Bioeng.* **1977**, *6*, 151.

DESIGN OF SELF-ASSEMBLING PEPTIDES AS CATALYST MIMETICS USING SYNTHETIC COMBINATORIAL LIBRARIES

S.E. Blondelle[1], E. Crooks[1], N. Reixach[1], and E. Pérez-Payá[1,2]

[1]*Torrey Pines Institute for Molecular Studies, 3550 General Atomics Ct, San Diego, CA 92121, USA,* [2]*Departament de Bioquímica i Biologia Molecular, Universitat de València, E-46100 Burjassot València, Spain*

1. Introduction

The *de novo* design of native protein-like structures may be envisioned as a two step process: the design of helical proteins that fold into the desired aggregational state and approximate tertiary structure, followed by a refinement of the hydrophobic core to achieve a well packed interior that adopts a unique rotameric state [1-2]. The introduction of even small modifications within the hydrophobic core of proteins usually affects their folding ability as well as packing stability [3-5]. However, a number of studies involving proteins with repacked cores indicate that such modified proteins can retain their biological function and that the nature of protein interiors may not be the sole features that contribute to conformational specificity [6-7]. A strategy for the design of novel functional proteins may therefore consist in designing a parent protein-like structure having a simplified hydrophobic core, and then optimizing the core structure by mutation/assay-based selection approaches. Combinatorial mutagenesis has been introduced to generate protein mutants with several amino acid replacements that exhibit similar or greater biological functions and/or structural features relative to the original protein sequence [8-10], as well as to engineer native protein-like structures [11-13]. Combinatorial chemical approaches have also now been applied to engineer artificial small proteins [14-17].

The design of artificial catalysts or enzymes represents an area of substantial

A. Aggeli et al. (eds.), Self-assembling Peptide Systems in Biology, Medicine and Engineering, 35–45.

emerging interest. We have therefore focused our efforts toward the design and construct of a uniquely folded tertiary structure, followed by the elaboration of this framework with appropriate catalytic groups. Our strategy uses self-assembling peptides for designing protein-like structures, and combinatorial chemistry approaches first to optimize the structural properties of our framework, and then its catalytic functions. We selected the decarboxylation of oxaloacetate as a catalytic model system based on the report by Benner and coworkers of a peptide that self-assembles into a four-α helix bundle and exhibits modest catalytic activity for this reaction [18].

2. Secondary Structure Optimization

2.1 LIBRARY DESIGN

Based on earlier studies on structure-activity relationships using basic polypeptides [19-20], an 18-mer peptide composed of leucine and lysine residues (YKLLKKLLKKLKKLLKKL-NH_2 - termed "YLK") was selected as the protein framework to build a mixture-based synthetic combinatorial library (SCL [16]). "YLK" folds from a random coil in benign buffer into an α-helix in lipid-like environments, and was found to be capable of self-aggregation, although lacking the specific interactions required to stabilize a unique conformation [19-20]. "YLK" aggregated forms possess hydrophobic cores mimicking the intrinsic characteristics of protein interiors. An initial SCL was built by randomizing four positions located on the hydrophilic face of the "YLK" amphipathic α-helix (i.e., replacing lysines at positions 6, 9, 13, and 16 [16]). This SCL was generated in the positional scanning format [21] and was composed of four separate positional SCLs. Each sublibrary was characterized by a single defined position ("O" position) and three mixture positions ("X" positions), and differs from one another by the position of the defined amino acid (TABLE 1A). The library contains a total of over 130,000 (19^4) "YLK" analogs (19 L-amino acids were used at each diversity position).

TABLE 1. "YLK"-based libraries

A	YKLLKOLLXKLKXLLXKL-NH_2	B	YKOLKWXLLKLKAXLEKL-NH_2
	YKLLKXLLOKLKXLLXKL-NH_2		YKXLKWOLLKLKAXLEKL-NH_2
	YKLLKXLLXKLKOLLXKL-NH_2		YKXLKWXLLKLKAOLEKL-NH_2
	YKLLKXLLXKLKXLLOKL-NH_2		

O is defined with one of the 19 naturally occurring L-amino acids (cysteine omitted), X is a close to equimolar mixture of the same 19 L-amino acids.

2.2. IDENTIFICATION OF OPTIMAL SEQUENCES

The "YLK" self-association process was found by circular dichroism (CD) studies to best fit a tetrameric concentration-dependence profile [16]. Based on a tetramer formation and assuming that the helix-coil transition is a totally cooperative two-state transition, the apparent free energy of helix formation (ΔG_f^{app}) of each peptide mixture was determined by measuring their CD spectra in the absence or presence of guanidinium chloride (Gdn.HCl). This resulted in four separate helical propensity scales from which residues were selected to generate individual sequences. These sequences can be separated into three groups: peptides expected to have a high, medium, or low level of α-helicity ("optimal", "mediocre", and "bad" peptides, respectively, TABLE 2). Thus, the "optimal" peptides were generated by incorporating residues having a high helical propensity in each of the four helical propensity scales, the "mediocre" peptides contain two residues with medium helical propensity and two with high helical propensity, either at the N- or C-terminal, and the "bad" peptide has the amino acid with the lowest helical propensity, proline, at three positions and leucine, a residue with medium helical propensity, at position 6. In agreement with the propensity scales, the optimal sequences were fully α-helical in benign buffer, the mediocre peptides were partially α-helical, and the negative control sequence was random (TABLE 2).

TABLE 2. Selected peptide sequences[a]

		% α-helix[b]	[Gdn.HCl]$_{50}$ (M)[c]
	"optimal" peptides		
[WLAE]	YKLLKWLLLKLKALLEKL-NH_2	93	5.2
[EWWM]	YKLLKELLWKLKWLLMKL-NH_2	84	4.9
[EAWR]	YKLLKELLAKLKWLLRKL-NH_2	82	4.4
[EAWD]	YKLLKELLAKLKWLLDKL-NH_2	80	4.7
	"mediocre" peptides		
[WLRI]	YKLLKWLLLKLKRLLIKL-NH_2	71	nd
[NKAE]	YKLLKNLLKKLKALLEKL-NH_2	67	3.4
	"bad" peptide		
[LPPP]	YKLLKLLLPKLKPLLPKL-NH_2	<10	nd

[a]The peptides are referenced by the amino acids at the mutation sites (shown in bold). [b]The CD spectra were recorded at 25°C, in benign buffer (5 mM MOPS, 240 mM NaCl, pH 7.4), at a peptide concentration of 70 μM. The % α-helicity has been calculated as $100[\theta]_{222}/^{max}[\theta]_{222}$. $^{max}[\theta]_{222}$ = -34,000 estimated from ellipticity of peptides with similar lengths [22]. [c]The denaturation studies were carried out in 5 mM MOPS, 240 mM NaCl, pH 7.4.

The stabilities of the peptides to Gdn.HCl denaturation were measured by monitoring the mean residue ellipticity at 222 nm as a function of denaturant concentration. The peptides unfolded cooperatively with varying midpoints ($[Gdn.HCl]_{50}$ - TABLE 2). As anticipated, the optimal sequences were found to be more stable toward chemical denaturation than the mediocre sequences and parent peptide, with [WLAE] showing the highest stability. Interestingly, the near-UV signal in the CD spectra of [EAWR] was lost at a low concentration in denaturant, a condition under which the peptide secondary structure remains intact, while the near UV CD signal of [WLAE] was lost at the same Gdn.HCl concentration as its CD signal in the far UV region. These results indicate the accumulation of an intermediate state at equilibrium at a given Gdn.HCl concentration for [EAWR], but the unfolding process of [WLAE] only involves the fully folded and the fully unfolded states. Furthermore, the thermal denaturation profile for [WLAE], [EAWD], and [EAWR] reflected the occurrence of a transition from a folded state to an unfolded state, with a clearer denaturation transition observed in the presence of 3.6 M Gdn.HCl. The differences in association stability seen between this series of peptides are noteworthy since, upon self-association,

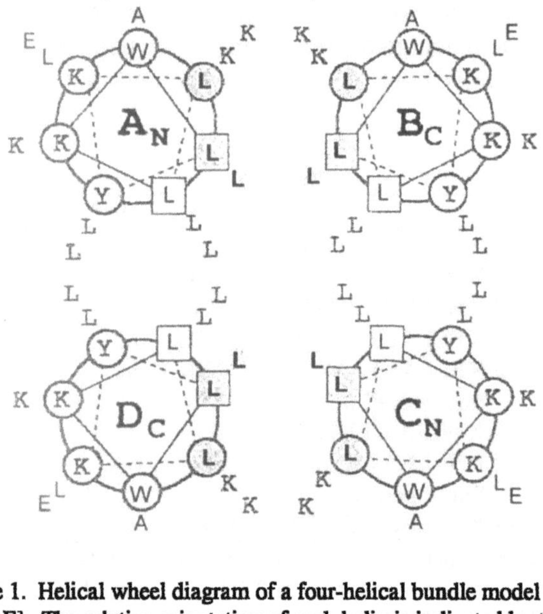

Figure 1. Helical wheel diagram of a four-helical bundle model for [WLAE]. The relative orientation of each helix is indicated by the subindex. The leucine residues corresponding to the diversity positions in the second SCL are shaded.

all peptides have the same hydrophobic core composed solely of leucine residues.

3. Hydrophobic Core Optimization

3.1. LIBRARY DESIGN

Based on the structural studies described above, [WLAE] was selected as a new library scaffold to refine the hydrophobic core and achieve a well

TABLE 3. Structural properties of the mixtures from each sublibrary exhibiting the highest helicity in benign buffer[a]

Defined residue[b]	$[\theta]_{222}$ (deg cm^2 dmol^{-1})	$[\text{Gdn.HCl}]_{50}$ (M)	slope
Position 3			
W	-17,180	3.4	-0.29
L	-14,900	3.2	-0.32
F	-14,110	3.1	-0.42
D	-13,940	2.1	-0.36
Y	-13,880	2.8	-0.34
S	-13,440	2.6	-0.44
Position 7			
L	-20,020	3.8	-0.41
W	-18,740	3.1	-0.29
I	-17,780	3.6	-0.38
F	-17,640	3.5	-0.45
M	-16,700	2.6	-0.45
V	-16,630	2.8	-0.40
Y	-15,990	2.7	-0.50
Position 14			
W	-19,210	4.5	-0.22
L	-19,210	3.9	-0.39
F	-19,150	3.2	-0.42
I	-18,620	3.2	-0.32
M	-18,370	3.1	-0.45
Y	-17,180	2.8	-0.42

[a]The CD and denaturation studies were performed at 25 °C in 5 mM MOPS, 240 mM NaCl, pH 7.4. [b]The residues selected for deconvolution are in bold.

packed protein interior. This second SCL was constructed in a positional scanning format by introducing three diversity positions replacing leucine residues at positions 3, 7, and 14 (TABLE 1B). As shown in Figure 1, if one assumes the folding into a tetrameric aggregate, the diversity positions are located in the middle of the hydrophobic core.

3.2. STRUCTURAL EVALUATION

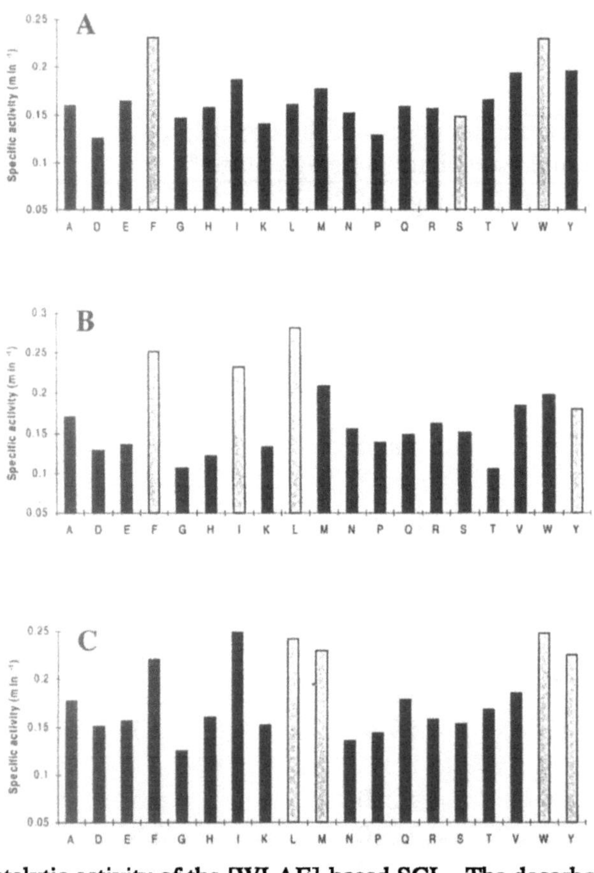

Figure 2. Catalytic activity of the [WLAE]-based SCL. The decarboxylation of oxaloacetate was followed by the loss of absorbance at 280 nm, arising from the enol of oxaloacetate. The specific activity was defined as the change in oxaloacetate concentration as a function of reaction time. Each peptide mixture was screened at 0.2 mM with 11.4 mM oxaloacetate in phosphate buffered saline. Each bar represents the activity of the mixture defined with the amino acid labeled on the x-axis. Each graph represents one sublibrary, i.e., defined residue at (**A**) position 3, (**B**) position 7, and (**C**) position 14. The mixtures selected for deconvolution are shown in grey.

As for the initial SCL, the ability of each peptide mixture to fold into an α-helix in benign buffer as well as their stability toward Gdn.HCl denaturation were determined by CD spectroscopy (TABLE 3). In contrast to the other two sublibraries, little variations in ellipticity at 222 nm were observed when the defined position was replacing leucine-3 (i.e., mixtures from the first sublibrary). Thus, the ranges in mean residue ellipticities $[\theta]_{222}$ were 7,030, 15,176, and 16,007 deg.cm^2.dmol^{-1}, for the sublibraries having position 3, 7, and 14 defined, respectively. These results suggest a small effect of the residues located at the N-terminal region on the aggregate core packing. On the other hand, significant differences in chemical stability were observed between the mixtures within the three sublibraries.

3.3. FUNCTIONAL STUDIES

As mentioned earlier, the last step of the design of artificial functional proteins consists in elaborating the protein framework with appropriate functional groups. The framework [WLAE] was found earlier to exhibit modest catalytic activity for the decarboxylation of oxaloacetate [23]. To validate our combinatorial library-based strategy, each mixture of the [WLAE]-based SCL was assayed for its ability to catalyze the decarboxylation of oxaloacetate [16,18,23]. It is noteworthy that a number of mixtures showed higher activity than the mixture defined with a leucine residue at position 3 (Fig. 2). This suggests the presence of a peptide within the library having higher catalytic activity than the peptide scaffold.

4. Identification of Catalytic Mimetics

Mixtures were then selected based on high ellipticity in benign buffer, high stability toward Gdn.HCl denaturation (i.e., high $[\text{Gdn.HCl}]_{50}$ values) and cooperativity of the denaturation profile reflected by a high absolute value of the slope (TABLE 3), as well as on high specific activity for the decarboxylation of oxaloacetate (Fig. 2). Peptides representing combinations of the selected residues from each sublibrary were then prepared and their structural and functional properties were evaluated.

The folding abilities and stabilities toward chemical and thermal denaturation were determined for each peptide. Only slight increases in both chemical and thermal stabilities were observed for a number of the peptides relative to the peptide scaffold [WLAE]. For instance, the most stable peptide was

YKWLKWLLLKLKAILEKL-NH_2 ($[Gdn.HCl]_{50}$ = 6.1 M, slope = -0.65). However, all of the peptides had a lower helical content as observed by CD spectroscopy. One should keep in mind that this initial deconvolution process was based on single sequence self-assembly. However, the optimal conformation may likely result from inter-molecular assemblies, which may occur when the peptides are present in a mixture making up the library, and be responsible for the conformational behavior seen at the mixture level. The conformational properties of various combinations of the existing peptides at equimolar ratios, as well as a deconvolution by omission strategies, will be explored at a later stage to address this issue.

When screened for catalytic activity in the decarboxylation of oxaloacetate, a 2-fold increase in specific activity was observed for the peptides relative to [WLAE] (TABLE 4). In agreement with the specific activities, the most active peptides [FYM] and [SIY] of this new series showed a Michaelis-Menten saturation kinetic curve corresponding to a k_{obs} value of 7.67 $M^{-1}s^{-1}$ and 3.09 $M^{-1}s^{-1}$, respectively. The catalytic constant of [FYM] is 3-fold greater than the catalytic constant of the peptide having the optimal

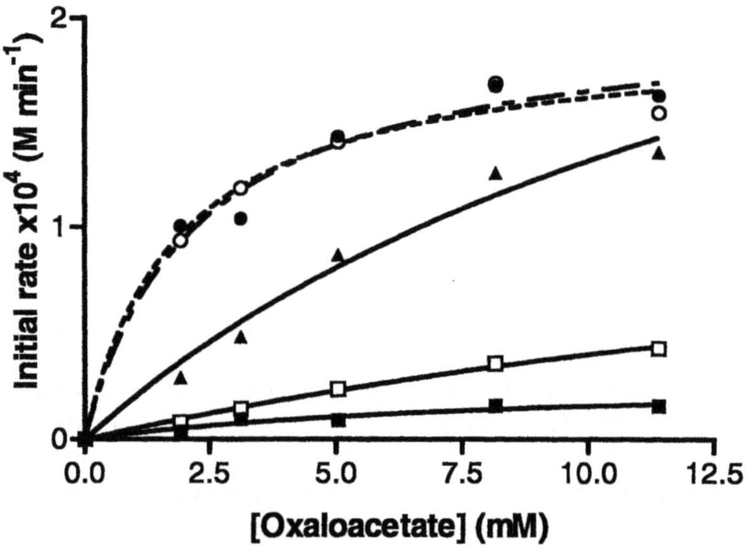

Figure. 3. Michaelis-Menten kinetics. They were performed in PBS buffer at 25°C by measuring the catalytic activity of each peptide at 0.2 mM over a range of substrate concentrations. The initial rates are plotted for [FYM] (- —- ●), [SIY] (- - - ○), [WLAE] (—— ▲), oxaldie-1 (—— □), and spontaneous hydrolysis (—— ■).

hydrophilic face [WLAE], and 10-fold greater relative to the peptide described by Benner et al. [18], oxaldie-1 (Fig. 3). Furthermore, a turn-over rate of about 150 to 200 was found for [SIY] and [FYM], and both peptides were recovered by HPLC and mass spectral analyses following 24h incubation with the substrate. As mentioned above, inter-molecular assemblies may lead to higher catalytic activity, and will be studied in the future.

TABLE 4. Catalytic activities of the designed peptides[a]

		Specific activity (min^{-1})[b]
[FYM]	YKFLKWYLLKLKAMLEKL-NH_2	0.67
[SIY]	YKSLKWILLKLKAYLEKL-NH_2	0.67
[FIY]	YKFLKWILLKLKAYLEKL-NH_2	0.64
[FYY]	YKFLKWYLLKLKAYLEKL-NH_2	0.60
[FIM]	YKFLKWILLKLKAMLEKL-NH_2	0.55
[WLAE]	YKLLKWLLLKLKALLEKL-NH_2	0.42
[EAWR]	YKLLKELLAKLKWLLRKL-NH_2	0.32
[YLK]	YKLLKKLLKKLKKLLKKL-NH_2	0.20
oxaldie-1	LAKLLKALAKLLKK-NH_2	0.13

[a]The peptides are referenced by the amino acids at the mutation sites (shown in bold). [b]The specific activities were determined as described for Figure 2.

5. Conclusion

Peptides that have the potential to form amphipathic α-helical structures in many instances aggregate in a highly cooperative manner in aqueous solution to form a hydrophobic core through strong interaction by the hydrophobic domains, and represent useful model systems for the study of protein folding and stability. For instance, although the packing of amino acid side chains in the interior of proteins contributes to the protein conformational specificity, comparative studies using protein variants containing repacked cores suggested that this feature alone cannot account for conformational specificity [6,7]. In the present studies, using self-aggregating peptides, we have demonstrated that the nature of the residues located on the hydrophilic face of an α-helix may also have a significant effect on the packing properties of α-helical bundles. Indeed, multiple substitutions on the hydrophilic face of an amphipathic peptide, i.e., face exposed to the solvent, was found to result in a clear change in its folding ability, as well as stability. Thus, a productive surface optimization on a simple peptide framework performed by combinatorial chemistry directed by structure selection procedures has resulted in a first series of self-aggregating peptides having enhanced folding ability and stability relative to the original framework. A second optimization by combinatorial approaches allowed the incorporation of functional groups into the core of the self-aggregating peptide framework. Thus, a second series of peptides were designed exhibiting catalytic activities for the decarboxylation of oxaloacetate. These results further support the utility of self-assembling peptides as a protein framework when combined with combinatorial chemistry approaches for the design of artificial functional proteins, such as catalytic mimetics. The optimal peptides [SIY] and [FYM] behaved as true catalysts, which implies a correct substrate recognition and binding, a transition state stabilization, and the final release of reaction products. While 1,000-fold less efficient than the natural enzyme, it is noteworthy that, in contrast to the natural enzyme, the catalytic activities of the identified peptides do not require the presence of a metal ion as a cofactor. Finally, the activities obtained for these catalytic self-assembling peptides are in the same order of magnitude as the activity levels first reported for larger molecules such as catalytic antibodies.

References

1 McGregor, M. J., Islam, S. A., and Sternberg, M. J.: Analysis of the relationship between side-chain conformation and secondary structure in globular proteins. *J. Mol. Biol.* **198** (1987), 295-310.

2 Bryson, J. W., Desjarlais, J. R., Handel, T. M., and DeGrado, W. F.: From coiled coils to small globular proteins: design of a native-like three-helix bundle. *Protein Sci.* **7** (1998), 1404-1414.

3 Eriksson, A. E., Baase, W. A., Zhang, X. J., Heinz, D. W., Blaber, M., Baldwin, E. P., and Matthews, B. W.: Response of a protein structure to cavity-creating mutations and its relation to the hydrophobic effect. *Science* **255** (1992), 178-183.

4 Sandberg, W. S. and Terwilliger, T. C.: Influence of interior packing and hydrophobicity on the stability of a protein. *Science* **245** (1989), 54-57.

5 Bowie, J. U., Reidhaar-Olson, J. F., Lim, W. A., and Sauer, R. T.: Deciphering the message in protein sequences: tolerance to amino acid substitutions. *Science* **247** (1990), 1306-1310.

6 Axe, D. D., Foster, N. W., and Fersht, A. R.: Active barnase variants with completely random hydrophobic cores. *Proc. Natl. Acad. Sci. USA* **93** (1996), 5590-5594.

7 Vetter, I. R., Baase, W. A., Heinz, D. W., Xiong, J. P., Snow, S., and Matthews, B. W.: Protein structural plasticity exemplified by insertion and deletion mutants in T4 lysozyme. *Protein Sci.* **5** (1996), 2399-2415.

8 Zhong, G., Smith, G. P., Berry, J., and Brunham, R. C.: Conformational mimicry of a chlamydial neutralization epitope on filamentous phage. *J. Biol. Chem.* **269** (1994), 24183-24188.

9 Wu, H., Yang, W.-P., and Barbas, C. F., III: Building zinc fingers by selection: Toward a

therapeutic application. *Proc. Natl. Acad. Sci. USA* **92** (1995), 344-348.
10 Gu, H., Kim, D., and Baker, D.: Contrasting roles for symmetrically disposed β-turns in the folding of a small protein. *J. Mol. Biol.* **274** (1997), 588-596.
11 Davidson, A. R. and Sauer, R. T.: Folded proteins occur frequently in libraries of random amino acid sequences. *Proc. Natl. Acad. Sci. USA* **91** (1994), 2146-2150.
12 Doi, N. and Yanagawa, H.: Screening of conformationally constrained random polypeptide libraries displayed on a protein scaffold. *Cell. Mol. Life Sci.* **54** (1998), 394-404.
13 Rojas, N. R. L., Kamtekar, S., Simons, C. T., Mclean, J. E., Vogel, K. M., Spiro, T. G., Farid, R. S., and Hecht, M. H.: De novo heme proteins from designed combinatorial libraries. *Protein Sci.* **6** (1997), 2512-2524.
14 Blondelle, S. E., Takahashi, E., Houghten, R. A., and Pérez-Payá, E.: Design of conformationally defined combinatorial libraries. In *Peptides 94: Proceedings of the 23rd European Peptide Symposium*; Maia, H. L. S., Ed.; ESCOM: Leiden, 1995, pp. 85-86.
15 Bianchi, E., Folgori, A., Wallace, A., Nicotra, M., Acali, S., Phalipon, A., Barbato, G., Bazzo, R., Cortese, R., Felici, F., and Pessi, A.: A conformationally homogeneous combinatorial peptide library. *J. Mol. Biol.* **247** (1995), 154-160.
16 Pérez-Payá, E., Houghten, R. A., and Blondelle, S. E.: Functionalized protein-like structures from conformationally defined synthetic combinatorial libraries. *J. Biol. Chem.* **271** (1996), 4120-4126.
17 Suzuki, N. and Fujii, I.: Design and characterization o intramolecular antiparallel coiled coils as a scaffold for conformationally defined synthetic combinatorial libraries. In *Peptide Chemistry 1996*; Kitada, C., Ed.; Protein Research Foundation: Osaka, 1997, pp.437-440.
18 Johnsson, K., Allemann, R. K., Widmer, H., and Benner, S. A.: Synthesis, structure and activity of artificial, rationally designed catalytic polypeptides. *Nature* **365** (1993), 530-532.
19 Blondelle, S. E. and Houghten, R. A.: Design of Model peptides having potent antimicrobial activities. *Biochemistry* **31** (1992), 12688-12694.
20 Blondelle, S. E., Ostresh, J. M., Houghten, R. A., and Pérez-Payá, E.: Induced conformational states of amphipathic peptides in aqueous/lipid environments. *Biophys. J.* **68** (1995), 351-359.
21 Pinilla, C., Appel, J. R., Blanc, P., and Houghten, R. A. Rapid identification of high affinity peptide ligands using positional scanning synthetic peptide combinatorial libraries. *Biotechniques* **13** (1992), 901-905.
22 Scholtz, J. M., Qian, H., York, E. J., Stewart, J. M., and Baldwin, R. L.: Parameters of helix-coil transition theory for alanine-based peptides of varying chain lengths in water. *Biopolymers* **31** (1991), 1463-1470.
23 Blondelle, S. E., Takahashi, E., Robles, B. L., Houghten, R. A., and Pérez-Payá, E.: Design of catalytic mimics for the decarboxylation of oxaloacetate using synthetic combinatorial libraries. In *Innovation and Perspectives in Solid Phase Synthesis & Combinatorial Libraries*; Epton, R., Ed.; Mayflower Sci.: Birmingham, UK, 1998, pp.163-166.

THERMODYNAMICS OF PROTEIN-PROTEIN AND PEPTIDE INTERACTIONS

ALAN COOPER
Chemistry Department, Glasgow University
Glasgow G12 8QQ, Scotland, UK

1. Summary

Direct measurement of thermodynamic parameters for association/dissociation of proteins and peptides in solution is now routine using isothermal microcalorimetry (ITC) titration and dilution techniques. Examples are given of measurements involving membrane receptors (colicin/porin), protein subunit assemblies (insulin, pyruvate dehydrogenase), vancomycin-peptide and synthetic receptors, and peptide "tapes". Limited kinetic information may also be obtained from calorimetric experiments. Although measurement of thermal data is relatively straightforward, interpretation and rationalisation of the parameters is frustrated by entropy-enthalpy compensation effects. Large variations in enthalpy (ΔH) and entropy (ΔS) of association frequently compensate to give significantly smaller changes in free energy (ΔG), indeed this familiar and ubiquitous homeostatic effect may be of considerable biological advantage in stabilising self-assembling systems against mutation or environmental changes. Nevertheless, the molecular basis for the phenomenon remains something of a mystery. Partly it arises because of the limited ΔG "window" available experimentally, which implies some large-scale apparent correlation of ΔH and ΔS, but this begs the question why ΔH and ΔS might vary so much in the first place. Temperature dependencies may be rationalised in terms of large heat capacity (ΔC_p) effects, which themselves are a manifestation of the thermodynamics of systems comprising multiple, weak (< kT) interactions. Macromolecular conformational dynamics and "quantum confinement" effects can also contribute.

A. Aggeli et al. (eds.), Self-assembling Peptide Systems in Biology, Medicine and Engineering, 47–64.

2. Introduction

There are various ways in which thermodynamic information about peptide and protein interactions may be obtained experimentally, many of them rather indirect. But recent advances in direct calorimetric techniques now make microcalorimetry the method of choice, both for absolute thermodynamic determinations and for simpler analytical purposes. I will concentrate here on the simple, direct microcalorimetric methods now routinely available, and will illustrate them using some examples of peptide-peptide and protein-protein interactions taken from current work in my group and with collaborators. Calorimetry is particularly appropriate here, since 1999 marks the 200th anniversary of the death of Joseph Black, who spent much of his early life in Glasgow devising the first calorimetric methods for studying chemical and physical processes. Black was particularly interested in the phase transitions of alcohol/water mixtures - what we now understand as cooperative hydrogen-bonded lattices; indeed... *"He waited with impatience for the winter"* in Glasgow so that he could pursue his experiments on freezing and melting that led to the concept of latent heat of fusion [1]. This gives a satisfying historical relevance to our current work here.

3. Thermodynamics and Microcalorimetry

3.1 Background Thermodynamics

In summary, thermodynamic equilibrium is usually expressed in terms of the change in Gibbs free energy, $\Delta G = \Delta H - T.\Delta S$; which represents the trade-off between two normally opposing factors: the need to maximise molecular attractions or bonds, as measured by the enthalpy change (ΔH), offset by the equally important entropic term (ΔS) representing thermal disruption of bonds or Brownian motion at the molecular level. The Gibbs free energy change for any process under standard conditions (ΔG^0) is just another way of writing the equilibrium constant (K) for the process, using the standard relationship $\Delta G^0 = -RT.\ln K$; and both entropy and entropy terms are just different integral functions of the more fundamental quantity, the heat capacity of the system or, for our purposes here, the change in heat capacity (ΔC_p) during any process:

$$\Delta H = \int_0^T \Delta C_p \, .dT \quad + \quad \Delta H(0)$$

$$\Delta S = \int_0^T (\Delta C_p / T).dT$$

where ΔH(0) is the enthalpy change for the process at 0K. These expressions show why calorimetry, involving the direct measurement of enthalpy and heat capacity changes, is the preferred technique for establishing absolute thermodynamic data.

3.2 DSC - The "Melting" and Aggregation of a Protein

Differential scanning calorimetry (DSC) measures directly the heat capacity and enthalpy changes during temperature changes and is now particularly suited to the study of temperature-induced changes in dilute solution [2,3]. Figure 1 shows DSC data for the thermal unfolding of a globular protein in aqueous solution.under near-physiological conditions. This is typical of the behaviour

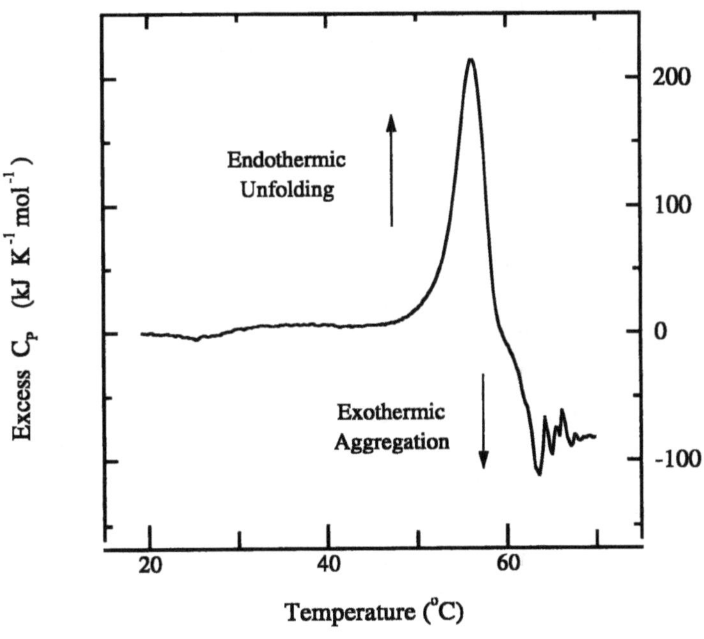

Figure 1: Typical DSC data for the thermal unfolding of a globular protein in aqueous solution

seen for many proteins. The initial uptake of heat (endotherm) corresponds to the enthalpy of unfolding (or "latent heat" of folding) of the protein as it undergoes a cooperative transition to a less-ordered, unfolded state. The finite size of the protein molecule means that the transition, even though fully cooperative, is relatively broad - a fundamental characteristic of phase transitions in mesoscopic systems [2,4], but otherwise the transition is similar to what one might expect for the melting of an ordered system, including the increase in heat capacity baseline (positive ΔC_p) following the transition. However, unfolded protein is sticky stuff, and exothermic aggregation frequently follows, and is usually irreversible. (This unfolded protein aggregate has been shown in various laboratories, including our own, to often show characteristics found also in amyloid-like systems.)

3.3 Isothermal Titration Microcalorimetry (ITC)

More direct, and generally less ambiguous experimental data on intermolecular interactions may be obtained using isothermal titration calorimetry (ITC) methods, which involve monitoring the absolute heat energy (enthalpy) changes resulting from sequential additions of one component to the other [3]. Basically two kinds of experiment are feasible here. Most commonly, simple mixing or addition experiments are used to study binding of one species to another. More recently, we have developed dilution techniques that allow us to measure the dissociation of pre-formed complexes. Examples of both approaches are given here.

Specific peptide-peptide interactions, such as the non-covalent interaction between the antibiotic vancomycin and bacterial cell-wall peptide mimics (Figure 2), can be characterized by straightforward titration experiments [3,5], and such methods have been used recently to investigate potential new antibiotics and synthetic receptors produced by combinatorial chemistry techniques [6]. Such experiments can also be used to probe similar ligand binding systems: enzyme-substrate, enzyme-inhibitor, antibody-antigen, drug-receptor, protein-protein and protein-DNA complexes, for example. This illustrates the power of ITC not only for absolute thermodynamic measurements but also as a non-invasive general-purpose analytical technique.

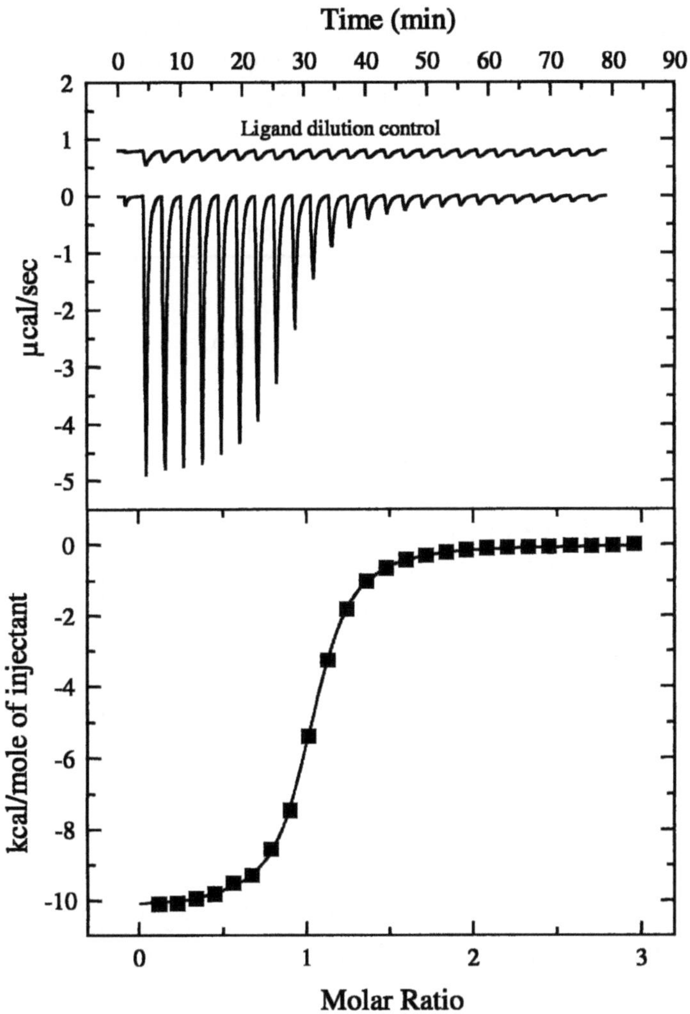

Figure 2: Typical ITC data illustrating peptide-peptide interactions. Sequential 10 µl additions of diacetyl-Lys-D-Ala-D-Ala (1.68mM) into vancomycin (0.12 mM, 1.4 ml ITC cell volume), pH 7, 25 °C. Each heat pulse (upper panel) corresponds to the exothermic binding of peptide ligand to the antibiotic. The integrated heat pulses (lower panel), corrected for control heats of dilution and concentration normalized, fit to a standard differential binding curve with n = 0.99 binding sites, enthalpy of binding, $\Delta H = -10.24$ kcal mol^{-1}, and association constant $K = 5.5 \times 10^5$ M^{-1}

Concentration dependence of ITC binding data sometimes indicates the presence of additional aggregation effects in binding systems, but this can be more easily measured in some instances by calorimetric dilution experiments.

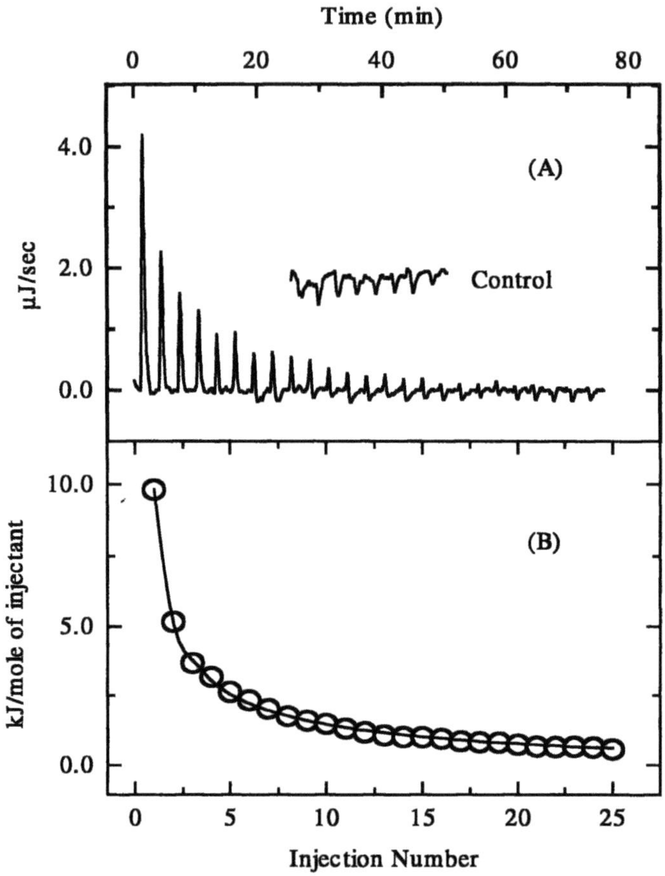

Figure 3: Typical ITC dilution data for the dissociation of insulin dimers in solution. (A) Endothermic heat pulses for 10 μl injections of insulin solution into buffer. (B) Integrated heat data fit to a dimer dissociation model with $K_{diss} = 12$ μM and $\Delta H_{diss} = 41$ kJ mol^{-1} . (See [7] for details.)

Injections of a small volume of dimeric or oligomeric peptide or protein into a larger volume of buffer (insulin, for example - Figure 3) results in dissociation of a fraction of the oligomers, usually with an endothermic heat uptake corresponding to the disruption of the specific protein-protein interface.

Subsequent injections give progressively smaller heat pulses as the protein concentration gradually builds up in the ITC cell, and further dissociation is suppressed. The shapes of these ITC dilution curves (Fig. 2B) can be fit to standard dimerization or polymerization models, and appropriate thermodynamic parameters derived. This method has been particularly effective in investigating the effects of cyclodextrins - cyclic oligosaccharides with a weak affinity for non-polar groups and amino acid residues - on the dimer dissociation of insulin, as indicated in Figure 4 [7]. Such techniques may well be effective in controlling or directing self assembly in assorted peptide systems.

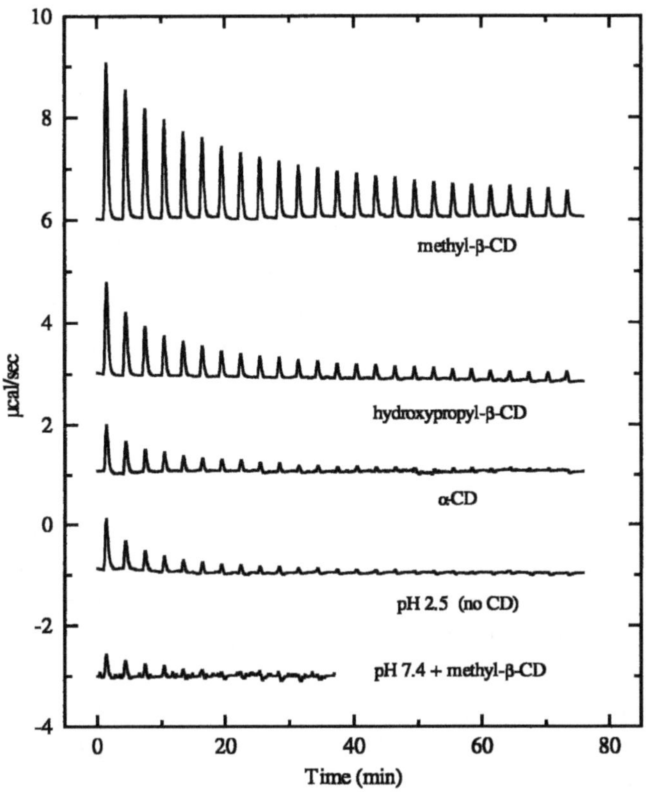

Figure 4: Effects of cyclodextrins on insulin dimer dissociation (see [7] for details).

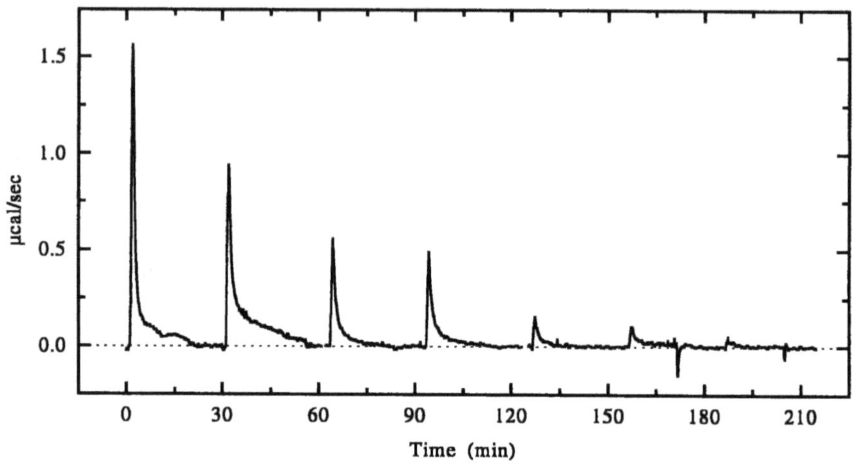

Figure 5: Example of calorimetric dilution experiments showing endothermic dissociation of synthetic peptide "tapes" (SOMS - Leeds) [DN1] = 550 μM in water. Sequential 50μl injections (dilutions) into water 50 seconds per injection, 25-30 minutes between injections

Of particular relevance to the main topic here, preliminary calorimetric dilution experiments on the dissociation of self-assembled peptide tapes are illustrated in Figure 5. Again dilution results in endothermic heat pulses but, in contrast to the insulin and related studies described above, where dissociation is essentially instantaneous on the calorimetric timescale, the endothermic effect continues for 30 minutes or more - presumably reflecting the very slow dissociation/re-arrangement kinetics in these systems. Some, but not all these kinetic effects can also be followed by other techniques such as circular dichroism (CD) and fluorescence, and the comparison between the different techniques will be instructive. Here calorimetry is used to give not only thermodynamic information about the nature of the interactions stabilising the peptide assemblies, but also unexpected kinetic data as well.

More complex biomolecular assemblies may also be examined by calorimetric titration methods you yield thermodynamic data for interactions at larger protein interfaces. Experiments involving membrane receptors, based on the colicin-porin system [8,9] illustrated in Figure 6, yield data for a range of mutants (see Table 1) which - at least in principle - should allow dissection of the various contributions to the interactions. However, interpretation is severely hampered by entropy-enthalpy correlation effects (see below).

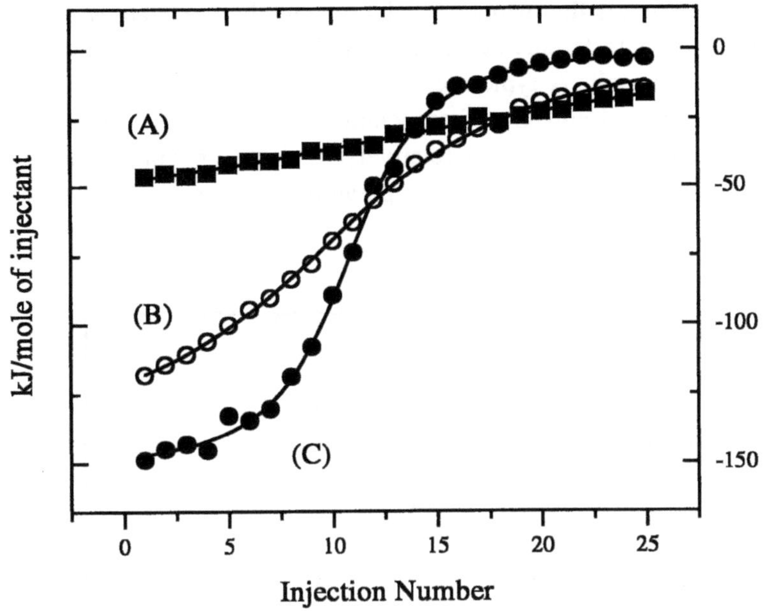

Figure 6: Examples of calorimetric titration isotherms for binding of different colicin N constructs to TolA-II-III (porin) receptor domains. Note the similar binding affinities, but widely different enthalpies (see [8,9] for details).

TABLE 1. Summary of thermodynamic data for binding of colicin constructs to assorted membrane receptor (TolA) components [9]

TolA + colicin fragments and domain mutants	K_d μM	ΔH^o kJ mol^{-1}	ΔG^o kJ mol^{-1}	ΔS^o J K^{-1} mol^{-1}
TolAII-III + colicin N	18.2	-64.5	-27.1	-125.6
TolAII-III + TR domain	8.1	-98.3	-29.1	-232.2
TolAII-III + T domain	1.4	-110.5	-33.4	-258.6
TolAIII + T domain	1.0	-103.0	-34.2	-230.6
T(H57A,S58A) + TolAII-III	9.8	-162.6	-28.6	-449.5
T(D59A) + TolAII-III	49.7	-85.7	-24.6	-205.1
T(G60A) + TolII-III	18.4	-115.7	-27.0	-297.4
T(S61A) + TolAII-III	3.2	-94.2	-31.4	-210.8
T(Δ26-30) + TolAII-III	1.0	-127.8	-34.2	-313.8

Equally disconcerting is the manner in which some thermodynamic parameters can vary radically with temperature, in a way that makes interpretation at the molecular level difficult. A typical example is shown in Figure 7, where the measured enthalpy and entropy of association of a particular protein-protein complex even change sign over a relatively modest temperature range, with association being endothermic at low temperature and exothermic at higher T.

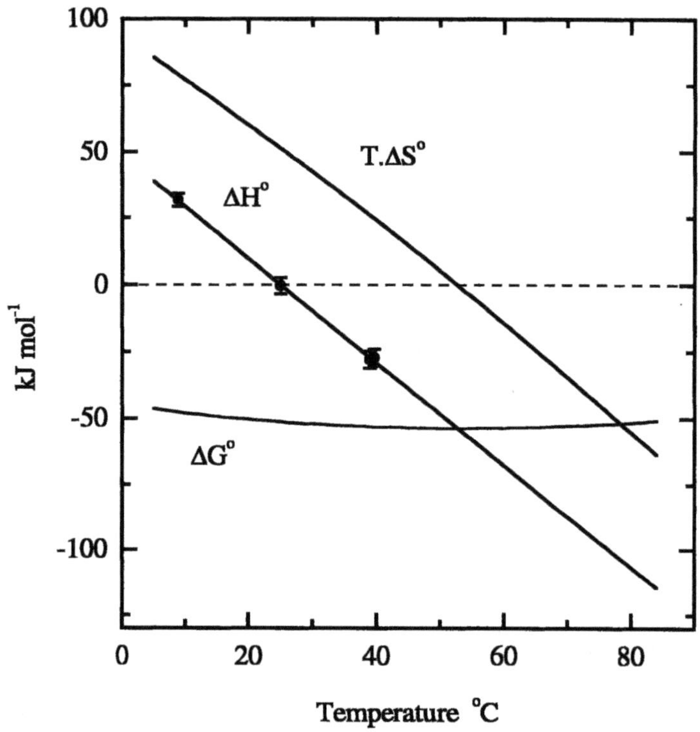

Figure 7: Entropy-enthalpy compensation in the pyruvate dehydrogenase multi-enzyme complex. E3:di-domain interaction [10].

This example, taken from ongoing work on subunit interactions in the pyruvate dehydrogenase multi-enzyme complex of a thermophillic organism [10], is typical of the sort of behaviour increasingly observed, especially in protein and peptide complexes involving a multiplicity of weak non-covalent interactions [4,11].

4. Enthalpy-Entropy Compensation

As shown in some of the examples above, and as also seen in many other systems involving non-covalent interactions, wide variations in absolute values of enthalpy (ΔH) and entropy (ΔS) of molecular association - sometimes involving even changes in sign - frequently compensate to give relatively much smaller variations in free energy (ΔG). This ΔH/ΔS compensation effect has been remarked upon many times in the past, and numerous explanations have been proposed. I will just concentrate here on more recent observations.

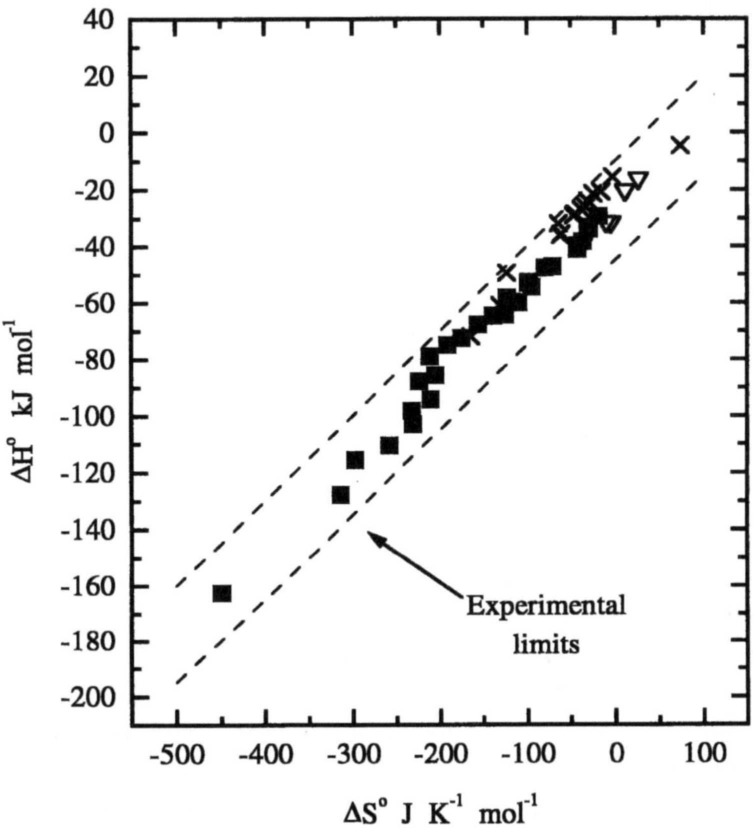

Figure 8: Combined enthalpy *versus* entropy plot for a range of apparently unrelated and chemically distinct systems, including vancomycin-peptides, insulin dimerization, colicin-porin and protein-protein subunit interactions. All data are taken from collaborative ITC studies in the Glasgow group.

4.1 Free Energy Windows

One remarkable feature of enthalpy-entropy correlation is that it appears to be an almost universal property of weakly interacting systems, with correlations apparent even between chemically distinct and totally unrelated species. For example, Figure 8 compares the experimental ΔH and ΔS values for an arbitrary selection of association reactions, measured over several years in Glasgow, using different ITC instruments, under different conditions, with peptides and proteins of differing sizes and structures. The linear correlation between ΔH and ΔS is clear, and the slope of the line (a "characteristic temperature") is close to 300 K. This is puzzling, since at first sight there should be no reason why such distinct systems should be so correlated. Even the colicin-porin data from Table 1, which form part of Figure 8, include interactions between protein fragments and mutants of various sizes, which should not necessarily bear any direct relationship to each other.

But, upon consideration, this correlation between enthalpies and entropies of association in diverse systems is merely a reflection of the limited Gibbs free energy window accessible for both experimental and functional reasons. Consider the following. Elementary thermodynamic expressions for Gibbs free energy,

$$\Delta G^\circ = -RT.\ln K = \Delta H^\circ - T.\Delta S^\circ$$

rearrange to give:

$$\Delta H^\circ = \Delta G^\circ + T.\Delta S^\circ$$

So, for <u>fixed</u> ΔG° , enthalpy/entropy plots <u>will</u> be linear with slope = T, where T is the temperature of the measurement, regardless of the system. Now, of course ΔG^0 is not necessarily fixed but will normally be confined to a relatively small range of values. For example, for instrumental reasons the range of K values normally accessible by ITC measurements is roughly in the range $K = 10^2 - 10^8\ M^{-1}$. This corresponds to an observable ΔG° range (or "window") of around = -10 to -45 kJ mol^{-1}. Other experimental methods will have similar limitations. Moreover, for functional and dynamic flexibility in biomolecular interactions, K values in most systems will be confined to a similar region. Consequently, although not necessarily constant, the range of observable ΔG° values is experimentally limited and may give rise to apparently linear plots. So if, for whatever reason, ΔH and/or T.ΔS are large compared to ΔG, then plots of ΔH versus ΔS may appear linear, with slope ≈ T. This is what is illustrated in Figure 8, where the dashed lines indicate the limits imposed by the free energy window estimated above.

4.2 Temperature and Heat Capacity (ΔC_p) Effects

The large temperature dependencies of ΔH and ΔS, such as seen in Figure 7, are simple thermodynamic consequences of a significant heat capacity change, or ΔC_p effect, associated with the interaction. Regardless of the origin of this ΔC_p, simple algebraic considerations show how this can lead to entropy-enthalpy compensation effects with change in temperature [5]. Recall that the variation of ΔH and ΔS with respect to some arbitrary reference temperature T_{ref} is given by the integral relationships:

$$\Delta H(T) = \Delta H(T_{ref}) + \int_{Tref}^{T} \Delta C_p .dT$$

and

$$\Delta S(T) = \Delta S(T_{ref}) + \int_{Tref}^{T} (\Delta C_p /T).dT$$

If ΔC_p is constant, independent of temperature (not necessarily true, but usually a reasonable approximation over a limited temperature range), then integration gives:

$$\Delta H(T) = \Delta H(T_{ref}) + \Delta C_p .(T - T_{ref})$$

$$\Delta S(T) = \Delta S(T_{ref}) + \Delta C_p .\ln(T/T_{ref})$$

For small changes in temperature with respect to absolute T_{ref}, $\delta T = T - T_{ref}$, these become:

$$\Delta H(T) = \Delta H(T_{ref}) + \Delta C_p .\delta T$$

$$\Delta S(T) = \Delta S(T_{ref}) + \Delta C_p .\ln(1 + \delta T/T_{ref}) \approx \Delta S(T_{ref}) + \Delta C_p .\delta T/T_{ref}$$

using the approximation $\ln(1 + x) \approx x$, for $x<<1$. Consequently, to the extent that this approximation is valid:

$$\Delta G(T) = \Delta H - T.\Delta S \approx \Delta H(T_{ref}) - T.\Delta S(T_{ref}) - \Delta C_p .\delta T^2/ T_{ref} = \Delta G(T_{ref})$$

to first order in δT. Moreover, over a limited temperature range for which this approximation is valid:

$$\Delta H(T) \approx \Delta H(T_{ref}) + T_{ref}.(\Delta S - \Delta S(T_{ref}))$$

so that a plot of ΔH versus ΔS would appear linear with slope T_{ref}.

4.3 "Thermodynamic Homeostasis"

Regardless of the molecular basis for entropy-enthalpy compensation, there is one interesting thermodynamic consequence that may be of some biological significance. This is illustrated in Figure 9, comparing the observed equilibrium constant for protein-protein complexation (K_{ass}), as a function of temperature, with what might be the behaviour in the absence of the ΔC_p-induced compensation effect. As we have seen, the large temperature dependencies of

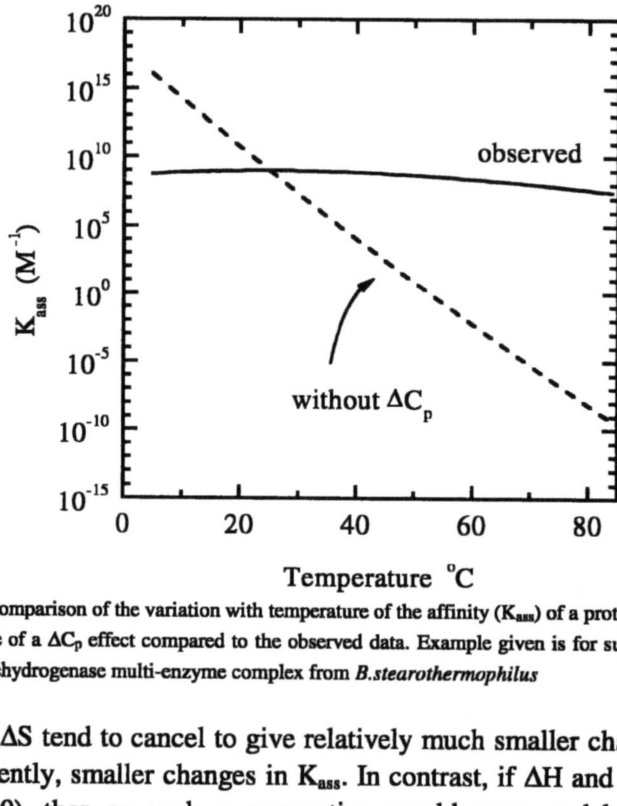

Figure 9: Comparison of the variation with temperature of the affinity (K_{ass}) of a protein-protein interaction in the absence of a ΔC_p effect compared to the observed data. Example given is for subunit interactions in the pyruvate dehydrogenase multi-enzyme complex from *B.stearothermophilus*

ΔH and ΔS tend to cancel to give relatively much smaller changes in ΔG^0 and, consequently, smaller changes in K_{ass}. In contrast, if ΔH and ΔS were constant (ΔC_p = 0), then no such compensation would occur and both ΔG^0 and K_{ass} would vary much more rapidly with T. Bearing in mind that biological systems depend on a very delicate balance of interactions, and that the extent of such interaction is determined by K and ΔG, rather than by ΔH and ΔS, this "thermodynamic homeostasis" would allow a developing or evolving organism to withstand much greater fluctuations in environmental conditions than might otherwise be the case. Such homeostasis might also be a useful design feature in synthetic self-assembling systems.

4.4 Hydrogen-bonded Networks

Large changes in heat capacity (ΔC_p) have long been regarded as the characteristic thermodynamic signature of hydrophobic interactions. But quite similar effects arise quite generally in order-disorder transitions in homogeneous systems, particularly those comprising hydrogen-bonded networks (Figure 10), and this may have significance for our understanding of

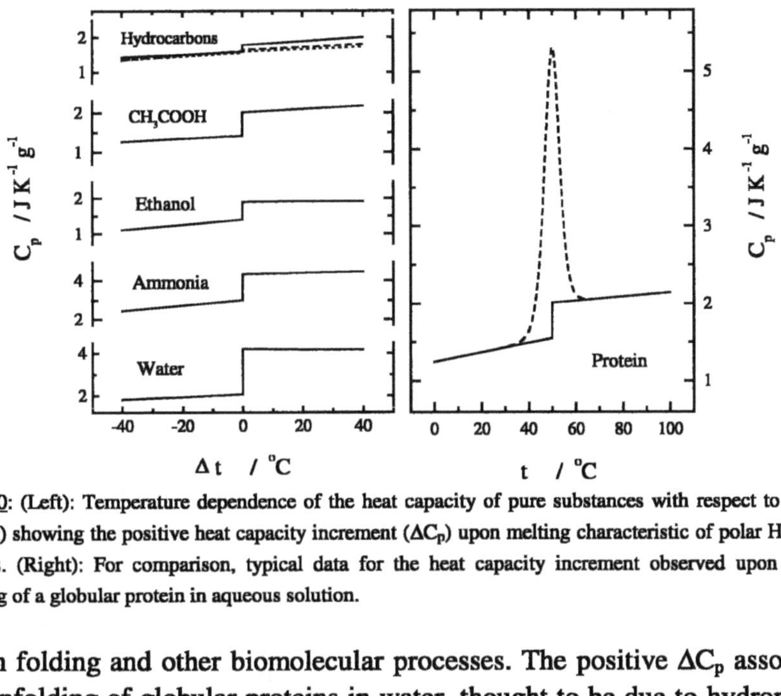

Figure 10: (Left): Temperature dependence of the heat capacity of pure substances with respect to melting point (Δt) showing the positive heat capacity increment (ΔC_p) upon melting characteristic of polar H-bonded networks. (Right): For comparison, typical data for the heat capacity increment observed upon thermal unfolding of a globular protein in aqueous solution.

protein folding and other biomolecular processes. The positive ΔC_p associated with unfolding of globular proteins in water, thought to be due to hydrophobic interactions, is also typical of the values found for the melting of crystalline solids, where the effect is greatest for the melting of polar compounds, including pure water. This suggests an alternative model of protein folding and peptide assembly based on the thermodynamics of phase transitions in hydrogen-bonded networks. Folded globular proteins or peptide assemblies may be viewed as cooperatively ordered hydrogen-bonded networks floating in an aqueous environment of less-well-ordered H-bonds, in which the degree of hydrogen bonding decreases with increasing temperature. The enthalpy of melting of the protein consequently increases with temperature. A simple statistical model [Cooper, in preparation], based on the overall number of protein and solvent hydrogen bonds in folded and unfolded states, shows how

ΔC_p from this source could match the hydrophobic contribution. The basic elements of this hypothesis are sketched below.

As a consequence of cooperativity, all (weak) peptide-peptide (P-P) hydrogen bond or other interactions contacts are intact in the complex:

Not all such interactions are replaced by water H-bonds when the complex dissociates or unfolds:

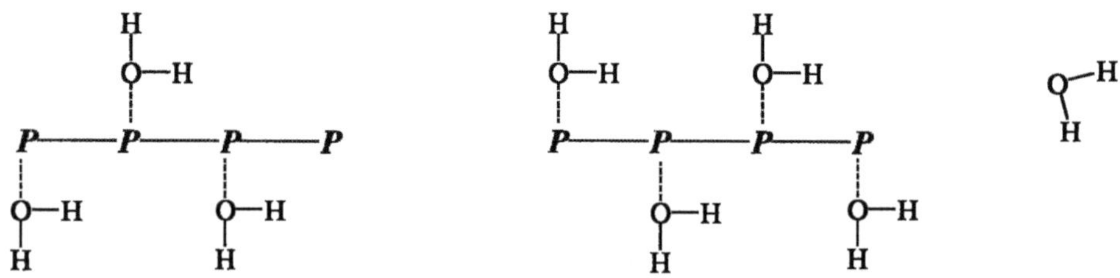

At higher temperatures there will be fewer water-peptide hydrogen bonds made to the dissociated or unfolded complex, so the enthalpy of unfolding or dissociation will be more positive:

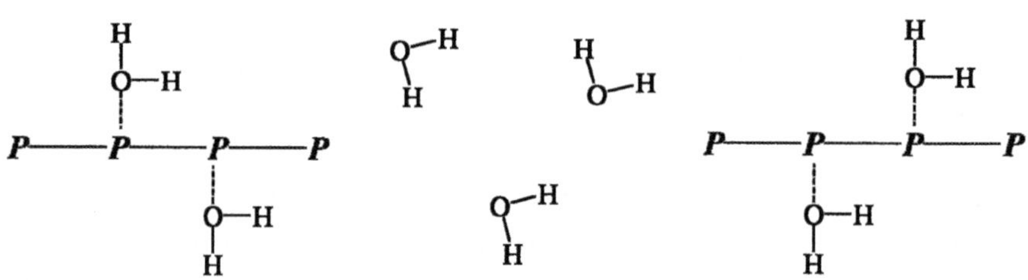

Since the overall number of H-bonds broken will increase with temperature, this will contribute to the increase in ΔH with temperature and to the positive ΔC_p . This confirms the growing view that the thermodynamics of protein folding, and other interactions in aqueous systems, is best described in terms of a mixture of polar and non-polar effects in which no one contribution is necessarily dominant.

4.5 Quantum Confinement

Any molecular association or assembly process involves the confinement or conformational restriction of atoms or groups, and this can give rise to unexpected effects at the quantum mechanical level [Cooper, in preparation]. Quantum effects of molecular confinement on thermodynamic parameters in weakly interacting systems may be summarised as follows:

- Over and above any other interactions, simple confinement of any particle (molecule) is an endothermic process. This is a quantum effect for which there is no macroscopic interpretation: it is simply a consequence of "zero point energy", arising from the Heisenberg Uncertainty Principle.

- The effects are normally negligible (in macroscopic or covalent systems), but become comparable to kT (i.e. comparable to non-covalent interactions) when freedom of movement at the molecular level drops below about 0.5 Å

- There is a trade-off between tightness of fit (favourable) and zero point energy (unfavourable).

- $\Delta H_{transfer}$(large cavity → smaller cavity) depends on temperature, i.e. it shows a ΔC_p effect.

- For "loosely-fitting" cavities, ΔC_p is small (and slightly positive). For "tightly-fitting" cavities, ΔC_p becomes negative.

- In the limit, for really tight fit (≈ 0.1 Å), $\Delta C_p \rightarrow -3R/2 \approx -12.5$ J K^{-1} mol^{-1} (per "degree of freedom")

This is comparable to experimental values such as: ΔC_p (protein folding) approximately -50 J K^{-1} mol^{-1} per residue (each with several degrees of freedom), or the ΔC_p associated with protein-protein interaction, burial of non-polar surface area, of around -1.34 J K^{-1} mol^{-1} / Å^2 ≡ -13 J K^{-1} mol^{-1} per water molecule Consequently it may be important to consider quantum confinement effects when designing self-assembling and other related systems.

5. Acknowledgements

The Biological Microcalorimetry Facility in Glasgow is funded jointly by BBSRC and EPSRC. Some of the experimental data presented here arise from collaborative projects with Dr. Jeremy Lakey (Newcastle), Professor N. Boden (SOMS, Leeds), Professor Richard Perham (Cambridge), and members of their groups.

6. References

1. Simpson, A.D.C. (1982). *"Joseph Black 1728-1799: a commemorative symposium"* (Royal Scottish Museum, Edinburgh, 1982); further historical details may be viewed at: http://www.chem.gla.ac.uk/~alanc/dept/black.htm

2. Cooper, A. (1999). *Thermodynamics of protein folding and stability*. "Protein: A Comprehensive Treatise", Volume 2, pp. 217-270. (Editor: Geoffrey Allen. JAI Press Inc., Stamford CT, 1999)

3. Cooper, A. (1997). *Microcalorimetry of protein-protein interactions*. Methods in Molecular Biology Vol.88: Protein Targeting Protocols, pp.11-22 (ed. Roger A. Clegg, Humana Press 1997).

4. Cooper, A. (1999) *Thermodynamic analysis of biomolecular interactions*. Curr.Opinion Chem.Biol. 3, 557-563.

5. McPhail, D. & Cooper, A. (1997). *Thermodynamics and Kinetics of Dissociation of Ligand-Induced Dimers of Vancomycin Antibiotics*. J.Chem.Soc. Faraday Trans. 93, 2283-2289.

6. Xu, R., Greiveldinger, G., Marenus, L.E., Cooper, A. & Ellman, J.A. (1999). *Combinatorial library approach for the identification of synthetic receptors targeting vancomycin resistant bacteria.* J.Am.Chem.Soc. 121, 4898-4899.

7. Lovatt, M., Cooper, A. & Camilleri, P. (1996). *Energetics of cyclodextrin-induced dissociation of insulin.* Eur.Biophys.J. 24, 354-357.

8. Evans, L.J.A., Cooper, A. & Lakey, J.H. (1996). *Direct measurement of the association of a protein with a family of membrane receptors*. J.Mol.Biol. 255, 559-563.

9. Raggett, E.M., Bainbridge, G., Evans, L.J.A., Cooper, A. & Lakey, J.H. (1998). *Discovery of critical tol A-binding residues in the bacteriocidal toxin colicin N: a biophysical approach.* Molecular Microbiology, 28, 1335-1343.

10. Jung, H.I., Bowden, S.J., Cooper, A, Kalia, Y.N. & Perham, R.N. (in preparation). *Entropy-driven protein-protein interactions in the assembly of the pyruvate dehydrogenase multienzyme complex from Bacillus stearothermophilus*

11. Dunitz, J.D. (1995) *Win some, lose some: Enthalpy-entropy compensation in weak intermolecular interactions.* Chemistry & Biology 2, 709-712.

THE MECHANISM OF AMYLOID FORMATION AND ITS LINKS TO HUMAN DISEASE AND BIOLOGICAL EVOLUTION

CHRISTOPHER M. DOBSON
Oxford Centre for Molecular Sciences, University of Oxford, New Chemistry Laboratory, South Parks Road, Oxford OX1 3QT, UK.

We have in our bodies approaching 100,000 different types of protein. These are engaged in promoting or controlling virtually every event on which our lives depend, and a vast machinery of translation and transcription has developed within cells for their manufacture. In recent years we have become familiar with the intricate and elegant forms that these molecules adopt in their functional native states as a result of the successful application of all the weapons in the structural biologist's arsenal. Moreover, we have begun to learn a great deal about the manner in which these structures are attained through the complex process of protein folding whether this takes place in the laboratory or within the natural environment of the living cell [1-3]. It is also increasingly clear that biological systems have evolved elaborate procedures to ensure that proteins fold correctly, or, if they do not, that they are discovered and degraded before any serious harm can ensue to the host organism [4].

Protein misfolding is linked to disease

Despite these controls a range of debilitating human diseases is associated with protein misfolding events that result in the malfunctioning of the cellular machinery [5]. Cystic fibrosis is one example, where mutations in the gene encoding a crucial transport protein result in its failure to fold correctly, and hence to be secreted in the quantity required for proper function. Other diseases, including some types of familial emphysema, result from mutations that result in improper trafficking of proteins to the sites where they are needed. Most attention recently has, however, been focused on a group of diseases where proteins or fragments of proteins convert from their normally soluble forms to insoluble fibrils or plaques which accumulate in a variety of organs including the liver, spleen and brain [6-9], Figure 1. The final forms of these aggregates often have a well-defined fibrillar nature, known as amyloid, giving rise to the use of the term amyloidosis to describe many of the clinical conditions with which they are associated.

This group of diseases, of which nearly twenty have been described, includes Alzheimer's and Parkinson's diseases, the spongiform encephalopathies such as Creutzfeldt-Jakob disease, type II diabetes, and a range of less well-known but often equally serious conditions such as fatal familial insomnia [6]. These diseases can be sporadic, inherited or even infectious, and are often manifest only late in life. Each disease is associated with a particular protein, and aggregates of these proteins are

A. Aggeli et al. (eds.), Self-assembling Peptide Systems in Biology, Medicine and Engineering, 65–74.

thought to be directly or indirectly the origin of the pathological conditions associated with the disease in question. In some cases the quantity of material involved is enormous, with several kilograms of protein being deposited in some manifestations of systemic amyloidosis. Remarkably, despite the range of proteins involved in these diseases, which include several well-known proteins such as lysozyme, transthyretin and the prions, all of which have unique and characteristic native folds, the fibrils in which they are found in the disease states are remarkably similar in their overall appearance [11].

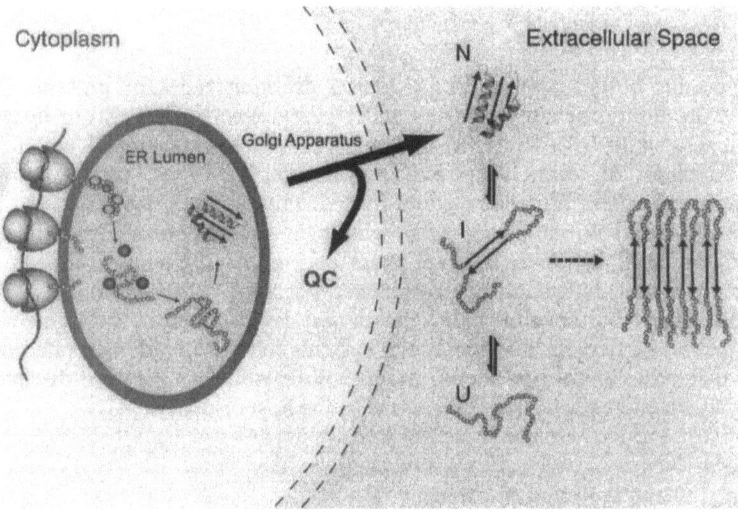

Figure 1. Schematic representation of the possible mechanism of formation of amyloid fibrils by a globular protein. After synthesis on the ribosome it is assumed that the protein folds in the endoplasmic reticulum, aided by molecular chaperones that deter aggregation of incompletely folded species. The correctly folded protein is secreted in from the cell, and functions normally in its extracellular environment. Under some conditions the protein unfolds, at least partially, and becomes aggregation prone. This can result in the formation of fibrils and other aggregates that accumulate in tissue. It is likely that small aggregates, as well as the highly organised fibrils and plaques, can give rise to pathological conditions in at least some cases. N, I and U refer to native, partially folded (intermediate) and unfolded states of the protein. QC refers to the quality control mechanism that prevents incompletely folded proteins being secreted from the endoplasmic reticulum (ER) [10]

Soluble proteins convert into aggregates under denaturing conditions

Studies of the mechanism of the conversion of the normally soluble proteins into amyloid fibrils have been aided considerably because, in many cases, the structural transitions of the disease-associated molecules can be reproduced under laboratory conditions. In order to achieve this, a common procedure has been to expose the folded proteins to mildly denaturing conditions, such as low pH or elevated temperatures. Those transitions that have been studied in detail have generally revealed the existence of intermediates prior to the formation of well-defined aggregates, and some have shown a clear kinetic lag phase before rapid development of fibrils occurs. This behaviour has analogies to the more familiar process of crystal growth or to polymer gelation events, and is thought to be associated with the need to develop "nuclei" or "seeds", small aggregates from which larger molecular assemblies can grow [12]. Indeed, the ability to enhance growth of fibrils by seeding is a possible mechanism for the rapid progression of many of these diseases following onset, and indeed for the infectivity of the spongiform encephalopathies associated with the prion proteins [13, 14].

Despite the association of the prions with the spongiform encephalopathies, no recombinant prion protein has yet been shown to be able to induce infectivity in animals and hence provide the ultimate proof of the "protein only" hypothesis of the prion diseases [15, 16]. Moreover, production of the aggregated ("scrapie") form from the soluble ("cellular") form of the prion protein has proved to be extremely difficult, although some conversion to protease resistant material can be achieved by addition of extracts from infective tissue [17]. The formation of fibrillar material with the characteristic morphology of amyloid has, however, recently been reported from a large fragment of the human prion protein [18]. This was achieved by exposing the soluble form of the protein to rather extreme conditions, including reduction of the single disulphide bond that stabilises the native state. Interestingly, this conversion was associated with the presence of a soluble intermediate having predominantly β-structure, rather than the largely helical nature of the native protein. Amyloid fibrils have long been known to be rich in β-structure, giving rise to the well known α–helix to β-sheet transition characterising the conversion of cellular to scrapie forms of the prion protein, and indeed of the conversion to aggregates of those other proteins whose native folds are largely helical [11, 19].

Amyloid is a generic structural form of proteins

The involvement of only a handful of proteins in amyloid diseases has commonly been thought to be associated with some particular conformational character of the protein sequences involved. But recently compelling evidence has accumulated that the ability to form amyloid is not a peculiarity of this small group of proteins. A clue to this came fortuitously from an NMR study of the SH3 module of PI3 kinase which has no connection with any known disease [20]. At low pH, where the protein is at least partially unfolded, the solution was found to turn into a viscous gel after several hours. When this gel was examined by electron microscopy and other techniques it was found to contain well defined fibrils with all the characteristics of those found associated with the amyloid diseases (Figure 2). Similar fibrils have been observed to form when a domain of fibronectin was heated to temperatures above that required for unfolding [21] and indeed are now being reported for an increasing range of other proteins and fragments of proteins.

A recent study set out to test the proposition that amyloid could be a common state of proteins, by designing solution conditions to see if fibrils could be deliberately formed from a soluble protein not associated with any disease [22]. The method selected was to add trifluoroethanol, a solvent known to denature proteins but, unlike traditional denaturants such as urea, one which generates partially unfolded states where hydrogen bonds between peptide groups are still stable. After several days under the chosen conditions the protein involved, acylphosphatase, duly formed amyloid fibrils with all the characteristics of those associated with disease. Interestingly, there are quite a number of observations in the early literature of fibrous gels being found under a variety of non-physiological conditions for a range of different proteins [23]. Although the relationship of these to the fibrils found in disease was not recognised at the time, these findings reinforce the idea that the ability to form amyloid could be a generic property of polypeptide chains. This ability can readily be rationalised by the fact that the intermolecular bonds that stabilise this material involve the peptide backbone, which is common to all proteins.

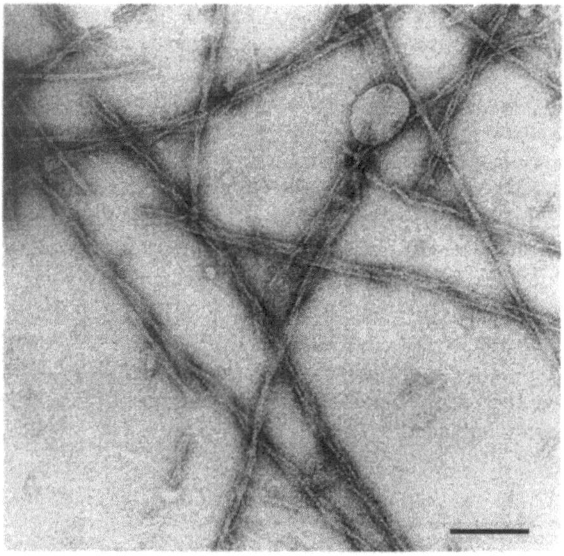

Figure 2. Electron micrograph of fibrils formed from an SH3 domain by incubation of a solution containing the protein at low pH [12]. Under these solution conditions the protein is partially unfolded and slowly aggregates to form a gel which contains the fibrils. Fibrils associated with the various amyloid diseases have a closely similar appearance to these fibrils formed under laboratory conditions [6]. The scale bar is 100 nm. Reproduced from Ref. 18. [Note: the term "amyloid" was originally used to describe the fibrils because their appearance was considered to be like that of starch (amylose). The traditional test for amyloid in tissue involves the observation of a red shift in the light absorption of the dye Congo red, and of a characteristic green birefringence under polarized light. Both these effects are attributable to the interaction of the dye molecule with the regularly ordered protein chains. The detection of the characteristic "cross-β" structure in X-ray diffraction and typical morphology in the EM micrographs is now generally considered essential if aggregates are to be described as amyloid. The term "prion" refers to a "proteinaceous infective particle" and is usually restricted to a protein associated with a phenomenon transmissible from one organism to another.

The degree of regularity of the fibrils grown under carefully controlled laboratory conditions can be greater than that of those commonly observed *in vivo*. Indeed, careful growth of the SH3 fibrils over a period of weeks resulted in material sufficiently well ordered to be studied in detail by cryo-electron microscopy [24]. Image reconstruction techniques devised by Saibil and her colleagues show clearly that the fibrils consisted of protofilaments, containing specific segments of β-sheet, wound around each other in a helical array. The resolution of these studies is not yet sufficient to reveal directly the structure in molecular detail. X-ray fibre diffraction studies, however, show that the polypeptide chains adopt a "cross-β" structure in which hydrogen bonds are formed between polypeptide chains in directions parallel to the fibre axis [11]. Using this information, a model was built to suggest how the polypeptide chains are assembled in the protofilaments (Figure 3). This arrangement of the polypeptide chain, albeit no doubt with variations on a theme for different sequences, may represent a first glimpse of the common structure of this alternative highly organised form of protein molecules.

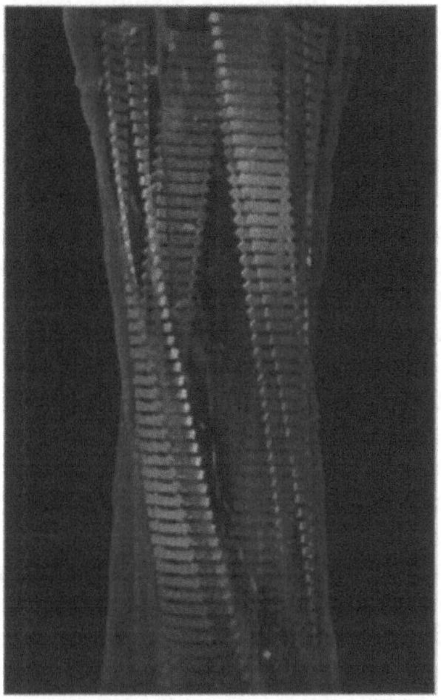

Figure 3. Molecular model of an amyloid fibril derived from cryo-EM analysis of fibrils grown from an SH3 domain. The fibril consists of four 'protofilaments' that twist around one another to form a hollow tube of diameter approximately 60 A [24]. The model shown here represents one way in which the regions of the polypeptide chain involved in β-sheet structure could be assembled within the fibrils. Figure courtesy of Helen Saibil.

Living systems avoid forming amyloid

If amyloid fibril formation is a generic property of polypeptide chains, why is its occurrence in biology restricted to a very small number of proteins? What prevents the rest of our proteins forming this material in our bodies? And why are aggregates normally seen only when the usual amino-acid sequence of the disease-related proteins is altered by mutation, or following infection, or in old age? These questions are particularly pertinent because, once formed, amyloid fibrils are essentially indestructible under physiological conditions, probably because of the large number of hydrogen bonds that must be disrupted to rescue the polypeptide chain from the aggregated state. Undoubtedly some amino acid sequences are more prone to aggregation than others, and some proteins are present *in vivo* at very much higher concentrations than others. But the fundamental answers to these questions must be that biology has found a way to avoid the formation of this material under normal physiological conditions. Many factors must be involved in this protective mechanism, but the selection of sequences during evolution that can fold efficiently to a globular form in which the polypeptide chain and the hydrophobic residues are sequestered in the interior is likely to be particularly important [22]. Thus, except when the protein is exposed to denaturing conditions, the peptide backbone is not accessible to form the interchain hydrogen bonds associated with amyloid fibrils (Figure 3).

One of the particularly interesting characteristics of proteins is that under normal conditions their native states are only marginally stable relative to denatured states. How then can the population of unfolded or partially folded species in a biological environment be low enough to avoid aggregation? The answer is likely to be associated with the cooperativity of the protein folding process [25]. The "two-state" nature of protein denaturation that is commonly observed means that no partially folded intermediate states are significantly populated, and indeed that all segments of the polypeptide chain are firmly locked into the close packed structure. Cooperativity of this type has long been associated with the existence of well-defined native proteins capable of carrying out their diverse functions, and of resisting degradation by proteolysis, but cooperativity may be equally crucial for the avoidance of aggregation. An additional and very important factor must be that the cellular environment has evolved in such a way that denaturation of proteins does not normally occur under conditions where the tendency of the unfolded chains to aggregate cannot be suppressed. In particular, pH and temperature are often carefully controlled, and molecular chaperones and degradation mechanisms are present to deal with the majority of aggregation-prone species.

The generic nature of amyloid formation therefore suggests that the existence of the diseases associated with this material results, at least in part, from the breakdown of the normal control and regulation mechanisms within a living organism. In support of this conclusion, conditions such as the low pH environments in endosomes associated with protein translocation or with lysosomes associated with protein degradation, have been implicated in these diseases [8]. Indeed, it has even been suggested that the prion proteins might experience environments in endosomes capable of reduction of the disulphide bond found to enhance the conversion to the fibrillar state [18]. Moreover, at least the majority of the mutations associated with the familial amyloid-related diseases are destabilising, and many give rise to intermediates which are indicative of the loss of the cooperativity associated with the wild-type protein [7]. Aggregates in Alzheimer's disease are associated with peptide fragments of a precursor protein that might result from the degradation of misfolded protein within the cellular environment [14]. Such peptides do not have a stable globular fold to protect them against aggregation.

New insights into evolutionary biology?

The conventional view of protein structures is that they have emerged under the pressure of the development of greater efficiency and new functionalities. Properties such as the cooperativity of their structures ensure that a single, well-defined structure exists in solution to allow efficient binding to other molecules and effective functional behaviour, for example catalysis of very specific chemical reactions. It is interesting to speculate that avoidance of aggregation, particularly to highly insoluble amyloid fibrils, is an equally important driving force in the evolutionary design of natural proteins. This is particularly significant in the context of the extraordinarily crowded nature of the cellular environment [26]. Interestingly, several proteins have been found to fold at rates close to the limits imposed by the physical principles that apply to all molecules [27]. This fits in with our preconceived notions that biology strives for efficiency, in the same way that at least some enzymes appear to have reached perfection in terms of the catalytic efficiency [28]. The ability of proteins to fold rapidly is not only an effective way of generating biological activity quickly, but is could be, in at least some cases, an evolutionary advantage in minimising the competing intermolecular processes of aggregation. It is interesting in this regard that the globular region of the prion protein has been found to fold rapidly in the absence of pathogenic mutations [29].

It appears, therefore, that biology has generally succeeded, through the coupled evolution of protein sequences and the environments in which they fold and function, in avoiding the formation of amyloid fibrils by the molecules selected to carry out the myriad functions of a living cell. But has the amyloid form of protein molecules evolved to be of any functional benefit in living systems? In general this seems unlikely, because of the intractable nature of the fibrils and the difficulty in controlling their growth once it is initiated. It is probable, however, that the major pathogenic species in at least some diseases are not the fibrils themselves but smaller aggregates that are precursors to their formation. If this is the case, then the formation of fibrils could in fact be a means of sequestering such species in a form in which they are less harmful (8). But recent findings suggest that the conversion of soluble proteins to fibrillar structures could be involved in biological processes of a rather different type from those associated with mammalian diseases.

Studies of *Saccharomyces cerevisiae* have resulted in the discovery of proteins which have infective properties related to those of the mammalian prions [30, 31]. These "yeast prions" have been found to convert *in vitro* into ordered aggregates with all the characteristics of amyloid fibrils, and to display classic seeding behaviour [32, 33]. A particularly remarkable aspect of this work, however, is that the ability to convert into the amyloid form can be passed from a parent to its offspring through the process of cell division. The yeast prions have therefore been described as non-chromosomal genetic material as they are the basis for inherited traits within the organism. This concept is revolutionary enough, but recently proteins with similar characteristics have been found in filamentous fungi. In this case, however, there is evidence that these prions might be involved in the mediation of the normal cellular functions in this organism, and not simply linked to disease [30, 31] If this is correct, it is one of the more extraordinary examples of a more general emerging concept that the control of folding events is intimately linked with the control of biological activity.

Opportunities for the Future

The discovery of mammalian prions has led to the hypothesis that infection can occur without the need for bacteria or viruses. The properties of the yeast and fungal prions now suggest that transfer of genetic information is possible without nucleic acids. This neatly turns the tables on RNA, where the discovery of ribozymes took the uniqueness of biological catalysis away from proteins [34]. It also perhaps gives food for thought about how information transfer could have taken place in early life forms in much the same way that the properties of ribozymes stimulated ideas about the control of chemical reactions in primitive organisms [35]. A better understanding of amyloid fibrils and the way they form may result in improved knowledge of the pathological conditions that lead to many of the diseases that are emerging in the aging population of the modern world, and of opportunities for their prevention and treatment. As well as assisting the development of pharmacological interventions designed to inhibit the nucleation and growth of pathological forms of proteins within an organism [36], there are prospects of therapeutic strategies based on the redesign of natural proteins to reduce their propensity to aggregate [37].

Research into the nature of amyloid fibrils and the mechanism of their formation is also likely to stimulate new ideas about the significance of the sequences that are emerging from the various genome projects. An important driving force in the evolution of protein sequences must be to maintain or improve their functional characteristics, but it must also be to ensure that they do not undergo inappropriate binding interactions in the complex and crowded milieu of the cell. Clearly the existence of molecular chaperones represents one important mechanism to avoid aggregation phenomena, the intrinsic properties of the sequences themselves must be the primary mechanism by which proteins are able to remain soluble in their biological environment. We have begun a systematic investigation using site-directed mutagenesis to explore the importance of individual residues of proteins in maintaining their resistance to aggregation [38]. As well of being of general significance in improving our understanding of protein structures, such studies may lead to an improved ability to design or redesign proteins, and to a deeper understanding of the fundamental driving forces behind biological evolution.

Finally, as in many areas of science, we undoubtedly have much to learn from biology. In the context of this article this fact is of particular interest in the discovery and development of novel materials and devices for future technological application. The amyloid structures formed from proteins are clearly natural and highly organised 'nanostructures' whose compositions are almost infinitely variable. The fibrils produced from SH3, for example, as illustrated in Figures 2 and 3, are tubular structures, and other forms of amyloid structure resemble the ribbons and tapes formed by the assemble of carefully designed peptide sequences [39]; and articles in this volume). Such materials may well turn out to have a wide range of novel and potentially important properties [40]. It would be an interesting twist of fate if the fibrillar structures that prove so deadly when their form in our bodies turn out to be of great value in the development of novel devices designed to make our lives more pleasant.

Acknowledgements

This article is a slightly modified version of a review originally published in Trends in Biochemical Sciences, 24, 329-332 (1999), and is reproduced with the generous permission of Elsevier Trends Journals, Cambridge UK. I acknowledge valuable discussions on this article with John Ellis and Carol Robinson. I am grateful to Jose Jimenez and Helen Saibil of Birkbeck College, London, for providing Figure 3, and to

Adam Rostom for producing Figure 1. This is a contribution from the Oxford Centre for Molecular Sciences which is funded by the BBSRC, EPSRC and MRC. The research of CMD is also supported by the Howard Hughes Medical Institute and the Wellcome Trust.

References

1. Fersht, A. R. (1999) *Structure and Mechanism in Protein Science* (W.H. Freeman, New York).
2. Dobson, C. M. & Ptits, O. B. (1999) The biological consequences of physical principles, *Current Opinion in Structural Biology* **9,** 89-91.
3. Radford, S. E. & Dobson, C. M. (1999) From computer simulations to human disease: Emerging themes in protein folding, *Cell* **97,** 291-298.
4. Suzuki, C. K., Rep, M., vanDijl, J. M., Suda, K., Grivell, L. A. & Schatz, G. (1997) ATP-dependent proteases that also chaperone protein biogenesis, *Trends in Biochemical Sciences* **22,** 118-123.
5. Thomas, P. J., Qu, B. H. & Pedersen, P. L. (1995) Defective protein-folding as a basis of human-disease, *Trends in Biochemical Sciences* **20,** 456-459.
6. Tan, S. Y. & Pepys, M. B. (1994) Amyloidosis, *Histopathology* **25,** 403-414.
7. Kelly, J. W. (1998) The alternative conformations of amyloidogenic proteins and their multi-step assembly pathways, *Current Opinion in Structural Biology* **8,** 101-106.
8. Lansbury, P. T. (1999) Evolution of amyloid: What normal protein folding may tell us about fibrillogenesis and disease, *Proceedings of the National Academy of Sciences of the United States of America* **96,** 3342-3344.
9. Perutz, M. F. (1999) Glutamine repeats and neurodegenerative diseases: molecular aspects, *Trends in Biochemical Sciences* **24,** 58-63.
10. Trombetta, E. S. & Helenius, A. (1998) Lectins as chaperones in glycoprotein folding, *Current Opinion in Structural Biology* **8,** 587-592.
11. Sunde, M. & Blake, C. (1997) The structure of amyloid fibrils by electron microscopy and X-ray diffraction, *Advances in Protein Chemistry* **50,** 123-159.
12. Mullins, J. (1993) *Crystallization* (Butterworth-Heineman, Oxford).
13. Horwich, A. L. & Weissman, J. S. (1997) Deadly conformations - Protein misfolding in prion disease, *Cell* **89,** 499-510.
14. Harper, J. D. & Lansbury, P. T. (1997) Models of amyloid seeding in Alzheimer's disease and scrapie: Mechanistic truths and physiological consequences of the time-dependent solubility of amyloid proteins, *Annual Review of Biochemistry* **66,** 385-407.
15. Prusiner, S. B. (1997) Prion diseases and the BSE crisis, *Science* **278,** 245-251.
16. Weissmann, C. (1999) Molecular Genetics of Transmissible Spongiform Encephalopathies, *Journal of Biological Chemistry* **274,** 3-6.
17. Caughey, B. & Chesebro, B. (1997) Prion protein and the transmissible spongiform encephalopathies, *Trends in Cell Biology* **7,** 56-62.
18. Jackson, G. S., Hosszu, L. L. P., Power, A., Hill, A. F., Kenney, J., Saibil, H., Craven, C. J., Waltho, J. P., Clarke, A. R. & Collinge, J. (1999) Reversible conversion of monomeric human prion protein between native and fibrillogenic conformations, *Science* **283,** 1935-1937.
19. Fink, A. L. (1988) Protein aggregation: folding aggregates, inclusion bodies and amyloid, *Fold. Des.* **3,** R9-R23.
20. Guijarro, J. I., Sunde, M., Jones, J. A., Campbell, I. D. & Dobson, C. M. (1998) Amyloid fibril formation by an SH3 domain, *Proceedings of the National Academy of Sciences of the United States of America* **95,** 4224-4228.
21. Litvinovich, S. V., Brew, S. A., Aota, S., Akiyama, S. K., Haudenschild, C. & Ingham, K. C. (1998) Formation of amyloid-like fibrils by self-association of a partially unfolded fibronectin type III module, *Journal of Molecular Biology* **280,** 245-258.
22. Chiti, F., Webster, P., Taddei, N., Clark, A., Stefani, M., Ramponi, G. & Dobson, C. M. (1999) Designing conditions for in vitro formation of amyloid protofilaments and fibrils, *Proceedings of the National Academy of Sciences of the United States of America* **96,** 3590-3594.
23. Clark, A. H., J.udge, F. J., Richards, J. B., Stubbs, J. M. and Suggett, A. (1981) Electron microscopy of network structures in thermally-induced globular protein gels, *Int. J. Pept. Protein Res.* **I17,** 380-392.

24. Jimenez, J. L., Guijarro, J. L., Orlova, E., Zurdo, J., Dobson, C. M., Sunde, M. & Saibil, H. R. (1999) Cryo-electron microscopy structure of an SH3 amyloid fibril and model of the molecular packing, *Embo Journal* **18**, 815-821.
25. Dobson, C. M. & Karplus, M. (1999) The fundamentals of protein folding: bringing together theory and experiment, *Current Opinion in Structural Biology* **9**, 92-101.
26. Luby-Phelps, K. (1994) Physical-properties of cytoplasm, *Current Opinion in Cell Biology* **6**, 3-9.
27. Eaton, W. A., Munoz, V., Thompson, P. A., Henry, E. R. & Hofrichter, J. (1998) Kinetics and dynamics of loops, alpha-helices, beta-hairpins, and fast-folding proteins, *Accounts of Chemical Research* **31**, 745-753.
28. Shakhnovich, E., Abkevich, V. & Ptitsyn, O. (1996) Conserved residues and the mechanism of protein folding, *Nature* **379**, 96-98.
29. Wildegger, G., Liemann, S. & Glockshuber, R. (1999) Extremely rapid folding of the C-terminal domain of the prion protein without kinetic intermediates, *Nature Structural Biology* **6**, 550-553.
30. Lindquist, S. (1997) Mad cows meet psi-chotic yeast: The expansion of the prion hypothesis, *Cell* **89**, 495-498.
31. Wickner, R. B., Edskes, H. K., Maddelein, M. L., Taylor, K. L. & Moriyama, H. (1999) Prions of yeast and fungi - Proteins as genetic material, *Journal of Biological Chemistry* **274**, 555-558.
32. Glover, J. R., Kowal, A. S., Schirmer, E. C., Patino, M. M., Liu, J. J. & Lindquist, S. (1997) Self-seeded fibers formed by Sup35, the protein determinant of [PSI+], a heritable prion-like factor of S-cerevisiae, *Cell* **89**, 811-819.
33. King, C. Y., Tittmann, P., Gross, H., Gebert, R., Aebi, M. & Wuthrich, K. (1997) Prion-inducing domain 2-114 of yeast Sup35 protein transforms in vitro into amyloid-like filaments, *Proceedings of the National Academy of Sciences of the United States of America* **94**, 6618-6622.
34. Cech, T. R. (1986) RNA as an enzyme, *Scientific American* **255**, 64 et seq.
35. Csermely, P. (1997) Proteins, RNAs and chaperones in enzyme evolution: A folding perspective, *Trends in Biochemical Sciences* **22**, 147-149.
36. Baures, P. W., Oza, V. B., Peterson, S. A. & Kelly, J. W. (1999) Synthesis and evaluation of inhibitors of transthyretin amyloid formation based on the non-steroidal anti-inflammatory drug, flufenamic acid, *Bioorganic & Medicinal Chemistry* **7**, 1339-1347.
37. Villegas, V., Zurdo, J., Filimonov, V. V., Aviles, F. X., Dobson, C. M. & Serrano, L. (2000) Protein engineering as a strategy to avoid formation of amyloid fibrils, *Protein Science, in press*.
38. Chiti, F., Capanni, C., Taddei, N., Stefani, M., Ramponi, G. & Dobson, C. M. Specific regions of a protein determine the kinetics of aggregation and amyloid formation, (*submitted for publication)*.
39. Aggeli, A., Bell, M., Boden, N., Keen, J. N., Knowles, P. F., McLeish, T. C. B., Pitkeathly, M. & Radford, S. E. (1997) Responsive gels formed by the spontaneous self-assembly of peptides into polymeric beta-sheet tapes *Nature* **386**, 259-262.
40. MacPhee, C. E. & Dobson, C. M. (2000) Chemical dissection and reassembly of amyloid fibrils formed by a peptide fragment of transthyretin, *Journal of Molecular Biology* **297**, 1203-1215.

TRANSGENIC PLANTS FOR LARGE SCALE PRODUCTION OF PEPTIDES AND PROTEINS

KLAUS DÜRING
MPB Cologne GmbH
Neurather Ring 1
51063 Köln
Germany
Tel. +49 / (0)221 / 64701 – 0
Fax +49 / (0)221 / 64701 – 99
e-mail: k.during@mpb-cologne.com
homepage: www.mpb-cologne.com

MPB Cologne GmbH is a young biotechnology company located in Cologne, Germany, the international center of excellence in transgenic plant science. Its main focus is development of transgenic plants as biofactories for a wide range of economically promising proteins for medical and technical applications. In addition, sophisticated expression of foreign proteins is used to engineer plants resistant to bacterial and fungal pathogens.

Plant storage organs as potato tubers or rape seeds are nature´s ideal biofactories for production of recombinant proteins in large amounts at very low cost, high efficiency and at highest environmental compatibility. Natural sources of energy are employed to convert basic building blocks into proteins of interest, facilitated by the highly elaborated protein biosynthesis machinery implemented in every plant cell.

Small peptides are an important class of proteins. Due to their small size they require special attention in developing biological production systems. Self-assembly peptides possess certain characteristics to be taken as the basis for methodological plant bioreactor development.

A. Aggeli et al. (eds.), Self-assembling Peptide Systems in Biology, Medicine and Engineering, 75–85.

Introduction

The first transformation of tobacco with a marker gene was reported in 1983 [1]. In 1988 Düring first succeeded to express and assemble a full-size monoclonal IgM antibody (comprising light and heavy chains) in transgenic tobacco [2, 3]. Both necessary genes had been introduced in one single *Agrobacterium* – transformation vector. Hiatt et al. [4] followed more than a year later with expression of an IgG antibody in transgenic tobacco using a slightly different method: they introduced the light and the heavy chain genes individually in separate vectors, selected for highly expressing regenerants which subsequently were crossed. In progeny plants assembled IgG antibody could be identified. Today, antibody production in transgenic plants, including recombinant antibodies, is an area of high commercial value.

Already for a longer time proteins can be produced in fermentation based systems, e.g. in bacterial cells, fungi, yeasts, insect cells or mammalian and human cell cultures. Despite the optimal biological presets available in human cells for production of human proteins, there are major drawbacks caused by low efficiency and high fermentation costs in human cell cultures. Transgenic animals and transgenic plants are actually emerging as new potent protein biofactories.

Plant storage organs as potato tubers or canola seeds are nature´s ideal biofactories for production of recombinant proteins in large amounts at very low cost, high efficiency and environmental compatibility. Natural sources of energy are employed to convert basic building blocks into proteins of interest, facilitated by the highly elaborated protein biosynthesis machinery implemented in every plant cell. In comparison to procaryotic and lower eucaryotic cell types animal and plant cells possess a biosynthetic machinery very similar to that of human cells. Therefore, more complex proteins can be produced in the more powerful plant and animal biofactory systems in future.

Advantages

Transgenic plants provide several advantages in comparison to fermentation based systems as well as to production of proteins in the milk of transgenic animals. All systems requiring fermentation need significant investments in technical facilities which are strongly increasing with scale-up. In addition, running costs for fermenters are high and, therefore, economical considerations are of importance during development of fermentation-based production systems for pharmaceutical and industrial proteins to be developed for the market.

A comprehensive comparison of important factors to be considered when choosing an appropriate production system for a specific protein are listed in the table below. Several factors clearly demonstrate the superiority of transgenic plants as biofactories over transgenic animals as direct competitors and fermentation-based systems. More complex proteins which can only be expressed at low rates in simple organisms require more potent eucaryotic cells for production. Transgenic plants evolve to be a highly potent opportunity for these purposes.

TABLE 1. Comparison of the transgenic plant bioreactor with other bioreactor systems

	Transgenic plants	*Transgenic animals*	*Mammalian cell culture*	*Insect cell culture*	*Fungi*	*Yeast*	*Bacteria*
Development time	intermediate	long	intermediate	intermediate	short	short	short
Probability of success and improvement	very high	high	limited	intermediate	limited	limited	limited
Production reproducibility	++	++	+++	+++	+++	+++	+++
Storage	very good	limited	limited	limited	limited	limited	limited
Scale-up: time	very quick	quick	intermediate	intermediate	intermediate	intermediate	intermediate
Scale-up: costs	very low	intermediate	high	high	high	high	high
Production volume	unlimited	unlimited	limited	limited	limited	limited	limited
Biological compatibility	very high	very high	high	high	high	intermediate	limited
Contamination with human pathogens	no	yes	yes	no	no	no	yes
Contamination with toxins	no	no	no	no	yes	no	yes

TABLE 2. Advantages of transgenic plant bioreactors compared to fermentation bioreactors

Plant production	Fermenter production
Subcellular compartment chloroplast does exist only in plants: specific high capacity for protein accumulation	Chloroplasts do exist neither in mammalian nor in bacterial cells
Due to biological conditions probability of success is very high; basically efficient production to be expected for most proteins	Due to biological conditions probability of success is intermediate to limited in all systems (because of diverse reasons); in cultures which can be simply handled in fermenters and are cost-effective to produce (e.g. bacteria) natural limitations exist for the protein production capacity; cultures with higher protein production capacity are more difficult to handle and are cost intensive (e.g. mammalian cultures)
Storage of plant material is technically established under industrial conditions -> prospective production and storage possible	Production in cell cultures -> no storage capability
Protein remains stable in production medium (e.g. tubers, seeds)	Protein has only limited storage stability in production medium
Purification of the protein can be postponed until sale	Purification of the protein must be performed immediately
Scale-up is extemely cost-effective as only seed has to be multiplied	Scale-up is costly because of biological reasons; the system must be newly adapted; technical prerequisites are very cost-intensive
Development time for scale-up is extremely low	Development time for scale-up is much higher
Production scale can be varied extremely easily, seeds can be stored at low cost for longer time periods; technology for long term storage established as well	Production scale must be thoroughly planned; sufficient fermenter capacity must be available
No contamination with human pathogens as these cannot replicate in plants	Dependent on the system contamination with human pathogens or toxins possible
Almost unlimited production volume, especially for large scale production (e.g. for technical applications)	Only limited production volumes

TABLE 3. Advantages of transgenic plant bioreactors compared to animal bioreactors:

Plant	Animal
Plant transformation is a highly efficient and robust routine technique on the basis of naturally occuring gene transfer	Transformation of animal cells is normally artificial and of low efficiency
A single differentiated plant cell (e.g. from a leaf) is competent for development of a fully new plant -> basis for simplicity of the technology	Requirement for complex technology
Hundreds of foreign proteins have been expressed in transgenic plants for research purposes	Development of transgenic animals is still very laborious and time consuming so that routine application is not yet possible in hundreds of labs; consequently much less experience available
Chloroplasts are special subcellular compartments existing only in plants; they have a specific high capacity for protein accumulation	Chloroplasts do not occur in mammalian cells
Development time much shorter than with animals	Development time is the highest of all bioreactor systems
Very high probability of success because of very promising conditions in all respects	Probability of success much lower than with plant transformation due to complexity and time requirements
Protein remains stable in the production medium under storage conditions in tubers or seeds	Protein has only limited storage stability in production medium (mostly milk)
Storage of plant material is technically established under industrial conditions -> prospective production and storage possible	Mostly production in milk -> no storage capacity
Purification of the protein can be postponed until sale	Purification of the protein must be performed immediately after production
Scale-up is a routine technology of plant breeders and growers (seed multiplication, identical progeny); time periods required are extremely low	Multiplication of a transgenic animal to establish a herd by identical offspring is still time consuming and problematic
Scale-up is extemely cost-effective as only seed has to be multiplied	Costs for scale-up are much higher than for seed multiplication
Production scale can be varied extremely easily, seeds can be stored at low cost for longer time periods; technology for long term storage established as well	Production scale must be planned thoroughly as multiplication is time consuming; a herd of transgenic animals must continuously be feeded
No contamination with human pathogens as these are not known to replicate in plants	Because of high degree of relationship almost the same contamination risks for human pathogens (viruses, toxins, etc.) exist as with humans themselves
Very low probability to find in plants proteins that share high homology with human proteins to be produced -> strong simplification of purification schemes, high purity of the produced protein rather easy to achieve	Due to high degree of relationship mammals often possess proteins with high homology to those human proteins to be produced which will be difficult to remove and, therefore, can be problematic minor components in the purified product (e.g. human serum albumin - bovine serum albumin)

By developing a proprietary set of plant bioreactor - type collections for production of the most important types of proteins, initially in potato and oilseed rape (canola) as important agricultural plant species, MPB Cologne GmbH creates an excellent basis for the market. Molecular farming not only opens a new era in production of highly valuable pharmaceutical proteins, but also enables economically feasible mass production of enzymes or new proteins as components of future biodegradable biosynthetic polymers.

Perspectives

Today, transformation of many plant species is already routine. Some important crop species are on the way to the market where the first transgenic products (e.g. cotton, corn, potato) already have been successfully established. For some crop species already more than half of the farmland has been planted with transgenic varieties. Protein expression has become a promising tool in plant biotechnology in recent years. Basic knowledge on efficient protein production has been gathered. MPB Cologne GmbH disposes of e.g. a routine expression system for recombinant scFv antibodies in transgenic potato tubers [5] and is developing a set of production factory systems for a diverse set of different protein types.

TABLE 4: Technological approaches for protein production in transgenic plants

Expression of foreign proteins in transgenic plants:

- in leaves or fruits
- in storage organs (tubers, seeds)

- accumulation in endoplasmic reticulum
- accumulation in extracellular space
- accumulation in chloroplasts
- accumulation as oleosin fusion proteins in oil-rich seeds (in the oil phase)

Optimization of plants for their general protein production capability as well as for specific proteins is under development. Specific promoters, expression and protein purification technologies and other factors will be prominent building blocks to further enhance yield of foreign proteins in transgenic plants.

Applications

Proteins can be used for a wide variety of applications. These include therapeutic as well as diagnostic proteins, biomedical proteins and proteins for technical applications. The power of the plant protein biosynthesis machinery prospects a promising future for transgenic plants as protein biofactories. Smaller amounts of protein can be produced in glasshouses whereas production at ton scale will be performed in the field and might eventually be linked to starch or oil production in some cases.

Specific requirements of e.g. therapeutic proteins have to be addressed appropriately. As outlined under „Advantages“ transgenic plants provide a plethora of advantages compared with transgenic animals which will make them a highly useful tool, also for therapeutic proteins, in future. First experiences have been gained and will quickly be extended. Regulatory guidelines for the use of recombinant proteins for therapeutic purposes are under development in the US with FDA similar to those for transgenic animal bioreactors and start to be developed in Canada as well.

Technical applications require less stringent production and purification conditions. On the other hand most often large scale production at ton scale has to be expected which makes transgenic plants the only technically and economically feasible production system.

MPB Cologne GmbH is engaged in all these fields of application and, besides developing a strategic patent portfolio, creates specific technologies and experiences for defined applications. Its activities concentrate on the core technology of sophisticated expression of foreign proteins in transgenic plants in order to develop optimized protein production technologies.

Antibodies: a valuable example for protein production in transgenic plants

Following the first reports on antibody production in transgenic plants [2-4] several laboratories engaged in further optimization of the basic science. A few years later, the group at Leicester University, UK, reported successful expression of a scFv recombinant antibody intracellularly [6] and extracellularly [7]. A group at Ghent University, Belgium, proved expression of Fab fragments as well [8]. A major breakthrough in proof-of-concept was achieved by Ma et al. [9] when they demonstrated that transgenic plants are able to synthesize and correctly assemble functional secreted (sIgA) antibodies. So far, recombinant sIgA antibodies can only be produced in transgenic plants. No other host is able to produce this complex protein type consisting of several heteromeric chains. Meanwhile, several other groups corroborated the validity of transgenic plants for expression and secretion of IgG antibodies [e.g. 10].

Further progress was achieved by a group at the Institute for Plant Genetics and Crop Plant Research (IPK) in Gatersleben, Germany, in expression of scFv by fusion to a

KDEL signal sequence causing subcellular localization in the endoplasmic reticulum [11, 12]. This group joined the one of Düring at that time to develop transgenic potato tubers as a biofactory for scFv antibodies. This was successfully achieved by routine expression levels in transgenic tubers at about 2% of total soluble protein. The produced antibody could be efficiently purified at analytical scale by metal affinity chromatography by binding to the integrated c-myc- or 6x HIS-tags [5]. Astonishing is the extraordinate stability of scFv antibodies in crude potato tuber extracts without any protease inhibitors. Similarly, a remarkable stability during storage of unprocessed tubers at 4°C assessed up to 1.5 years was demonstrated.

The scFv production technology in potato tubers is further developed at MPB Cologne GmbH as a commercially viable as well as a model system. Protein purification schemes have been developed relying on conventional chromatographic techniques instead of affinity chromatography enabling scale-up for industrial scale.

Small peptides: a challenging type of proteins to be produced in transgenic plants

Peptides are very small compared to most proteins. Therefore, they are especially sensitive to proteolytic degradation. This has to be carefully considered when small peptides shall be produced in foreign hosts such as bacteria, yeasts or plant cells. Peptides have been expressed rarely so far in transgenic plants, mostly with the aim to achieve enhanced resistance to plant pathogenic bacteria and fungi. The first one was cecropin which proved to be extremely protease sensitive [13]. Rational design generated an optimzed peptide which was more resistant to plant proteases [14].

The second peptide which should be produced in tobacco was $(GVGVP)_{121}$, forming a polymer. Only a low expression level could be achieved due to technical limitations [15].

Concluding, all expression-relevant factors should be optimized for development of plant biofactory systems for the production of small peptides (e.g. codon usage, destabilizing mRNA sequences, promoters, peptide targetting, etc.). In case of self-assembly peptides their inherent biophysical properties have to be considered and can be favourably exploited.

Application of peptides for plant resistance engineering

Another business area of MPB Cologne GmbH based on sophisticated protein expression in transgenic plants is resistance engineering towards bacterial and fungal plant pathogens. No appropriate chemical treatment methods are available against bacteria and conventional breeding was so far mostly unsuccessful to provide germplasm with enhanced resistance to bacteria. Large amounts of fungicides are used for treatment of fungal infestations as again most crop plants lack resistance towards major fungal pathogens. Therefore, conventional breeding needs to be complemented by genetic engineering which allows new target-directed resistance strategies for inclusion in conventional plant breeding strategies.

Monogenic resistance traits have proved not to be safe and durable enough. This experience has to be taken into account also when genetic engineering is used for development of novel resistance traits. MPB Cologne GmbH's aim is development of oligogenic resistance strategies which combine several independent modes of action. The group of Düring possesses long-standing experience with T4 lysozyme as a resistance factor [16, 17, 18]. A newly detected antimicrobial mechanism in lysozymes is the basis for optimization of the T4 lysozyme resistance trait followed up at MPB Cologne GmbH. Düring et al. [19] figured out that a short amphipathic peptide (13 amino acids) in the C-terminus of T4 lysozyme is solely responsible for its antimicrobial activity. Positive results in vitro as well as in vivo could be demonstrated for a broad spectrum of gram-negative and gram-positive bacteria as well as several plant pathogenic fungi. Again, a peptide, here embedded in the amino acid context of T4 lysozyme, mediates the essential function of this protein the enzymatic activity of which is not essential for cell killing. Hen egg white lysozyme was tested in parallel and revealed the same results. Further approaches using different modes of action are under development Goal is the short-term availability of oligogenic and oligomechanistic resistance gene constructs.

Summary

Transgenic plants have proved to be a highly promising biofactory for production of recombinant proteins. „Plantibodies" have been among the first to be extensively worked on and today constitute a competitive field of commercial interest, especially for therapeutic antibodies to be produced expectedly around 10-100kg per year for an individual antibody.

A first non-therapeutic tumor imaging antibody has been produced in transgenic maize by Monsanto in cooperation with NeoRx providing the antibody. First clinical trials have been successfully completed and orally communicated so far at several conferences. Another „plantibody" having been brought to successful clinical trials is an anti-caries active agent [20]. These two plant produced antibodies are the first examples of a new commercial technology with a huge economic potential. Transgenic plants as bioreactors will allow the development of a large number of potential therapeutic proteins into successful pharmaceuticals by enabling large scale production at low cost. Therefore, they will have significant implications on the future development of the pharmaceutical industry.

Furthermore, plant bioreators for production of a variety of different peptides are under development and will dramatically change the cost and availability of these new materials.

In addition, peptides also are important as active agents in plant resistance engineering

References

1. Herrera-Estrella L, Depicker A, Van Montagu M, Schell J (1983) Expression of chimaeric genes transferred into plant cells using a Ti-plasmid-derived vector. Nature 303: 209-213
2. Düring K (1988) Wundinduzierbare Expression und Sekretion von T4 Lysozym und monoklonalen Antikörpern in *Nicotiana tabacum*. Ph.D. thesis. Universität Köln
3. Düring K, Hippe S, Kreuzaler F, Schell J (1990) Synthesis and self-assembly of a functional monoclonal antibody in transgenic *Nicotiana tabacum*. Plant Mol Biol 15: 281-293
4. Hiatt A, Cafferkey R, Bowdish K (1989) Production of antibodies in transgenic plants. Nature 342: 76-78
5. Artsaenko O, Kettig B, Fiedler U, Conrad U, Düring K (1998) Potato tubers as a biofactory for recombinant antibodies. Mol Breed 4: 313-319
6. Owen M, Gandecha A, Cockburn B, Whitelam G (1992) Synthesis of a functional anti-phytochrome single-chain Fv protein in transgenic tobacco. Biotechnol 10: 790-794
7. Firek S, Draper J, Owen MRL, Gandecha A, Cockburn B, Whitelam GC (1993) Secretion of a functional single-chain Fv protein in transgenic tobacco plants and cell suspension cultures. Plant Mol Biol 23: 861-870
8. De Neve M, De Loose M, Jacobs A, Van Houdt H, Kaluza B, Weidle U, Van Montagu M, Depicker A (1993) Assembly of an antibody and its derived antibody fragment in *Nicotiana* and *Arabidopsis*. Transgen Res 2: 227-237
9. Ma JKC, Hiatt A, Hein M, Vine ND, Wang F, Stabila P, Vandolleweerd C, Mostov K, Lehner T (1995) Generation and assembly of secretory antibodies in plants. Science 268: 716-719
10. Van Engelen FA, Schouten A, Molthoff JW, Roosien J, Salinas J, Dirkse WG, Schots A, Bakker J, Gommers FJ, Jongsma MA, Bosch D, Stiekema WJ (1994) Coordinate expression of antibody subunit genes yields high levels of functional antibodies in roots of transgenic tobacco. Plant Mol Biol 26: 1701-1710
11. Artsaenko O, Peisker M, Zur Nieden U, Fiedler U, Weiler EW, Müntz K, Conrad U (1995) Expression of a single-chain Fv antibody against abscisic acid creates a wilty phenotype in transgenic tobacco. Plant J 8: 745-750
12. Fiedler U, Conrad U (1995) High-level production and long-term storage of engineered antibodies in transgenic tobacco seeds. Biotechnol 13: 1090-1093
13. Florack D., Allefs S., Bollen R., Bosch D., Visser B., Stiekema W. (1995) Expression of giant silkmoth cecropin B genes in tobacco. Transgenic Research 4: 132-141

14. Owens L.D., Heutte T.M. (1997) A single amino acid substitution in the antimicrobial defense protein cecropin B is associated with diminished degradation by leaf intercellular fluid. Molecular Plant - Microbe Interactions 10: 525-528

15. Zhang X., Urry D.W., Daniell, H. (1996) Expression of an environmentally friendly synthetic protein-based polymer gene in transgenic tobacco plants. Plant Cell Reports 16: 174-179

16. Düring K., Porsch P., Fladung M., Lörz, H. (1993) Transgenic potato plants resistant to the phytopathogenic bacterium *Erwinia carotovora*. Plant Journal 3: 587-598

17. Düring, K. (1993) Can lysozymes mediate antibacterial resistance in plants? Plant Molecular Biology 23: 209-214

18. Düring, K. (1996) Genetic engineering for resistance to bacteria in transgenic plants by introduction of foreign genes. Molecular Breeding 2: 297-305

19. Düring K., Porsch P., Mahn A., Brinkmann O., Gieffers W. (1999) The non-enzymatic microbicidal activity of lysozymes. FEBS Letters 449: 93-100

20. Ma JK, Hikmat BY, Wycoff K, Vine ND, Chargelegue D, Yu L, Hein MB, Lehner T (1998) Characterization of a recombinant plant monoclonal secretory antibody and preventive immunotherapy in humans. Nature Med 4: 601-606

ASSEMBLY MODULATION OF CHANNEL-FORMING PEPTIDES

S. FUTAKI

Institute for Chemical Research, Kyoto University

Uji, Kyoto 611-0011, Japan

1. Introduction

Natural ion channel proteins are highly exquisite molecules which consist of two main functional units, 'the pore' and 'the gate'. The pore is the pathway of ions going through the membrane, which decides the conductance and the ion selectivity of the channel. The gate is the mechanism that controls the open probability of the channel. Many channels open (or close) in accordance with stimulation from the outside, such as attachment of a specific ligand or change in the membrane potential. Many ion channel proteins are composed of multiple transmembrane and extramembrane segments. It is interesting to investigate how these peptide segments can discriminate each other in the presence of membranes to form their three dimensional structure and exert their sophisticated channel function. These problems are also interesting as examples of the molecular recognition or supramolecular structural formation under membrane-associated conditions.

To create artificial ion channel peptides with these functions is a challenge in the

A. Aggeli et al. (eds.), Self-assembling Peptide Systems in Biology, Medicine and Engineering, 87–104.

field of peptide engineering. These molecules can be candidates for artificial signal transmission devices. The direct understanding of the structure-function relationship of channel molecules is also expected through the design of channel molecules. It has been reported that amphiphilic helical peptides of ~20 amino acid residues self-assemble in the membrane to form ion-channels. It is amazing that the assembly of a mere ~20 amino acid peptide can manifest a significant function of ion channel proteins which are often composed of more than 1000 amino acid residues. However, self-assembly of channel-forming peptides often generates ion-channels with multiple conductance as the result of the random assembly of peptides. To refine the function of artificial ion-channels, it should be necessary to modulate the assembly state of these peptides. Here we will make a survey of the efforts to control the association of channel-forming peptides and we will also introduce our approach of assembly modulation of a channel-forming peptide antibiotic, alamethicin, through an extramembrane leucine-zipper segment.

1.1. ION-CHANNEL FORMATION BY THE SELF-ASSEMBLING α-HELICAL PEPTIDES

De novo design of artificial ion channels is one of the targets of peptide engineering. Lear et al. designed two amphiphilic peptides, H-$(LSSLLSL)_3$-NH_2 and H-$(LSLLLSL)_3$-NH_2, and found that the former peptide formed an ion-channel of similar ion-permeability and lifetime with the acetylcholine receptor ion-channel and that the latter formed a proton channel [1]. Oiki et al. also reported that a 23-residue peptide corresponding to one of the transmembrane segments of the nicotinic acetylcholine receptor (M2 in δ-subunit) exhibited features characteristic of the acetylcholine receptor channel, such as single channel conductance, channel life-time and cation specificity [2]. Following these pioneering works, various amphiphilic α-helical peptides were reported to have channel-forming activity [3-5]. Usually the helical structure of short peptides are not very stable, however, the assembly of amphiphilic helical peptides is stabilized by the membrane with amphiphilic amino acids surrounding the channel pore and hydrophobic amino acids facing the membrane. These peptide channels often lack well-organized mechanisms to control the association state of the peptides. Thus they often show multi-levels of conductance derived from an uncontrolled association number of channel-forming peptides. Inconsistency in the association number of the pore-forming peptides results in the failure of attaining a fine pore structure to exert a specific channel function. To improve the channel function, it would be necessary to think about controlling the association state of the transmembrane peptides.

1.2. USE OF TEMPLATES FOR THE ASSEMBLY CONTROL OF TRANSMEMBRANE PEPTIDES

One of the most practical approaches to control the association state (or association number) of channel forming transmembrane peptides would be to use a template. Montal et al. applied Mutter's concept of template assembled synthetic proteins (TASP) [6] to control the assembly of the channel-forming peptides corresponding to the M2 segment of the δ-subunit of the acetylcholine receptor [7]. There are many other reports that use Mutter templates to organize the channel peptide assembly [3-5, 8]. We have also devised a peptide template on which four peptide segments having different amino acid sequences can be introduced [9]. As an application of our template, four individual peptide segments corresponding to the S4 segments in repeat I-IV of the sodium channel were assembled on the template [10]. The four-helix-bundle protein thus obtained worked as an ion channel with rectification (Figure 1).

As an alternative template, Åkerfeldt et al. utilized a tetraphenylporphrin for

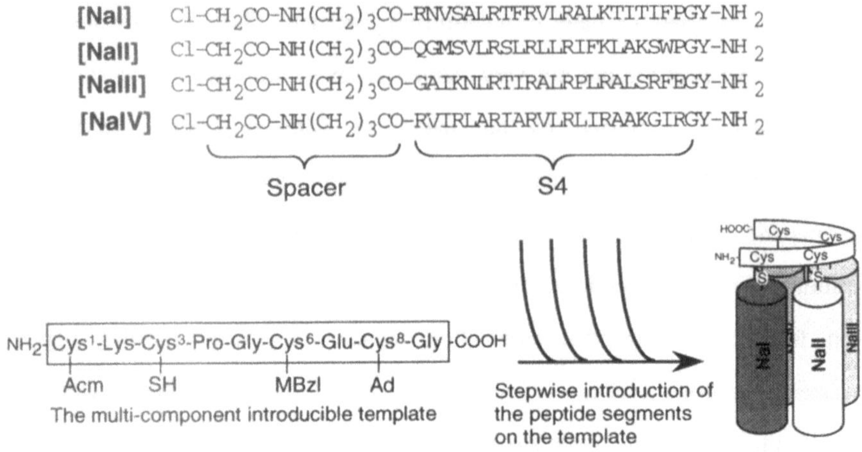

Figure 1. Assembly of four individual helices on the multi-component introducible template to give a channel protein.

assembling an analog of the above mentioned de novo designed transmembrane peptide, H-(LSLLLSL)$_3$-NH$_2$ [3, 11]. There are also reports on the employment of cyclic peptide templates containing *p*-aminobenzoic acid derivatives [12,13].

By tethering channel-forming peptide molecules on a template, convergence of the channel conductance and duration of the channel open states have often been observed. Without doubt, the template-using approaches are useful, but sometimes more than one level of conductance have been observed even for the channel formed by tethered molecules most likely due to the aggregation of the tethered molecules or the polymorphism in the conformation within a molecule [8, 9].

1.3. ASSEMBLY MODULATION THROUGH HYDROGEN BONDING WITHIN THE MEMBRANES

Other examples of assembly modulation are the ones attained with the help of possible hydrogen bonding within the membranes. A natural channel-forming peptide, gramicidin A, takes its characteristic β-helical structure. Two molecules of gramicidin A associate in the membrane to form a channel with their N-termini coming together, where ions go through the center of the helices. Attempts to endow more sophisticated functions to this cylindrical channel motif are reported [14, 15]. Artificial cyclic peptides designed by Ghadiri are also regarded to make assemblies in the membranes by hydrogen bonding to form a nano-tube structure [16].

2. Assembly control of transmembrane peptides through extramembrane peptide segments

We now turn our attention to natural membrane proteins. Natural membrane proteins often have large extramembrane segments. It is natural to think about the possibility of the contribution of extramembrane peptide segments as well as the transmembrane segments to their structural formation. If we can employ extramembrane segments to interact with a specific ligand that induces a conformational change in the transmembrane segments, or if the assembly of the transmembrane peptides can be modulated via the extramembrane segments with the help of some ligand, these concepts can be extended to the design of an artificial sensor and hybrid receptor (Figure 2). As our first step towards this goal, we have designed a hybrid peptide of alamethicin and a leucine-zipper segment as a model, and examined if the assembly of the transmembrane segment could be modulated by the extramembrane segment.

2.1. DESIGN OF THE ALAMETHICIN-LEUCINE ZIPPER HYBRID PEPTIDE

The design of the hybrid peptide (Alm-LeuZ) is shown in Figure 3. As the transmembrane segment, a segment derived from the peptide antibiotic, alamethicin,

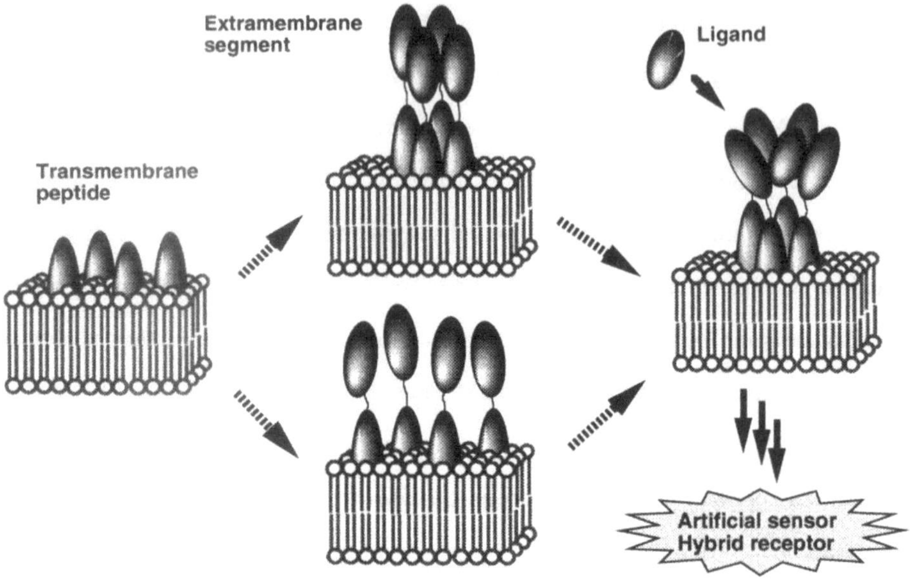

Figure 2. Extramembrane modulation of transmembrane peptide assembly

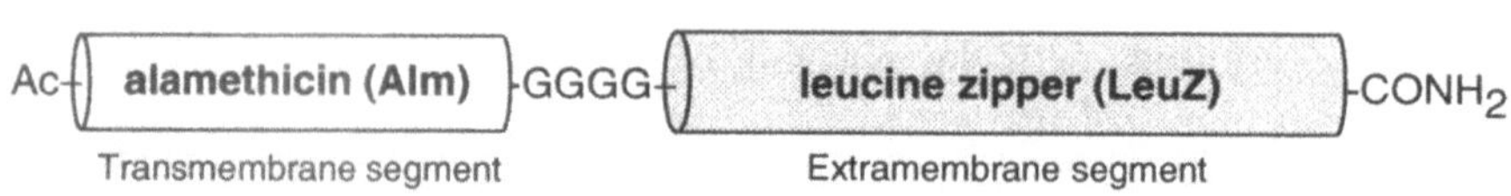

Ac-UPUAUAQUVUGLUPVUUEQF-GGGG-RXKQLEDKVEELLSKNYHLENEVARLKKLVGE-NH_2 (Alm-LeuZ)

Ac-RXKQLEDKVEELLSKNYHLENEVARLKKLVGE-NH_2 (LeuZ)

Ac-UPUAUAQUVUGLUPVUUEQF-NH_2 (Alamethicin amide)

Figure 3. Design of the alamethicin-leucine zipper hybrid peptide (Alm-LeuZ). U: α-aminoisobutyric acid (Aib); X: norleucine (Nle).

was placed on the N-terminus as a transmembrane segment. As an extramembrane segment, the leucine zipper sequence derived from a yeast transcription factor, GCN4, was placed on its C-terminus. A Gly tetramer linker was introduced as a linker to connect these segments. Met in the natural GCN4 segment was replaced with Nle to prevent oxidation. Alamethicin is a fungus derived peptide antibiotic, which is known as a typical ion-channel forming peptide [17]. Aib (α-aminoisobutyric acid) residues in the sequence help the peptide to form a helical structure. One reason why we chose alamethicin is that this peptide randomly assembles in the membrane to form channels with various open states. This characteristic would be preferable for the assessment of the contribution of extramembrane segments to the assembly control. Also, there are many reports on the nature of the alamethicin channel and we may utilize these data for the comparison with those obtained from the hybrid peptide. The GCN4 leucine-zipper segment is known to form a very stable α-helical coined-coil dimer in water [18]. The nature of the GCN4 leucine-zipper segment has also been extensively studied as one of the minimal protein folding motifs. We have introduced this segment to the C-terminus of alamethicin as a model of extramembrane segments.

2.2. SYNTHESIS AND CHARACTERIZATION OF THE HYBRID PEPTIDE

Peptide synthesis was conducted by Fmoc-solid-phase peptide synthesis on a Rink amide resin as previously reported (Figure 4) [19]. For the construction of a peptide segment corresponding to the leucine-zipper segment and $(Gly)_4$-linker, each amino acid was incorporated stepwise onto the resin using benzotriazole-1-yl-oxy-tris-pyrrolidino-phosphonium hexafluorophosphate (PyBOP) in the presence of 1-hydroxybenzotriazole (HOBt) and *N*-methylmorpholine (NMM) as a coupling system. For the construction of the alamethicin segment, this coupling system did not work due to the steric hindrance of the Aib residues. Therefore, we employed the amino acid

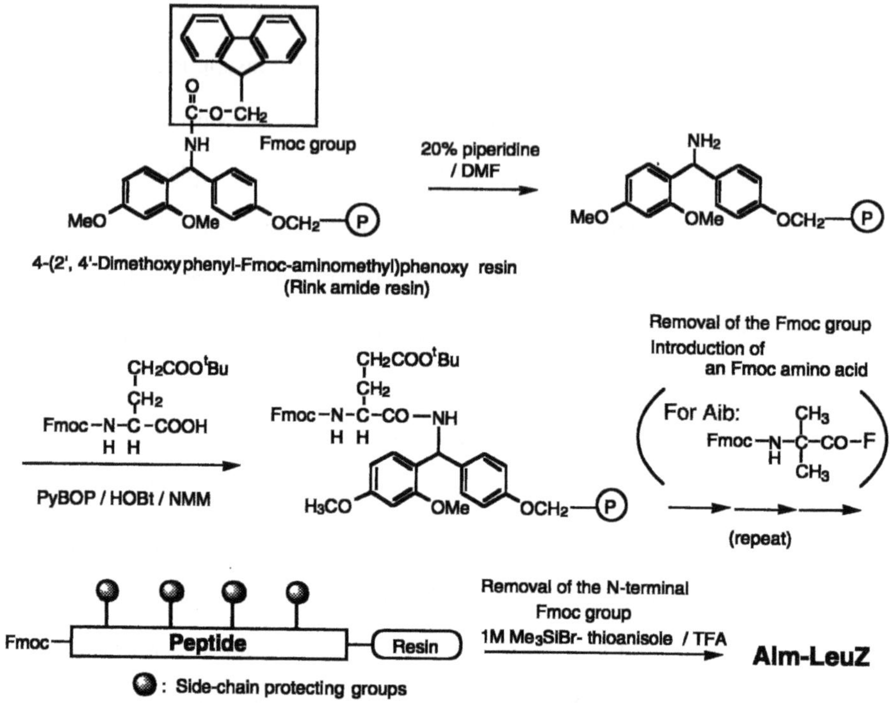

Figure 4. Preparation of the alamethicin-leucine zipper hybrid peptide (Alm-LeuZ).

fluoride activation method [20]. After the N-terminus Fmoc group was removed by 20% piperidine in DMF and the N-terminus acetylated, the peptide resin was deprotected using 1M trimethylsilylbromide-thioanisole in TFA in the presence of ethanedithiol and m-cresol [21]. HPLC purification of the sample afforded a pure 56-residue peptide. Fidelity of the product was ascertained by time of flight mass spectrometry (TOFMS). The CD spectrum of the hybrid peptide in the presence of liposomes made from egg yolk phosphatidylcholine in the presence of 5 mM 4-(2-hydroxyethyl)-1-piperazineethanesulfonic acid (HEPES) (pH 7.4) showed a double minima around 222 nm and 208 nm, suggesting that the hybrid peptide takes a helical structure in the presence of the membrane ($[\theta]_{222}$: -2.5×10^4 deg·cm^2/dmol; peptide concentration: 7 μM).

2.3. APPLICATION OF PLANAR LIPID BILAYER METHOD TO THE ASSESSMENT OF ASSEMBLY STATES OF TRANSMEMBRANE PEPTIDES

Assessment of the assembly states of the transmembrane peptides was conducted using the planar lipid bilayer method (Figure 5) [22]. Two teflon compartments are connected by a pinhole having a diameter of ~300 μm. Lipid bilayers were then formed and the peptide was introduced from one side of the membrane. In this experiment, these

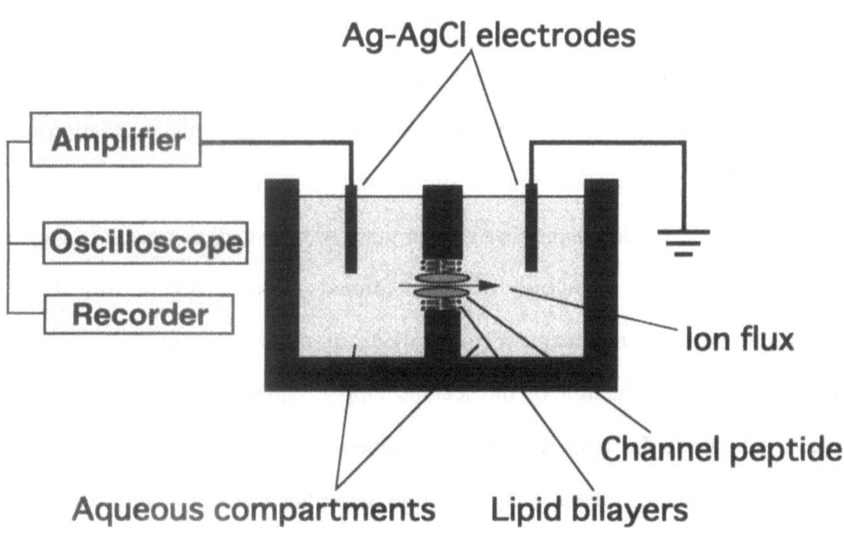

Figure 5. Experimental system for recording ion-channel currents through lipid-bilayers.

chambers were filled with 1M KCl as the electrolyte, and ion fluxes going through the membrane were monitored as a channel current. This system is equivalent to the patch-clamp system which enables one to observe the ion flux going through a single channel pore. A different channel pore structure would give a different ion-flux, which should reflect the channel current levels. On the other hand, if the assembly of the transmembrane peptides is modulated, convergence of the channel current levels would be expected. Thus, by monitoring the ion channel current, we may expect to monitor the assembling state of the transmembrane peptides in real time.

2.4. CONVERGENCE OF THE ASSOCIATION NUMBER OF ALAMETHICIN BY THE INTRODUCTION OF AN EXTRAMEMBRANE LEUCINE-ZIPPER SEGMENT

Figure 6a shows the ion channel current of alamethicin amide that corresponds to the N-terminus of the hybrid peptide. As are typical in alamethicin-related peptides, multi-levels of channel conductance, which reflect the presence of different association states, were observed for this peptide. On the other hand, a channel recording of Alm-LeuZ (50 nM) showed a predominant single open state (designated as open 1), suggesting that a unique number of hybrid peptides formed a pore (Figure 6b). Thus the association number of the alamethicin peptides in the membrane was successfully controlled by the introduction of the leucine zipper segment at the C-terminus. The channel conductance for open 1 was comparable to that obtained from the template-assembled alamethicin tetramer [13]. Thus the assembly was tentatively deduced as a tetramer. A higher current level designated as open 2 was also observed to a lesser frequency, probably due to the presence of an aggregation state of a higher association number or a different conformational state.

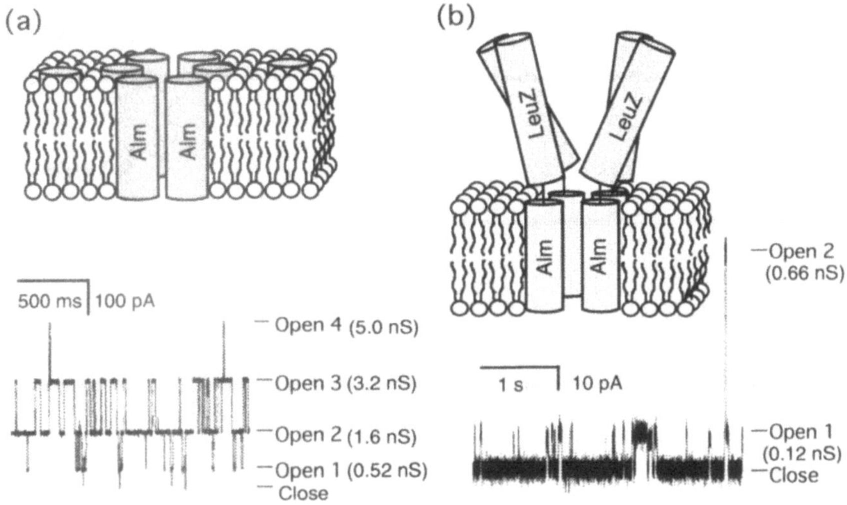

Figure 6. Schematic representation of possible assembly states and single channel recordings of alamethicin amide (10 nM) (a) and Alm-LeuZ (50 nM) (b) at +100 mV.

2.5. CHARACTERISTICS OF THE HYBRID PEPITDE CHANNEL

The assembly modulation was voltage dependent. The assembly was modulated up to 120 mV under the peptide concentration of 50 nM (Figure 7). At higher voltages, higher conductance states were often observed. Also, the modulation was concentration dependent. When the peptide concentration was 10 nM, the assembly was effectively modulated even at 140 mV. On the other hand, at 100 nM, the modulation becomes less efficient even at 80 mV (Figure 8).

The hybrid peptide was supposed to be predominantly introduced to the membrane from the N-terminal side of the molecule because only at positive voltages has a significant channel current been observed. The natural alamethicin channel shows only weak cation selectivity because its poorly organized pore structure resulting from

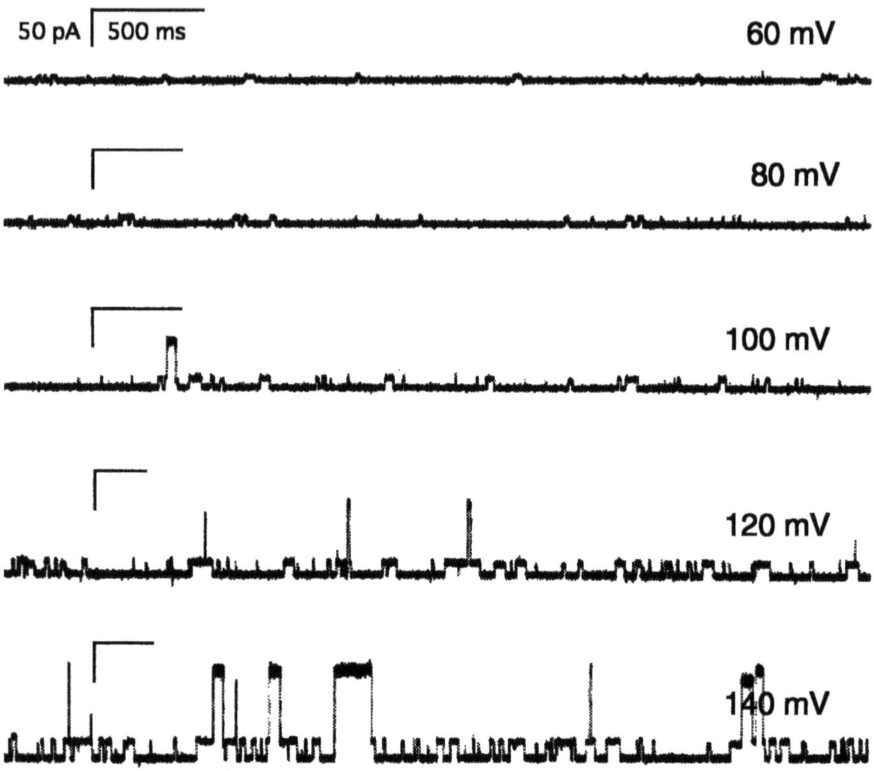

Figure 7. Effect of voltage on the modulation of the assembly of Alm-LeuZ (peptide concentration: 50 nM).

the frequent change in the association number is not able to precisely discriminate ions. By the introduction of the extramembrane segment, the pore structure became more organized and the cation selectivity of the channel became prominent. K^+ was calculated to be about 6 to 7 times more permeable than Cl^-.

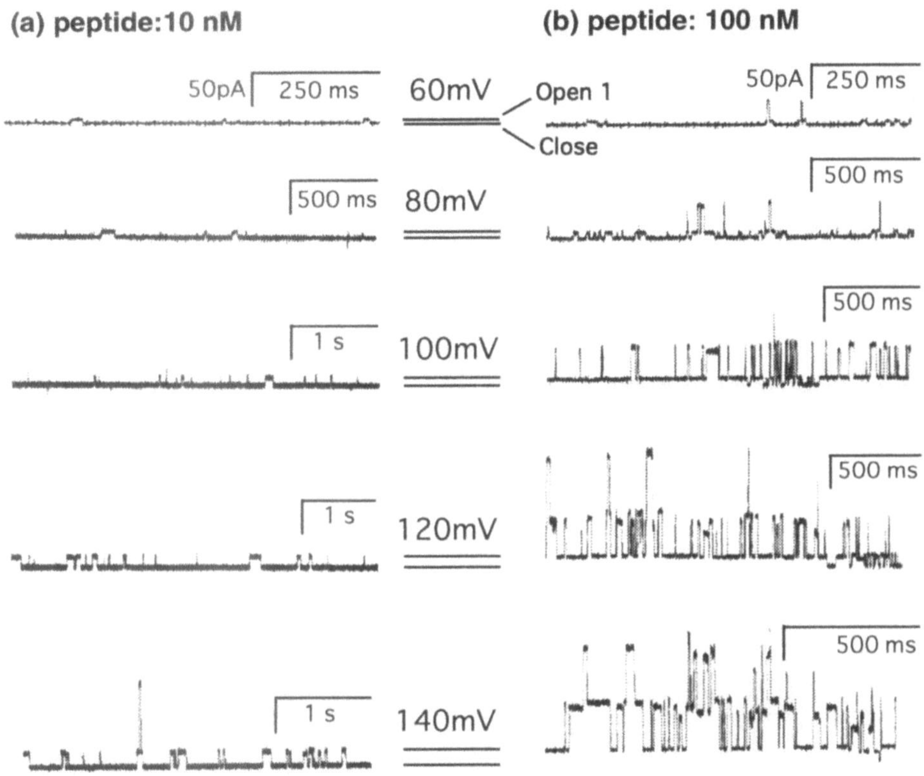

Figure 8. Effect of concentration on the assembly modulation of Alm-LeuZ.

2.6. PLAUSIBLE MECHANISM OF ASSEMBLY MODULATION

To learn more about the assembly modulation, we have synthesized two analogs (Figure 9). One is the alamethicin-[cFos]leucine zipper hybrid peptide (Alm-[cFos]LeuZ) that has the cFos-derived leucine zipper peptide instead of the GCN4-derived leucine zipper. cFos usually forms heterodimers with other proteins such as cJun and is not regarded to form stable homodimers. The other has an extramembrane

Ac-UPUAUAQUVUGLUPVUUEQF-GGGG-RXKQLEDKVEELLSKNYHLENEVARLKKLVGE-NH_2
(Alm-LeuZ)

Ac-UPUAUAQUVUGLUPVUUEQF-GGGG-LTDTLQAETDQLEDEKSALQTEIANLLKEKEKLEFILAAY-NH_2
(Alm-[cFos]LeuZ)

Ac-UPUAUAQUVUGLUPVUUEQF-GGGG-RXKQGEDKVEEGLSKNYHGENEVARGKKLVGE-NH_2
(Alm-[Gly]LeuZ)

Figure 9. Design of Alm-LeuZ analogs.

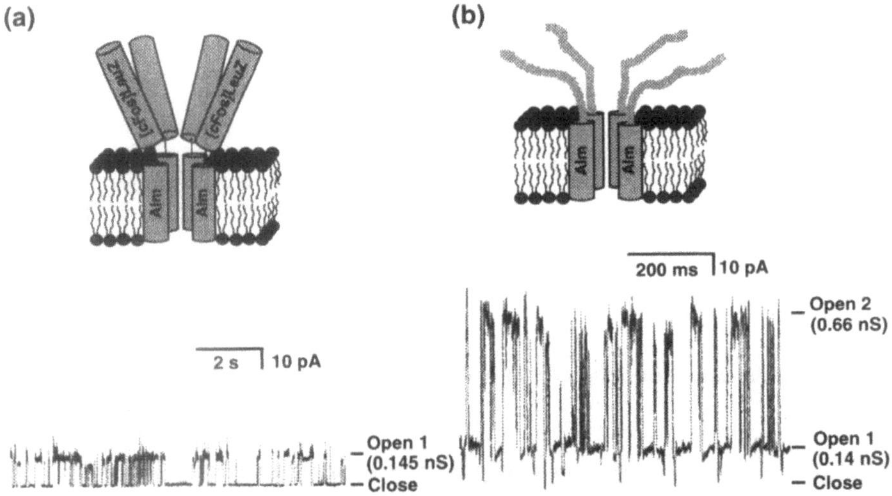

Figure 10. Schematic representation of possible assembly states and single channel recordings of Alm-[cFos]LeuZ (10 nM) (a) and Alm-[Gly]LeuZ (3 nM) (b) at +100 mV.

segment where leucines in the hepted repeat of the GCN4 leucine zipper were replaced with glycines, and this extramembrane segment is not expected to form a helical structure. Synthesis of these analogs was conducted basically in the same manner as already stated. In contrast with the CD spectrum of the GCN4 leucine zipper, the helical content of the cFos leucine zipper based on the molecular ellipticity at 222 nm is only one-third that of GCN4. In the case of the Gly-replaced leucine zipper, the peptide took an almost random structure. However, the results of the channel current measurements of these three peptides were very interesting. Even without a strong tendency for dimer formation, the observed channel conductance of the Alm-[cFos]LeuZ hybrid peptide was very comparable to that of the GCN4 hybrid peptide (Figure 10a). In the case of the Gly-substituted hybrid peptide, where the extramembrane segment took on a poor helical structure, the conductance level was significantly modulated as compared with those of natural alamethicin (Figure 10b). At the peptide concentration of 3μM, the observed conductance states, namely open 1 and open 2, are very comparable with those of the original hybrid peptide.

2.7. IMPLICATION OF THE RESULTS TO THE STRUCTURAL FORMATION OF NATURAL MEMBRANE PROTEINS

Of course further study is necessary, but these results may present a new concept of the structural formation of the membrane proteins. The importance of the mutual recognition has often been emphasized. However, as long as the transmembrane segments have a tendency to assemble as often seen in channel forming peptides such as alamethicin, steric hindrance between the extramembrane segment that prevents excess assembly of protein subunits may play an important role in the assembly control.

3. Conclusions

In conclusion, although within limited ranges of concentration and applied voltage, we have shown that the assembly of the alamethicin molecules was successfully modulated by the extramembrane leucine zipper sequence. As the assembly was modulated, ion-selectivity of the channel became prominent. We have also shed light on the effect of the bulkiness of the extramembrane segments on the assembly modulation. The results obtained in this study should provide useful information on the development of artificial sensors and receptors.

Acknowledgement

The author is grateful to Masayuki Fukuda and Kayoko Yamauchi for their efforts during the course of this study. This work was supported by Grant-in-Aids for Scientific Research from the Ministry of Education, Science and Culture of Japan, the Nissan Science Foundation and the Research Foundation for Pharmaceutical Sciences. The author is also grateful to Professors Yukio Sugiura, Kyoto University, and Mineo Niwa, The University of Tokushima, for their helpful discussions.

References

1. Lear, J.D., Wasserman, Z.R., and DeGrado, W.F.: Synthetic amphiphilic peptide models for protein ion channels, *Science* **40** (1988) , 1177-1181.
2. Oiki. S., Danho, W., Madison, V., and Montal, M.: M2δ, a candidate for the structure lining the ionic channel of the nicotinic cholinergic receptor, *Proc. Natl. Acad. Sci. U.S.A.* **85** (1988), 8703-8707.
3. Åkerfeldt, K.S., Lear, J.D., Wassernab, Z.R., Chung, L.A., and DeGrado, W.F.: Synthetic peptides as models for ion channel proteins, *Acc. Chem. Res.* **26** (1993), 191-197, and the references cited therein.
4. Montal, M.: Protein folds in channel structure, *Current Opp. Structral Biol.* **6** (1996) 499-510, and the references cited therein.

5. Futaki, S.: Peptide ion channels: Design and creation of function, *Biopolymers (Peptide Science)* **47** (1998), 75-81, and the references cited therein.

6. Mutter, M., Altmann, E., Altmann, K.-H., Hersperger, R., Koziej, P., Nebel, K., Tuchscherer, G., Vuilleumier, S., Gremlich, H.-U., and Müller, K.: Artificial folding units by assembly of amphiphilic secondary structures on a template, *Helv. Chim.* **71** (1988), 835-847.

7. Montal, M., Montal, M.S., and Tomich, J.M.: Synporins-synthetic proteins that emulate the pore structure of biological ionic channels, *Proc. Natl. Acad. Sci. U.S.A.* **87** (1990), 6929-33.

8. Grove, A., Mutter, M., Rivier, J.E., and Montal, M.: Template-assembled synthetic proteins designed to adopt a globular, four-helix-bundle conformation form ionic channels in lipid bilayers, *J. Am. Chem. Soc.* **115** (1993), 5919-5924.

9. Futaki, S., Aoki, M., Fukuda, M., Kondo, F., Niwa, M., Kitagawa, K., and Nakaya, Y.: Assembling of the four individual helices corresponding to the transmembrane segments (S4 in repeat I-IV) of the sodium channel, *Tetrahedron Lett.* **38** (1997), 7071-7074.

10. Futaki, S., Aoki, M., Ishikawa, T., Kondo, F., Asahara, T., Niwa, M., Nakaya, Y., Yagami, T., and Kitagawa, K.: Chemical ligation to obtain proteins comprising helices with individual amino acid sequences, *Bioorg. Med. Chem.* **7** (1999), 187-192.

11. Åkerfeldt, K.S., Kim, R.M., Camac, D., Groves, J.T., Lear, J.D., and DeGrado, W.F.: Tetraphilin: A four-helix proton channel built on a tetraphenylporphyrin framework, *J. Am. Chem. Soc.* **114** (1992), 9656-9657.

12. Ishida, H., Donowaki, K., Inoue, Y., Qi, Z., and Sokabe, M.: Synthesis and ion channel formation of novel cyclic peptides containing a non natural-amino acid, *Chem. Lett.* **1997**, 953-954.

13. Matsubara, A., Asami, K., Akagi, A., and Nishino, N.: Ion channels of cyclic template-assembled alamethicins that emulate the pore structure predicted by barrel-stave model, *J. Chem. Soc. Chem. Comunn.* **1996**, 2069-2070.

14. Oiki, S., Koeppe II, R.E., and Andersen, O.S.: Voltage-dependent gating of an asymmetric gramicidin channel, *Proc. Natl. Acad. Sci. U.S.A.* **92** (1995), 2121-2125.

15. Woolley, G.A., Jaikaran, A.S.I., Zhang, Z., and Peng, S.: Design of regulated ion channels using measurements of cis-trans isomerization in single molecules, *J. Am. Chem. Soc.* **117** (1995), 4448-4454.

16. Ghadiri, M.R., Granja, J.R., and Buehler, L.K.: Artificial transmembrane ion channels from self-assembling peptide nanotubes, *Nature* **369** (1994), 301-304.

17. Sanson, M.SP.: The biophysics of peptide models of ion channels, *Prog. Biphys. Molec. Biol.* **55** (1991), 139-235.

18. O'Shea, E.K., Rutkowski, R., and Kim, P.S.: Evidence that the leucine zipper is a coiled-coil, *Science* **243** (1989), 538-542.

19. Futaki, S., and Kitagawa, K.: Peptide-unit assembling via disulfide cross-linking: A versatile approach which enables the creation of artificial proteins comprising helices with different amino acid sequences, *Tetrahedron* **53** (1997), 7479-7492.

20. Kaduk, C., Wenschuh, H., Beyermann, Forner, K., Carpino, L.A. and Bienert, M.: Synthesis of Fmoc-amino acid fluorides via DAST, an alternative fluorinating agent, *Lett. Peptide Sci.* **2** (1995), 285-288.

21. Fujii, N., Otaka, A., Sugiyama, N., Hatano, M., and Yajima, H.: Evaluation of trimethylsilyl bromide as a hard-acid deprotecting reagent in peptide synthesis, *Chem. Pharm. Bull.* **35** (1987), 3880-3883.

22. Montal, M., and Mueller, P.: Formation of bimolecular membranes from lipid monolayers and a study of their electrical properties, *Proc. Natl. Acad. Sci., U.S.A* . **69** (1972), 3561-3566.

MOLECULAR CASTING OF INFECTIOUS AMYLOIDS, INORGANIC AND ORGANIC REPLICATION: NUCLEATION, CONFORMATIONAL CHANGE AND SELF-ASSEMBLY

D. C. GAJDUSEK
Department of Human Retrovirology
Academic Medical Center (AMC)
University of Amsterdam
1105 AZ Amsterdam
The Netherlands

1. Abstract

Kuru in New guinea, the first proven slow virus infection of man, led to the discovery of the transmittability of Creutzfeldt Jakob Disease (CJD) and of its clinical variants, Gertsmann-Sträusler Schanker disease (GSS) and fatal familial insomnia (FFI). It is recognized that many diseases of animals (scrapie, transmissible mink encephalopathy, wasting diseases of mule deer and Rocky Mountain elk, bovine spongiform encephalopathy (BSE), bovine and feline spongiform encephalopathy in many exotic zoo species) are all the same disorder with similar pathogenesis.

2. Introduction

Kuru is the source of our awareness of the infectious amyloids and recognition that nucleation of change to cross β-pleated conformation and amyloid fibril polymerization in a normal host protein could cause disease as well as the changes of normal aging, and thus the concept of the conformational diseases. It gave the first example of an infectious agent originating *de novo* by spontaneous generation in the host without a chain of lateral or vertical infection. This occurs as a rare stochastic event in which the host precursor protein changes to the β-pleated conformation of an infectious nucleant at the incidence of 1 per million per annum, the worldwide incidence of sporadic CJD. In the familial forms of CJD (or GSS and FFI) caused by point mutations with the amino acid change or octopeptide repeats, the likelihood of this spontaneous stochastic event is increased by six orders of magnitude [1-12].

Kuru caused by mortuary practice set the pattern for high-tech neo-cannibalism induced CJD as an iatragenic or dietary disease caused by an infectious nucleant, with high resistance to heat or antiseptics. It also demonstrated dependence of the incubation period on dose and genetic homozygosity of a silent polymorphism on codon 129 of the host precursor protein, which we also now see in new-variant CJD from BSE infection of man and iatrogenic CJD from CJD-contaminated human growth hormone [1-12].

A. Aggeli et al. (eds.), Self-assembling Peptide Systems in Biology, Medicine and Engineering, 105–112.

Nucleation of conformational change in host precursor proteins and their polymerization into fibrils is a general phenomenon in amyloidogenesis in a large group of diseases. All amyloid deposits in the genetically determined amyloidoses with point mutations in the precursor protein contain not only the mutated polypeptide, but also the normal unmutated polypeptide as well, in heterozygous subjects of these autosomal dominant diseases. The mutated polypeptide in cross β-pleated conformation has induced the patterning, or nucleated the β-pleated conformational change and the copolymerization into fibrils of the normal polypeptides. In fact, other heterologous amyloidogenic proteins in the milieu interieur are also often induced into β-pleated conformation, and thus contaminate the primary amyloid [1-12].

Most β-pleated fibrils will absorb and bind by van der Waal forces or H-bonds, a wide series of other proteins and polypeptides, because of close geometric matching. These include ubiquitin, chymotrypsin, light chains of gamma globulin, and oligomers of most of the glycosaminoglycans and proteoglycans. In fact, if the fibril polymerization occurs in a medium containing these, they will be copolymerized.

3. Fantasy of a "Virus" without Carbon Atoms

The process of nucleation or template-induction of conformational change of a normal host protein to a pathological disease-producing conformation, which autocatalytically initiates a cascade of such further transformations, underlies the pathogenesis of Kuru, Creutzfeldt-Jakob disease (CJD), scrapie and bovine spongiform encephalopathy (BSE). The cross-β-pleated conformation easily self-assembles into fibrils on further nucleation of the fibrilogenesis. In these diseases, the iatrogenic or experimentally induced infections across a species barrier depend on the extent of geometric matching between the precursor proteins of the different species. Strain specific behavior of homologous infectious nucleants within a single species is probably the result of specific strain conformations, each inducing different epitaxic pattern setting, again a matter of geometric matching. This strain-specific pattern setting could be at the initial conformational change to a β-pleated conformation or at the nucleation of polymerization to semi-crystalline arrays and fibril growth.

The forces involved in surmounting the energy barrier to the new conformation are surely van der Waal and perhaps also coulombic, as in other changes of state. The quantum dynamics of the transition to the lower energy-state of sheets or fibrils is not yet elucidated.

The crossed β-pleated conformation of all amyloids and probably also their fibril polymerization prevent proteases from disrupting most of their peptide bonds. When condensed into fibrils, the close association with H-bonding confers on the protein further added resistance to heat. However, the experience of many laboratories that a small fraction of scrapie infectivity may resist prolonged exposure to temperatures over 300°C indicates a heat resistance hardly conceivable for protein structures. This survival of extremely heat-resistant infectivity is typically of very low titer, below 10^{-1}, or with less than one infectious dose per milliliter from suspensions in which 10^7 to 10^{10} infectious doses per milliliter were initially present. Thus, only one infectious particle per 10^1 to 10^4 million survives prolonged exposure to such high temperatures. Even this should not be the case at 360°C.

I previously attributed this survivability of some infectivity for decades to the possibility, that an inorganic clay or silica or other ceramic footprint or fossil-like replica of the necessary molecular geometry has been made. They can then nucleate the conformational change of the scrapie precursor protein to its infectious isoform: $PrP^c \rightarrow PrP^{sc}$.

4. Material Science and Engineering of Self-Assembling Inorganic – Organic Complex Solids

Engineers in material science have been directing the structural and morphological pattern of inorganic crystals and of noncrystalline solids or ceramics by interaction of organic regularly self-assembled templates with inorganic minerals to produce high order co-organizational states of consolidated matter far beyond the unit cell and usual inorganic crystal structures. Local rather than general energy minimal determines the crystal structures [13-24].

Thus, they are making micromolecular sieves of zeolites, silica and of aluminum silicates, and of other minerals with pore size and connectivity determined by the organic templates. They form organic-inorganic complexes which solidify, and from which the organic molecules are later removed by calcination or by heating to 600°C, leaving a microporous or macroporous replica in silica or aluminum silicates. Biomineralization of a single cell organism and of bacterial fibrils has been achieved [13-24].

This is essentially a molecular tectonics of biomineralization or the production of footprints or fossils of the structural patterns of biological organization and of organic macromolecules.

5. Amyloid Enhancing Factors are Scrapie-like Agents

AA amyloidosis in man and animals is a natural concomitant of aging and also a progressive fatal disease under single gene control in hereditary Mediterranean fever. The precursor serum amyloid protein (SAA) from which the amyloid fibrils are formed is made in the liver. Since the 1950s, in order to study this amyloidosis in experimental animals, an acceleration of this phenomena of aging has been accomplished in rats and mice by provoking non-specific inflammation with the associated increase in the serum level of the SAA-protein precursor of the AA amyloid fibrils. This is usually produced by the intra-peritoneal injection of $AgNO_3$ or of heterologous casein. With the rise to very high levels of the SAA protein, the amyloid deposits appear earlier than normally, first in the lung and spleen and subsequently in other organs, as early as one month after $AgNO_3$ injections are started in the mouse, rather than at 2 years as in normal aging.

Organ extracts from old animals, that already formed AA in their tissues (but not from younger AA-amyloid-free mice) even at high dilutions, contain an amyloid enhancing (augmenting, inducing, accelerating, facilitating) factor which causes the primed animal to start depositing amyloid, usually first in the lungs, within one to two weeks, thus triggering the process to death weeks before it would otherwise have occurred. Purification of this amyloid enhancing factor (AEF) has been as elusive as that of the scrapie agent. Hol, Niewold et al demonstrated that microfibrils of AA amyloid are themselves a fibrilar AEF. Thus the fibrilar AEF nucleates the change of the normal serum SAA protein to the cross β-pleated conformation and its polymerization into amyloid fibrils as does the scrapie "virus" when in contact with its scrapie precursor protein, PrP^c. The new AA amyloid fibrils so formed have similar AEF activity or "infectivity" [25].

Per Westermark and his colleagues have achieved successful nucleation of AA-amyloid deposition using synthetic fibrils made from synthetic polypeptides of amyloidogenic sequences from the precursor protein of human pancreatic islet amyloid and from transthyretin and labeled these with ^{125}I. They have demonstrated AEF activity of these synthetic mice primed by $AgNO_3$ injections for early AA amyloid deposition, and caused its appearance several days earlier than anticipated by injections of small fibrils made from these synthetic peptides, but not with the monomeric form of these peptides. Only the small fibrils can either serve as seeds or niduses for nucleation of AA amyloid deposition [25].

These synthetic heteronucleants of amyloid deposition have thus been shown to behave like scrapie-like "viruses". Other such systems should be studied. All glycosaminoglycans and proteoglycans copolymerize with or absorb onto all β-pleated protein fibrils – as does Congo red – and they too might be expected in oligomeric form to serve as such AEF heteronucleants as they do for tropocologen gel formation [26]. Many amyloidogenic peptides will copolymerize into fibrils when other amyloidogenic peptides are forming fibrils in their presence.

Westermark's group has clearly demonstrated this, even with synthetic heteronucleants, rather than with homonucleant fibrils from AA amyloid (or SAA protein) sequences. Thus amyloid enhancing fibrils are scrapie-like "viruses"; conversely, scrapie "virus" (or prion) is an amyloid enhancing factor [25].

The process of heteronucleation by other amyloids resembles the heteronucleation by minerals of many proteins' crystallization which McPherson and Shlichta have investigated. The low levels of some infectivity often found to survive long exposure to temperatures over 300°C after the sudden reduction of infectivity titers by many $\log_{10}$ may well be accounted for by the formation of molecular casts or atomic ghost replicas. Although these agents are small, these heat-resistant low level infectivities are frequently observed. These infectivities may be caused, in part, by inorganic clay, ceramics, glass footprints, or fossil templates of infectious organic amyloid nucleants [27-28]. Thus I suggest the high temperature incineration of the synthetic polypeptide nucleating fibril amyloid enhancing factors of Per Westermark and his group to contain no further C-atoms (600°C) amd in the presence of silica and

Table 1. Various infectious agents

Virus: a submicroscopic parasite requiring for its replication the informational and energy systems of the host
Computer viruses
Industrial infectious nucleants
Ice-nine (Kurt Vonegut)
Viroids
Infectious amyloids (proteins): unconventional (atypical) viruses; prions.
Amyloid enhancing factors: amyloid accelerating, augmenting factors.
Synthetic amyloid heteronucleants
Molecular casts or fossils: atomic ghost replicas
Viruses from the inorganic world

Table 2. Synonymous terminology used in referring to conformational change and fibrils polymerization in amyloidogenesis

Nucleated change to β-pleated conformation
Nucleated induction
Nucleated polymerization
Nucleation of fibrilogenesis
Heterogeneous nucleation
Epitaxial nucleation (geometrical matching)
Epitaxic crystallization

Seeding
Nidus induced assembly
Nidus induced aggregation
Self-assembly

Conformational mimicry
Conformational templating
Conformational matching
Conformational imprinting
Amyloid enhancing factor – enhancement
Amyloid augmenting factor – augmentation
Amyloid facilitating factor – facilitation
Amyloid accelerating factor – acceleration
Amyloid inducing factor – induction

Instructional amyloidogenesis
Instructional fibrilogenesis
Heterodimer formation
One-dimensional crystallization
Ordered molecular packing
Ordered protein aggregation
Formation of ordered nuclei

surfactants to form such molecular casts on molds or atomic ghost replicas. If this is the case, we can really speak of the fantasy of a "virus" from the inorganic world.

References

1. Gajdusek, D.C. (1988) Transmissible and non-transmissible amyloidoses: autocatalytic post-translational conversion of host precursor proteins to β-pleated configurations. *J. of Neuroimm.* **20**, 95-110.

2. Guiroy, D.C. and Gajdusek, D.C. (1988) Fibril-derived amyloid enhancing factors as nucleating agents in Alzheimer's disease and transmissible virus dementia. In "Molecular Genetic Mechanisms in Neurodegenerative Disorders" P. Brown, L. Bolis and D.C. Gajdusek, editors. *Discussions in Neuroscience* **5**, 69-73.

3. Gajdusek, D.C. (1990) Fantasy of a "virus" from the inorganic world: pathogenesis of cerebral amyloidoses by polymer nucleating agents and/or "viruses". In: Neth R., Gallo R.C., Greaves M., Gaidicke G., Geha S., Mannweiller K., Ritter J., editors, "*Modern Trends in Human Leukemia VIII*," New York. Springer-Verlag, pp. 481-499.

4. Gajdusek, D.C. (1991) The transmissible amyloidoses: genetic control of spontaneous generation of infectious amyloid proteins by nucleation of configurational change in host precursors: Kuru-CJD-GSS-Scrapie-BSE. *European Journal of Epidemiology*, **7**, 567-577.

5. Gajdusek, D.C. (1991) Genetic control of *de novo* generation of infectious amyloids from human precursor proteins in Mediterranean Jews and European Slavs. Proceedings of the European Community Commission Symposium, November 12-14, Brussels. In: *Sub-acute Spongiform Encephalopathies, Current Topics of Veterinary and Animal Science.* Vol.55. R. Bradley, M Savey and B.A. Marchant, editors. Kluwer Academic Publishers, Dordrecht, the Netherlands. pp. 91-114.

6. Gajdusek, D.C. (1991) Infectious amyloidoses: transthyretin familial amyloidotic polyneuropathy as a paradigm for genetic control of spontaneous generation of transmissible amyloids by patterned configurational change of host precursors in CJD and other spongiform encephalopathies. In "*Molecular Neurovirology*," R.F. Roos, editor Humana Press, Totawa, N.J. pp. 1-20.

7. Gajdusek, D.C. (1994) Spontaneous generation of infectious nucleating amyloids in the transmissible and nontransmissible cerebral amyloidoses. *Molecular Neurobiology*. **8**, 1-13.

8. Gajdusek, D.C. (1994) Nucleation of amyloidogenesis in infectious and noninfectious amyloidoses of brain. *Annals of the New York Academy of Sciences* **724**, 173-190.

9. Gajdusek, D.C. (1994) Spontaneous generation of infectious amyloid nucleants in the transmissible and nontransmissible brain amyloidoses. In "*Advances in Research on Neurodegeneration. II: Etiopathogenesis*." Y. Mizuno, D.B. Calne, R. Horowski, editors. Birkhauser, Boston. pp. 187-207.

10. Gajdusek, D.C. (1994) Infectious and noninfectious amyloidoses of the brain: systemic amyloidoses as predictive models in transmissible and nontransmissible amyloidotic neurodegeneration of Creutzfeldt-Jakob disease and aging brain and Alzheimer's disease. Chapter 19 in "*Neurodegenerative Diseases*," D.B. Calne, editor. W.B. Saunders, Philadelphia, pp. 301-317.

11. Gajdusek, D.C. (1994) Genetic control of nucleation of precursors to infectious amyloids in the transmissible amyloidoses of man. *Brit. Med Bulletin*, **49**, 913-931.

12. Gajdusek, D.C. (1996) Infectious amyloids: subacute spongiform encephalopathies as transmissible cerebral amyloidoses. Chapter 91 in B.N. Fields, D.M. Knipe, P.M. Hensley et al, editors. "*Freid's Virology*" Third Edition. Lippincott-Raven Publishers, Philadelphia, pp. 2851-2900.

13. Burkett S.L. and Davis M.E. (1995) Mechanisms of structure direction in the synthesis of pure-silica zeolites. I. Synthesis of TPA/Si-ZSM-5, *Chem. Mater.* **7**, 920-928.

14. Burkett S.L. and Davis M.E. (1995) Mechanisms of structure direction in the synthesis of pure-silica zeolites. II. Hydrophobic hydration and structural specificity. *Chem. Mater.* **7**, 1453-1463.

15. Burkett S.L. and Davis M.E. (1995) Towards the rational design and synthesis of zeolites. *Mater. Res. Soc. Symp. Proc.*, **371**, 3-14.

16. Burkett S.L. and Davis M.E. (1996) Synthetic mechanisms and strategies for zeolite synthesis in *Comprehensive Supramolecular Chemistry*, Volume 7, Chapter 16. J.L. Atwood, et al., Eds., Oxford, Pergamon/Elsevier.

17. Burkett S.L. and Davis M.E. (1996) Spatial organization and patterning of gold nanoparticles on self-assembled biolipid tubular templates. *Chem. Commun.*, 321-322.

18. Burkett S.L., Sims S.D. and Mann S. (1996) Synthesis of hybrid inorganic-organic mesoporous silica by co-condensation of siloxane and organosiloxane precursors. *Chem. Commun.*, 1367-1368.

19. Davis M.E., Chen C.-Y., Burkett S.L. and Lobo R.F. (1994) Synthesis of (alumino)silicate materials using organic molecules and self-assembled organic aggregates as structure-directing agents. *Mater. Res. Soc. Symp. Proc.*, **346**, 831-842.

20. Davis M.E. and Burkett S.L. (1995) Towards the rational design and synthesis of microporous and mesoporous silica-containing materials. *Zeoraito* **12**, 3333-47.

21. Davis S.A., Burkett S.L., Mendelson N.H. and Mann, S. (1997) Bacterial templating of ordered macrostructures in silica and silica-surfactant mesophases. *Nature* **385**, 420-423.

22. Mann S., Davis S.L., Fowler C.E., Mendelson N.H., Sims S.D., Walsh D. and Whilton N.T. (1997) Sol-gel synthesis of organized matter. *Chem. Mater.* **9**, 2300-2310.

23. Sims S.D., Burkett S.L. and Mann S. (1996) The synthesis and characterization of ordered mesoporous materials incorporating organosiloxanes. *Mater. Res. Soc. Symp. Proc.* **431**, 77-82.

24. Whilton N.T., Burkett S.L. and Mann S. (1998) Hybrid lamellar nanocomposites based on organically-functionalized magnesium phyllosilicate clays with interlayer reactivity, *J. Mater. Chem.* **8**, 927-1932.

25. Johan, K., Westermark G., Engström U., Gustavsson A., Hultman P., Westermark P. (1998) Acceleration of amyloid protein A amyloidosis by amyloid-like synthetic fibrils. *Proc. Nat. Acad. Sci. USA* **95**, 2558-2563.

26. Obrink, B. (1972) The influence of glycosaminaglycans on the formation of fibers from monomeric tropocollagen *in vitro*. *Eur. J. of Biochem.* **84**, 129-137.

27. McPherson A., Shlichta P. (1988) Heterogeneous and epitaxial nucleation of protein crystals on mineral surfaces. *Science* **239**, 385-387.

28. Shlichta P. (1991) Heterogeneous/epitaxial (HE) nucleation of protein crystals in relation to the formation of amyloid fibrils. In "*Regulation and Genetic control of brain amyloid.*" D.C. Gajdusek, K. Bayreuther and P. Brown, editors. *Brain Research Reviews* **16**, 105-106.

29. Weiss, A. (1981) Replication and evolution in inorganic systems. *Angev. Chem.* **20**, 850-860.

STRUCTURE AND STABILISATION OF SELF-ASSEMBLING PEPTIDE FILAMENTS.

NICHOLAS J. GAY AND MARTYN SYMMONS
Department of Biochemistry
University of Cambridge
CAMBRIDGE, CB2 1GA
UK

MARISA MARTIN-FERNANDEZ AND GARETH JONES
CLRC Daresbury Laboratory
WARRINGTON, WA4 4AD
UK

Abstract

The formation of filaments and fibrils by specific peptide substrates is widespread in nature and underlies the pathology of diseases such as Alzheimer's and Creutzfelt Jacob (CJD). These processes result from a conformational switch which allows polymerisation of the protein usually into highly stable β sheet structures. Here we describe the properties of a peptide derived from a 24 amino acid motif called a leucine rich repeat which undergoes such a structural transition. We show that a conserved amide sidechain is required for filament formation and contrast this with the role played by these residues in the native structures of LRR proteins. Interestingly, mutations involving amide residues are often associated with amyloid disease, including predispositions to sporadic CJD. We also describe studies of early events in filament formation by both the LRR peptide and a derivative of Alzheimer β protein 1-42 using the technique of time resolved fluorescent anisotropy.

1. Introduction

The leucine rich repeat is a widespread motif in proteins, being found in extracellular, membrane cytoplasmic and nuclear proteins (Buchanan and Gay, 1996). They are about 24 amino acids long and are found in blocks of

A. Aggeli et al. (eds.), Self-assembling Peptide Systems in Biology, Medicine and Engineering, 113–125.

up to about 40 consecutive tandem repeats. In general terms they function to mediate protein-protein interactions for example in cell adhesion and signal transduction. One important class of LRR proteins are the Toll transmembrane signalling receptors. Originally identified as part of the dorso-ventral patterning system of the *Drosophila* embryo, Toll-like receptors also fulfil roles in innate immune responses to bacterial and fungal pathogens in both invertebrates and humans. Human Toll like receptors are thought to be involved also in the development of disease states such as septic shock and rheumatoid arthritis. At present the structure of only two LRR proteins are known, the cytoplasmic ribonuclease inhibitor (which has atypical repeats) (Kobe and Deisenhofer, 1993; Kobe and Deisenhofer, 1995) and the U2SnRNPA', a ribonucleoprotein component of the spliceosomal apparatus (Price *et al.* 1998). As shown in Fig 1, the latter repeats form a coiled coil consisting of a short β strand, a turn and a region of variable secondary structure. The consensus sequence for the typical repeat is:

P X X X **F** X X **L** X X **L** X X **L** X **L** X X **N** X **I** X X **L**

The conserved asparagine residue is nearly invariant in LRRs and forms part of the β turn linking the two secondary structure elements where it adopts an unusual left handed α-helical conformation (Fig. 1) (see Srinivasan *et al.*, 1994).

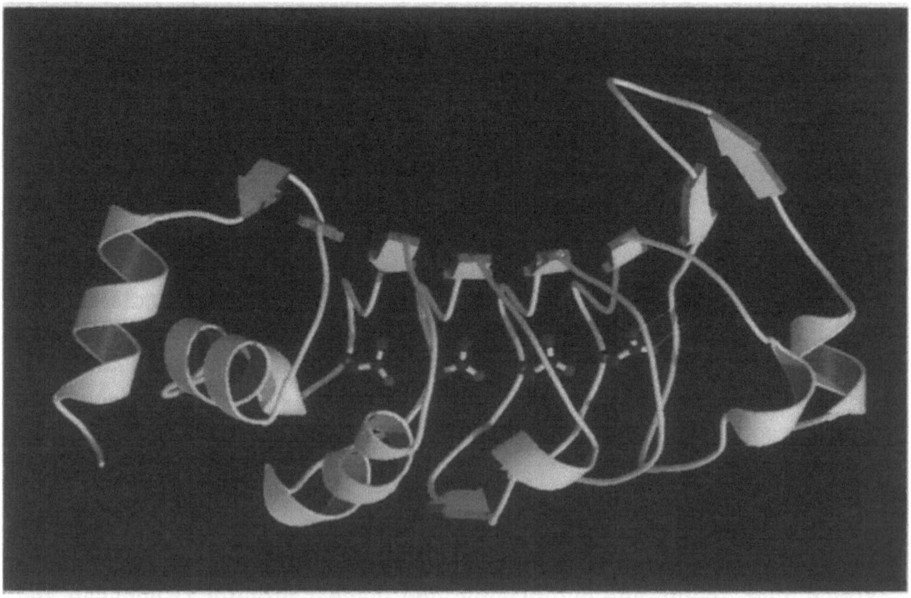

Figure 1. The five typical LRRs of U2snRNPA' The LRRs are presented as ribbon diagrams and the positions of the conserved asparagine sidechains are indicated by space filling models. Taken from Price *et al.*, 1997.

2. A single unit LRR from the Toll receptor forms spontaneously into filaments.

Some years ago we synthesised a peptide derived from *Drosophila* Toll receptor of the following sequence: PANLLTDMRNLSHLELRANIEEM (LRRN). This peptide when dissolved at pH 7.5 formed into a viscous gel (Gay *et al.* 1991; Symmons *et al.*, 1997). This suggested that peptide monomers were polymerising into extended filaments similar to those formed by Alzheimer and prion peptides. Examination of the gels using electron microscopy confirmed this. Using the technique of rotary shadowing the filaments are seen to be interleaved with each other and the accumulation of stain suggests that each strand is a tape-like structure similar to those formed by other designed synthetic peptides (Aggeli *et al.* 1997) (Fig. 2). Assembly of the filaments is accompanied by a change in the circular dichroism spectra of the peptide solution from random coil to β-sheet structure (Fig 3), suggesting that polymer formation involves a structural transition of individual peptide subunits in an end-dependent manner. We have further analysed the structure of LRR filaments by X-ray

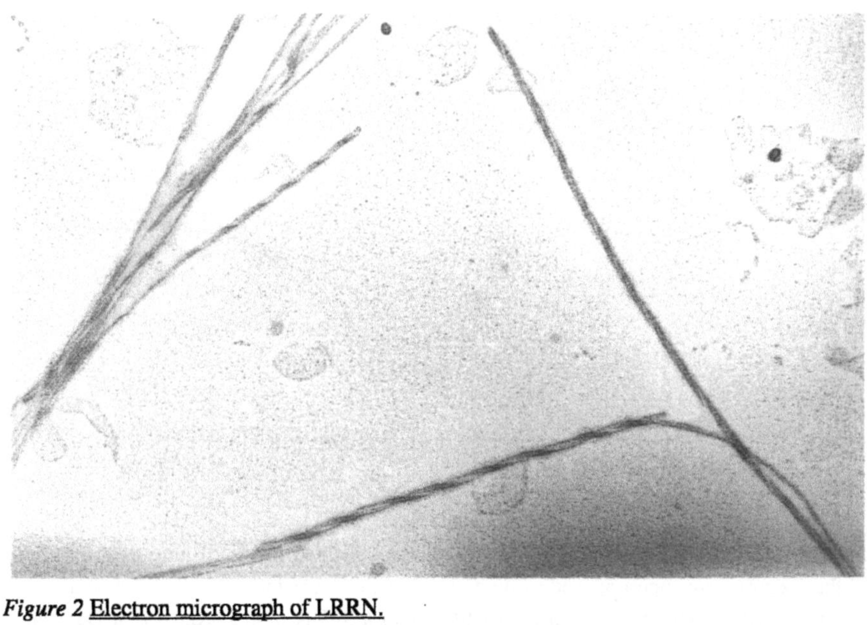

Figure 2 Electron micrograph of LRRN.
LRRN fibres (x54700). Scale bar indicates 100nm Samples of peptide were polymerised as described (see Symmons *et al.* 1997) and then diluted 1/50 in 10mM sodium phopshate pH 7.8. Sample was stained by the rotary shadowing method. Samples were examined with a Philips EM300 instrument.

fibre diffraction studies of oriented, dried fibres. The filaments show 'cross β' diffraction with predominant reflections at 4.76Å on the meridian (Fig 4), a pattern first described in 1968 by Geddes *et. al.* This shows that the hydrogen bonds of the β-sheets are oriented parallel to the filament axis. Thus LRRN filaments are composed of subunits with an extended polypeptide in contrast to the folded structure adopted by native LRRs in proteins. These structural properties of LRRN are very similar to those of fibrils formed by both Alzheimer β 1-42 peptide and of prion protein PrP peptides (Inouye *et al.*, 1993; Kirschner *et al.* 1986; Kirschner *et al.*, 1987; Nguyen *et al.*, 1995).

3. The conserved amide residue of LRRN plays a critical role in filament polymerisation.

In order to determine whether the conserved asparagine residue in LRRN was important for filament formation we synthesised two derivatives which placed another amide, glutamine (LRRQ), and aspartic acid (LRRD)

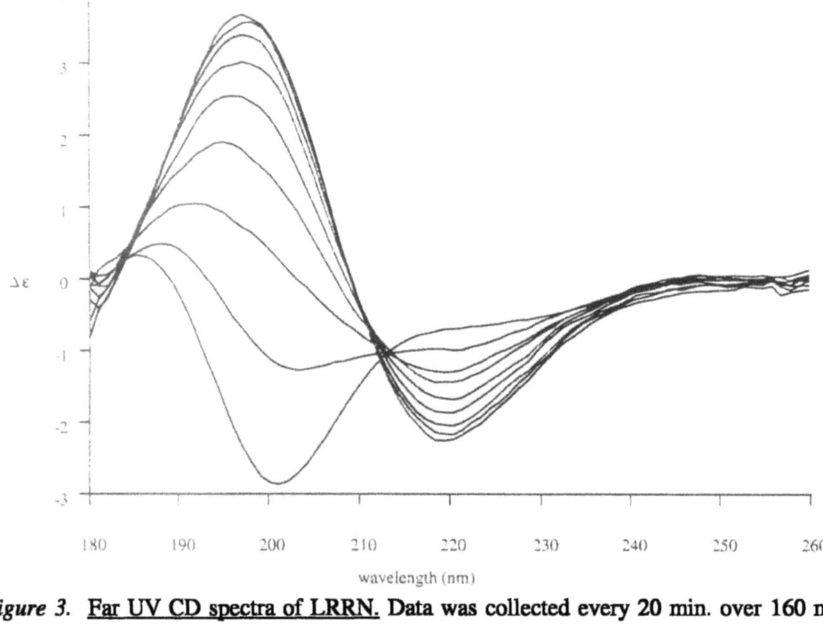

Figure 3. Far UV CD spectra of LRRN. Data was collected every 20 min. over 160 min. and the spectra were superimposed to show the change in structure. CD measurements were taken using UV radiation from station 3.1 SRS with a SEYA monochromator at Daresbury Laboratory, CLRC, Daresbury, Warrington, Cheshire UK. In these experiments the peptide was dissolved at 1 mg/ml directly in 10 mM sodium phosphate buffer (pH 7.8) and allowed to polymerise over time in a 0.1 mm path length cuvette. The spectra are derived from single scans subtracted from a buffer baseline and are slightly smoothed.

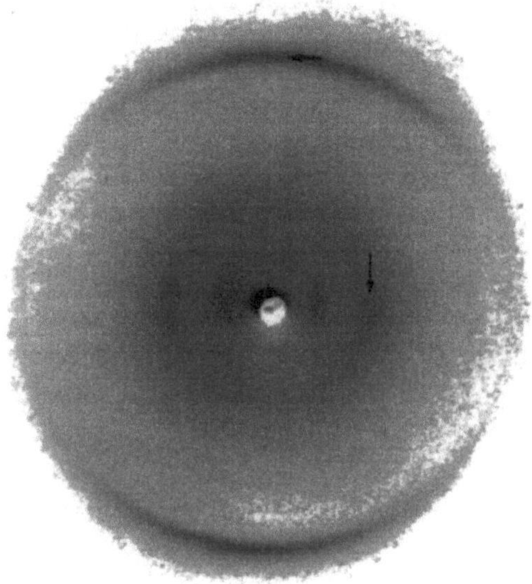

Figure 4. Fibre diffraction studies of LRRN filaments. LRRN peptide was dissolved in water at 10 mg/ml and polymerisation initiated by addition of phosphate buffer (10mM final concentration) to give a pH in the range 7.0 to 8.2. 10μl aliquots were immediately suspended vertically between glass-rods aligned with the magnetic field of a 9T superconducting magnet. The solutions dried into fibres over 30 hrs at 7°C and 92% R.H. When dry they were removed and stored at 75% R.H. Fibres formed at pH7.4 were transparent and showed blue birefringence and these were chosen for fibre diffraction studies. The pH 7.0 fibres were transparent but had red birefringence and gave diffraction patterns suggesting greater disorientation. Fibres from pH 8.2 were not transparent and gave much weaker birefringence. Shown here is the diffraction pattern obtained from a pH 7.4 fibre mounted vertically in the Keele fibre camera (0.2 mm collimator, 75% R.H.) on Station 7.2 SRS Daresbury (λ = 1.488 Å). The pattern was recorded on a MAR research 120mm image plate (20 min exposure). A calibration ring at 3.136 Å was obtained by dusting the fibre with silicon powder. The pattern was converted to CCP13 image format using the program CONV (Denny, 1993) displayed and calibrated using the program FIX. This allowed the fibre-to-detector distance, centre, rotation, and tilt to be estimated. It was also used to measure the position of intensity maxima in reciprocal space (Denny, 1993).

Large arrow indicates the 4.76 Å meridional intensity. Small arrow indicates broad intensity in the 11 Å region. Other less intense maxima are observable at 22 Å on the equator and in the 4.2 Å region slightly off the meridian. No intense features were observed close to the central maximum, and no distinct feature is observable on the meridian corresponding to a spacing of 9.4 Å. Crystalline spots at outer corners are portions of calibration ring.

at that position. LRRQ was still able to form filaments and these were ultrastructurally indistinguishable from those of LRRN (Fig 2), although the kinetics of formation was slower. By contrast LRRD was unable to polymerise even at very high concentrations (ca 10mM). This suggests that formation of the filaments requires an amide residue at the conserved position and that sidechain-sidechain or sidechain mainchain interactions stabilise the filaments. It is interesting that glutamine can substitute in the filaments because in both LRRs and the related coiled-coil motif called the β-helix (Yoder and Jurnak, 1995) asparagine seems to be required. This may reflect either steric contraints in these native structures not present in the peptide filaments or the pseudo-achirality of asparagine, a feature which favours formation of the left handed α-helical conformation (Srinivasan *et al* 1994).

4. Predispositions to amyloid disease often involve mutation to amide residues.

The properties of LRRN peptide described here suggest that stabilization caused by inter-molecular interactions between side chains may be important in the formation of amyloid deposits in neurodegenerative diseases. It is interesting to note that mutations causing pre-disposition to such diseases frequently involve amino acid changes to glutamine or asparagine. A major inherited form of Creuzfelt Jacob disease (CJD) and the related condition fatal familial insomnia is caused by a change of an aspartic acid residue to asparagine (D178N) (Goldfarb *et al.*, 1992). Furthermore, a natural variant of sheep prion protein which when homozygous confers a predisposition to the development of scrapie is a change from arginine to glutamine in the same region of the molecule (R171Q) (Westaway *et al.*, 1994). Another striking correlation is found with a polymorphism in the prion protein (Q219E/K) carried in heterozygous form by 6% of the Japanese population. An analysis of 85 cases of CJD dementia showed that none had this polymorphism but, by contrast, 6 of 43 patients with non-CJD dementia did carry the mutation (Shibuya *et al*, 1998). Thus Q219E/K probably confers resistance to the development of sporadic CJD. The amyloid plaques of Alzheimer's disease are composed of a 42 amino acid peptide processed from amyloid precursor protein. A related familial condition, cerebral haemorrhage with amyloidosis (Dutch form), is caused by mutation of aspartate 22 in the β-peptide to glutamine. This change enhances the rate of formation and the stability of fibrils in vitro (Soto *et al.*, 1995). The Icelandic form of this disease is caused by a leucine to glutamine change in a different protein cystatin C (L58Q) (Ghiso *et al.*, 1986). A third familial amyloidosis is caused by a mutation in gelsolin, again resulting in an additional amide

(D187N). A 30 amino acid peptide which contains the mutation forms into fibrils 9 times faster than an equivalent wild type peptide (Maury and Nurmiaho, 1992). Finally, accelerated senescence in the SAM mouse is caused by a proline to glutamine change (P5Q) in the serum apolipoprotein. This condition is characterised by generalised accumulation of amyloid fibrils consisting of the mutant apolipoprotein (Kunisada *et al.*, 1986).

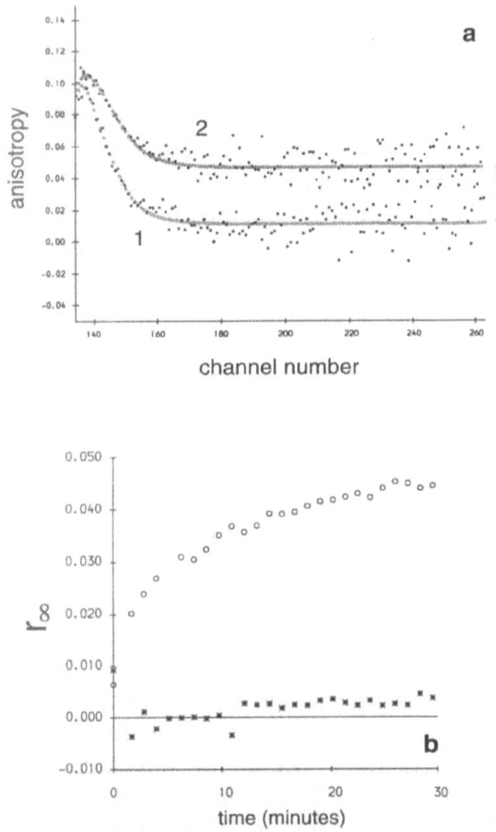

Figure 5. **(a)** Fluorescence anisotropy decay of Alz peptide (β1-28F3W) (75 μl peptide (814μM stock) in 215 μl dH_2O). 1) at time zero before adding 10 μl Tris; 2) 30 minutes after adding 10 μl Tris. The grey lines are the best fit obtained with the exponential law (Lakowicz, 1983): $r(t) = (r_0 - r_\infty) \exp(-t/\phi) + r_\infty$, where r_0 is the initial anisotropy, ϕ the rotational correlation time and r_∞ is the asymptotic anisotropy. The value of ϕ in (1) is 0.55 ns, and in (2) is 0.50 ns. The time calibration is 0.0788 ns/channel. The arrow represents the difference between the values of r_∞ before and after the polymerisation reaction. **(b)** Time course of r_∞ derived from successive measurements of the anisotropy decay. The time resolution is 60 seconds. (circles) Alz peptide (75 μl peptide in 215 dH_2O)- 10 μl of Tris were added at time zero; (asterisks) Alz peptide (18.75 μl in 280 μl dH_2O) – 2.5 μl Tris added at time zero.

In another group of inherited neurodegenerative diseases exemplified by Huntington's Chorea, the age of onset and severity of the disease is correlated with the presence of expanded poly-glutamine repeats at the C-terminal of the susceptibility protein. Perutz *et al.* (1994) have shown that polyglutamine peptides can polymerise into rod shaped structures made of β-sheets. These are stabilized by 'glutamine zippers' and it is suggested that such a polymerisation reaction underlies the disease process.

5. How can an we study early events in filament formation?

It is of great importance to study the initiation of polymer formation not only in the context of diseases such as Alzheimer's but also in designed peptide tapes. Nucleation of polymeric filaments is thought to require the slow production of an initiator chemically distinct from the mature polymer subunit and subsequent rapid polymerisation (Lansbury and Jarrett, 1993). The nature of these nucleation precursors is of profound importance because compounds that inhibit their formation could be used as therapies that prevent commencement or progression of amyloid diseases. These initiators are intrinsically difficult to study because like transition states in enzyme mediated catalysis they are probably chemically unstable and therefore highly transient. In view of this we have been developing biophysical methods that may allow precursors to be studied directly. In these experiments, quantitative determination of the kinetics of peptide nucleation were based on the measurement of time-resolved fluorescence anisotropy decays (Chuang and Eisenthal, 1972; Weber, 1971; Lakowicz, 1983). The anisotropy decay is defined as the difference between the rate of the fluorescence decay of a fluorescent species in the directions parallel and perpendicular to the polarisation of the exciting light. Anisotropy in the fluorescence emission results from photoselection of fluorophores on excitation with linearly polarised light, as more efficient interactions occur for molecules with dipolar axes parallel to the direction of polarisation (Lakowicz, 1983). Immediately after excitation, fluorescence photons are emitted still polarised along the direction of polarisation of the exciting light. At a later time during the fluorescence decay the fluorescent molecular species has had time to rotate in space and consequently the emission becomes depolarised. The rate of anisotropy decay can therefore report on the rotational correlation time of a fluorophore.

Time-resolved decays of tryptophan anisotropy have been used to report on the kinetics of peptide nucleation. The lifetime of the emission of tryptophan can range from 0.3 - 7.4 ns depending on the polarity of the environment, whereas the segmental motion of a freely rotating tryptophan

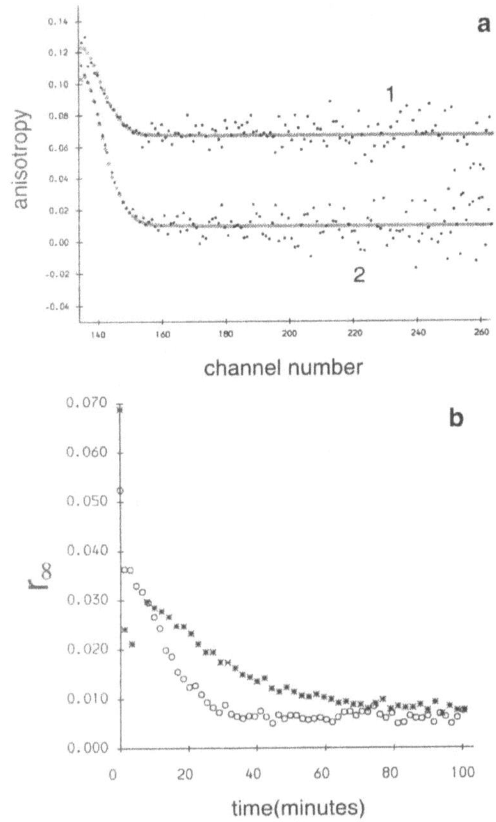

Figure 6. **(a)** Fluorescence anisotropy decay of LRRN peptide (L4W) (50 μl peptide (10mg ml^{-1})in 250μl dH_2O). 1) at time zero before adding 5 μl Tris; 2) 100 minutes after adding 5 μl Tris. The grey lines are the best fit obtained with the exponential law (as in Figure 1). The value of ϕ in (1) is 0.28 ns and in (2) is 0.23 ns. The time calibration is 0.0788 ns/channel. The arrow represents the difference between the values of r_∞ before and after the reaction. **(b)** Time course of r_∞ derived from successive measurements of the anisotropy decay. The time resolution is 100 seconds (circles) LRRN peptide (50 μl peptide in 250 μl dH_2O)- 2.5 μl of Tris were added at time zero; (asterisks) LRRN peptide (100 μl peptide in 195 μl dH_2O) – 5 μl Tris added at time zero.

side chain in water would be of the order of 10 ps (Diggle *et al*, 1995). These fluorescence decay lifetimes are in a similar time range to the rotational correlation time of peptides smaller than about 50 kDa. Therefore, the value of the anisotropy decays to zero within the lifetime of the probe for peptides smaller that about 50 kDa because during the lifetime of their fluorescence the tryptophans in the peptides can complete angular displacements in all directions. For larger peptides the anisotropy does not decay to zero but to an asymptotic value (r_∞). As the peptide increases in size the value of r_∞ will increase accordingly, and therefore the measurement of r_∞ can be used to determine the time course of peptide nucleation.

5. 1 TIME-RESOLVED ANISOTROPY DECAY COLLECTION.

Anisotropy decays were obtained by measuring the components of the fluorescence decay parallel [$I_{par}(t)$] and perpendicular [$I_{per}(t)$] to the excitation light: $r(t) = (I_{par}-I_{per}) / (I_{par}+2I_{per})$. The excitation light was vertically polarised and fluorescence decays were collected at right angles in station 13.1b of the synchrotron radiation source (SRS) at Daresbury Laboratory using the time-correlated single photon counting method (O'Connor *et al.*, 1979). The SRS provides light pulses that have a full width half-maximum of 160 ps at a repetition rate of 3.125 MHz. Continual measurement of anisotropy decays has been performed with a data acquisition system previously described (Martin-Fernandez *et al.* 1996; 1998). The wavelength of the excitation was 295 with a band pass of 5 nm, and the emission was collected above 330 nm.

6. References

Aggeli, A., Bell, M., Boden, N., Keen, J.N., Knowles, P.F., McLeish, T.C.B., Pitkeathly, M. and Radford, S.E. (1997) Responsive gels formed by spontaneous self assembly of peptidesinto polymeric beta-sheet tapes. *Nature* **386**, 259-262.

Buchanan, S.G.S.C. and Gay, N.J. (1996) Structural and functional diversity in the leucine rich repeat family of proteins. *Prog. Biophys. Mol. Biol.* **65**, 1-44..

Chuang, T.J. and Eisenthal, K.B. (1972) Theory of fluorescence depolarisation by anisotropic rotational diffusion. *J. Chem. Phys.* **57**, 5094-5097.

Denny, R. C. (1993) The CCP13 Newsletter No. 2 (Daresbury Laboratory, Warrington U.K.)

Diggle, C., Hutton, M., Jones, G.R., Thomas, C.M., and Jackson, J.B. (1995) Properties Of The Soluble Polypeptide Of The Proton-Translocating Transhydrogenase From Rhodospirillum-Rubrum Obtained By Expression In Escherichia-Coli. *European Journal of Biochemistry* **228**,719. .

Fraser, P.E. Nguyen, J. T., Surewicz E. K and Kirschner, D. A. (1991) pH dependent structural transitions of Alzheimer amyloid peptides. *Biophys. J.* **60**, 1190-1201.

Gay, N.J., Packman, L.C., Weldon, M.A.and Barna, J.C.J. (1991) A leucine rich repeat peptide derived from the *Drosophila* Toll receptor forms extended filaments with a β-sheet structure. *FEBS Lett.* **291**, 87-91.

Geddes, A. J., Parker, K. D. Atkins, E. D. T., and Beighton, E. (1968) "Cross-β" conformation in proteins *J. Mol. Biol.* **32**, 343-358.

Ghiso, J., Jensson, O. and Frangione, B. (1986) Amyloid fibrils in hereditary haemorrhage with amyloidosis of Icelandic type is a variant of γ trace basic protein (cystatin C). *Proc. Natl. Acad. Sci. (USA)* **83**, 2974-2978.

Goldfarb, L.G., Petersen, R.B., Tabaton, M., Brown, P., Leblanc, A.C., Montagna, P., Cortelli, P., Julien, J., Vital, C., Pendlebury, W.W., Haltia, M., Wills, P.R., Hauw, J.J., McKeever, P.E., Monari, L., Schrank, B., Swergold, G.D., Autiliogambetti, L., Gajdusek, D.C., Lugaresi, E. and Gambetti, P. (1992) Fatal familial insomnia and familial Creutzfeldt-Jakob disease - disease phenotype determined by a DNA polymorphism. *Science* **258**, 806-808.

Inouye, H., Fraser, P.E. and Kirschner, D.A. (1993) Structure of β-crystalline assemblies formed by Alzheimer β–amyloid protein analogues: analysis by X-ray diffraction. *Biophys. J.* **64**, 502-519.

Kirschner, D. A. Abraham, C. and Selkoe, D. J. (1986) X-ray diffraction from interneuronal paired helical filaments and extraneuronal amyloid fibers in Alzheimer disease indicates a cross-β conformation. *Proc Natl Acad Sci USA* **83**, 503-507.

Kirschner, D. A. Inouye, H., Duffy, L. K., Sinclair, A., Lind, M. and Selkoe, D. J. (1987) Synthetic homologues to protein from Alzheimer disease forms amyloid-like fibrils in vitro. *Proc Natl Acad Sci USA* **84** 6953-6957.

Kobe, B. and Deisenhofer, J. (1993) Crystal structure of porcine ribonuclease inhibitor, α protein with leucine rich repeats. *Nature* **366**, 751-756.

Kobe, B. and Deisenhofer, J. (1995) Proteins with leucine rich repeats. *Curr. Opin. Struct. Biol.* **5**, 409-416.

Kunisada, K., Higuchi, K., Shin-ichi, A. and Yamagishi, H. (1986) Molecular cloning and nucleotide sequencing of cDNA for murine senile amyloid protein: nucleotide substitutions found in apolipoprotein A-II cDNA of senescence accelerated mouse. *Nucl. Acids. Res.* **14**, 5929-5939.

Lakowicz, J.R., *Principles of Fluorescence Spectrocopy* (Plenum, New York, 1983).

Lansbury, P.J. and Jarret, J.T. (1993) Seeding one-dimensional crystallization of amyloid: a pathogenic mechanism in Alzheimer's disease and scrapie. *Cell* **73**, 1055-1058.

Maury, C.P.J. and Nurmiaho, E.L. (1992) Creation of amyloid fibrils by mutant asn 187 gelsolin peptides. *Biochem. Biophys. Res. Comm.* **183**, 227-231.

Martin-Fernandez, M.L., Tobin, M.J., Clarke, D.T., Gregory, C.M. and Jones, G.R. (1996) A high sensitivity time-resolved microfluorimeter for real-time cell biology. *Rev. Sci. Ins.* **67**, 3716-3721.

Martin-Fernandez, M.L., Tobin, M.J., Clarke, D.T., Gregory, C.M. and Jones, G.R. (1998) Sub-nanosecond polarised microfluorimetry in the time domain: An instrument for studying receptor trafficking in live cells. *Rev. Sci. Ins.* **69**, 540-543.

Nguyen, J.T., Inouye, H., Baldwin, M.A., Fletterick, R.J., Cohen, F.E., Prusiner, S.B. and Kirschner, D.A. (1995) X-ray diffraction of scrapie prion rods and PrP peptides. *J. Mol. Biol.* **252**, 412-422.

O'Connor D.V.O., Ware, W.R., and Andre, J.C. (1979) Deconvolution of fluorescence decay curves. A critical comparison of techniques. *J. Phys. Chem.* **83**,1333-1343.

Perutz, M.F., Johnson, T., Suzuki, M. and Finch, J. (1994) Glutamine repeats as polar zippers: their possible role in inherited neurodegenerative diseases. *Proc. Natl. Acad. Sci. USA* **91**, 5355-5358.

Price, S.R., Evens, P.R. and Nagai, K. (1998) Crystal structure of the spliceosomal U2B" - U2A' protein complex bound to a fragment of U2 small nuclear RNA. *Nature* **394** 645-650.

Shibuya, S., Higuchi, J. and Shin, R.W. (1998) Codon 219 lys allele is not found in sporadic CJD. *Annals Neurol.* **43**, 826-828.

Soto, C., Castano, E.M., Frangione, B. and Inestrosa, N.C. (1995) The α-helical to β-strand transition in the amino terminal fragment of amyloid β-peptide modulates amyloid formation. *J. Biol. Chem.* **270**, 3063-3067.

Srinivasan, N., Anaradha, V.S., Ramakrishnan, C., Sowdhamini, R. and Balaram, P. (1994) Conformational characteristics of asparaginyl residues in proteins. *Int. J. Pept. Prot. Res.* **44**, 112-122.

Symmons, M.F., Buchanan, S.G., Clark, D.T., Jones, G. R. and Gay, N.J. (1997) X-ray diffraction and far UV CD studies of filaments formed by a leucine rich repeat peptide: structural similarity to the amyloid fibrils of prions and Alzheimer's disease β protein. *FEBS Lett.* **291**, 87-91.

Weber, G. (1971) Theory of fluorescence depolarisation by anisotropic Brownian rotations. Discontinuous distribution approach. *J. Chem. Phys.* **55**, 2399-2407.

Westaway, D., Zuliani, V., Cooper, C.M., DaCosta, M., Neuman, S., Jenny, A.L., Detwiler, L. and Prusiner, S.B.(1994) Homozygosity for prion protein alleles encoding glutamine 171 render sheep susceptible to natural scrapie. *Genes Develop.* **8**, 959-969.

Yoder, F. and Jurnak, F. (1995) The parallel β-helix and other coiled folds. *FASEB J.* **9**, 335-342.

DESIGNED COMBINATORIAL LIBRARIES OF NOVEL AMYLOID-LIKE PROTEINS

M.H. HECHT, M.W. WEST, J. PATTERSON, J.D. MANCIAS, J.R. BEASLEY, B.M. BROOME AND W. WANG
Department of Chemistry
Princeton University, Princeton NJ, 08544 USA
hecht@princeton.edu

1. Abstract

The deposition of amyloid is associated with several neurodegenerative diseases including Alzheimer's disease and the prion diseases. To probe the relationship between amino acid sequence and the propensity to form amyloid, we studied a combinatorial library of sequences designed *de novo.* All sequences in the library were designed to share an identical pattern of alternating polar and nonpolar residues as would be consistent with the formation of amphiphilic β-sheet structure. While the polar/nonpolar patterning was identical for all members of the library, the precise identities of these side chains were not constrained and were varied combinatorially. The *de novo* sequences were expressed in bacteria and purified. Biophysical characterization revealed that the proteins self-assemble into large oligomers visible by electron microscopy as amyloid-like fibrils. Like natural amyloid, the *de novo* fibrils are composed of β-sheet secondary structure and bind the diagnostic dye, Congo Red. Thus, an alternating pattern of polar and nonpolar residues is sufficient to cause designed sequences to assemble into amyloid-like fibrils. This finding prompted us to question the distribution of alternating patterns in the sequences of natural proteins. Analysis of a large database of natural protein sequences revealed that alternating patterns occur less frequently than other patterns with similar compositions. The under-representation of alternating patterns in natural proteins, coupled with the observation that such patterns promote assembly of amyloid-like structures in *de novo* proteins, suggests that sequences of alternating polar and nonpolar amino acids are inherently amyloidogenic and consequently have been disfavored by evolutionary selection.

2. Introduction

Amyloid deposits are associated with several neurodegenerative diseases including Alzheimer's disease, senile systemic amyloidosis, and spongiform encephalopathies (e.g. 'mad cow' disease) (1-10). The primary component of amyloid differs from one disease to another: In Alzheimer's it is the 42 residue Aβ peptide; in senile systemic amyloidosis it is the transthyretin protein; and in spongiform encephalopathy it is the prion protein (2-5). Despite substantial differences in both sequence and length (40 - 250 residues), these diverse proteins assemble into amyloid structures that are

A. Aggeli et al. (eds.), Self-assembling Peptide Systems in Biology, Medicine and Engineering, 127–138.

remarkably similar to one another. They all form fibrils composed of β-strands running perpendicular to the fibril axis.

What molecular determinants predispose a protein to form amyloid? The extreme dissimilarity between the various amyloidogenic sequences, coupled with the unavailability of simple model systems, has limited our understanding of the sequence determinants of amyloidogenesis. However, the similarity among the structures of the different amyloids suggests they may share unifying structural determinants. Here we describe an assessment of these determinants through the study of a combinatorial library of protein sequences designed de novo. Because of their combinatorial origins, the sequences in the library differ significantly from one another. Yet, because the combinatorial diversity was constrained by elements of rational design, all sequences in the library share essential global features. Structural and biophysical characterization of the *de novo* proteins suggests that these global features can suffice to predispose amino acid sequences to form amyloid fibrils.

3. Results

3.1 DESIGN OF NOVEL SEQUENCES

3.1.1. *Polar/Nonpolar Patterning of β-Strands*

The structures of natural amyloid fibrils are composed of β-strands running perpendicular to the long axis of the fibril ("cross-β" structure) (7-10). Therefore, for model proteins to mimic amyloid, their sequences must be designed to form multimeric β-strands. Previous research in our laboratory demonstrated that for peptides destined to assemble into multimers, secondary structure is dictated by the periodicity of polar and nonpolar residues in the linear sequence (11). When the periodicity of the linear sequence matches the repeat pattern of a particular secondary structure, the peptide will form that structure -- irrespective of the intrinsic propensities of the individual amino acids (11). For example, a sequence of polar (O) and nonpolar (●) amino acids with the pattern ●●OO●●OO●OO●●OO● has a nonpolar residue every 3 or 4 positions. This matches the α-helical repeat of 3.6 residues/turn. Hence, peptides with this sequence pattern form amphiphilic α-helices that self-assemble into bundles and bury the nonpolar faces of the individual helices in the core of the bundle (11). Conversely, sequences with the alternating pattern ●O●O●O●O● have a periodicity of 2. This matches the structural repeat of β-strands with successive side chains pointing up-down-up-down-etc. Such sequences form amphiphilic β-strands that bury their nonpolar faces by aggregating into large β-sheet structures (11, 12). These earlier findings with short peptides suggest that *de novo* proteins designed to contain segments of alternating polar/nonpolar periodicity might be predisposed to oligomerize into cross-β structures.

We designed a library of *de novo* proteins such that all sequences were constrained to include this β-type pattern of alternating polar and nonpolar residues (13, 14). Combinatorial diversity was incorporated into the library by allowing polar residues to be His, Lys, Asn, Asp, Gln or Glu; and nonpolar residues to be Leu, Ile, Val or Phe. These combinatorial sets of amino acids were encoded by libraries of synthetic genes in which polar residues are encoded by the degenerate DNA codon XAX, and nonpolar residues by the degenerate DNA codon XTX (where X represents a combinatorial mixture of bases).

3.1.2. *Length of β-Strands*
The amyloid fibrils found in nature are typically composed of two or more protofilaments, each with a diameter of approximately 30 Å (7, 9). This diameter presumably includes the cross β-strand itself and the turns connecting successive β-strands. Thus the β-strands (not including turns) presumably span 20 Å to 25 Å. As β-structure has a translation of 3.2 Å per residue for parallel strands and 3.4 Å per residue for anti-parallel strands (15), approximately seven residues are required to span this distance.

3.1.3. *Relative Abundance of Polar and Nonpolar Residues*
β-strands composed of 7 alternating polar and nonpolar residues could contain either an excess of polar or an excess of nonpolar residues. Both options were considered. However, earlier work with model peptides showed that alternating sequences with an excess of nonpolar residues form intractable gels (12). Therefore the O●O●O●O pattern was chosen, with 4 polar (O) and 3 nonpolar (●) residues per β-strand (14).

3.1.4. *Parallel Versus Anti-Parallel β-Strands*
Structural studies of amyloid composed of peptides derived from the Alzheimer's Aβ sequence have found evidence for both parallel and antiparallel β-structure (7, 16). In our designed library, anti-parallel β–strands were chosen (14) because they can be linked by short turns or loops. Parallel strands were avoided because long connections between strands are difficult to design.

3.1.5. *Topology: Up-Down Versus Greek Key*
Anti-parallel β-strands can be hydrogen-bonded either to β-strands adjacent in the primary sequence (up-down topology) or to β-strands distant in the primary sequence (Greek key topology) (17). For any particular amino acid sequence, the choice between these pairings is influenced by the side chains on the respective strands (18). Since different amyloid proteins have different sequences, natural amyloid structures probably include contributions from both topologies. Likewise, because each member of our combinatorial library of *de novo* proteins has a different sequence, we expect some will favor up-down topology, while others will favor Greek key. The combinatorial underpinnings of the binary code strategy allow many different side chain combinations, and therefore both topologies are possible.

3.1.6. *Length of Turns*
Because the turns between β-strands must accommodate either up-down or Greek key topology, they must be long enough to join strands that are not neighbors in the 3-dimensional structure. Hence, they must be longer than 2 residues. Yet, they must also be short enough to favor turns rather than extensive loops or meanders. Turns of 3 or 5 residues were ruled out by considerations of 'negative design' (19). We wished not only to favor the desired structure -- amphiphilic β-strands separated by chain reversals -- but also to disfavor undesired structures, such as long uninterrupted amphiphilic β-strands running the length of a fibril. A sequence with an odd number of turn residues (e.g. O●O●O●O-t-t-t-O●O●O●O) could maintain alternating periodicity, but a sequence with an even number (e.g. O●O●O●O-t-t-t-t-O●O●O●O) cannot; the turn residues interrupt the periodicity. Therefore, we designed 4-residue turns (14).

3.1.7. *Sequences of Turns*
The detailed structure of amyloid has not been determined. Therefore, it is not known what types of turns link successive strands in the cross-β structure. In designing our library, we chose combinatorial turn sequences that allow various types of turns (e.g. I, I', II, II' etc.). The combinatorial diversity of the designed turn sequences was constrained to residues known to have high turn potentials at *all* positions *irrespective* of position or type. The residues having the highest overall turn potential are Gly, Asn, Asp, Pro, and Ser (20). Of these, we excluded Pro because its potential depends on its specific position in particular turn types. Based on these considerations, the turn sequences were combinatorially varied among Gly, Asn, Asp, and Ser. This mixture is encoded by the degenerate DNA codon RRT ('R' represents A or G).

3.1.8. *Length of Protein Sequence*
Amyloid structures from different diseases are composed of polypeptides of different lengths (2-8). For example, the dominant sequence in Alzheimer's plaque is only 42 residues, whereas the sequence of transthyretin in senile systemic amyloidosis is 127 residues. Thus it is clear that sequences of various lengths can assemble into amyloid. We chose to design sequences that would be short enough to facilitate straightforward gene construction, but long enough to resemble independent protein domains. The current library was designed to encode sequences encompassing 6 β-strands (7 residues each) punctuated by 5 turns (4 residues each). A schematic diagram of this pattern is shown in figure 1.

3.2 CONSTRUCTION OF A LIBRARY OF SYNTHETIC GENES

A diverse library of synthetic genes was constructed such that each member of the library encodes a different amino acid sequence, yet all sequences conform to the designed polar/nonpolar pattern shown in figure 1. The library of genes was constructed by assembling four synthetic oligonucleotides in which most codons were combinatorially degenerate (14). These degenerate codons encode amino acid sequences consistent with the alternating polar/nonpolar pattern of the β-strands (O●O●O●O), as well as combinatorially diverse residues in the first and fourth turns. Within each β-strand, the polar residues (His, Lys, Asn, Asp, Gln or Glu) were encoded by the degenerate DNA codon VAW, while nonpolar residues (Leu, Ile, Val or Phe) were encoded by NTT ('V' represents G, C or A; 'W' represents A or T; and 'N' represents any base). Within the combinatorial turn sequences, the residues Gly, Asn, Asp or Ser were encoded by the degenerate codon RRT ('R' represents G or A). The sequences of the second, third, and fifth turns were not varied, but were held constant to facilitate annealing and priming for the enzymatic synthesis of complementary strands.

Figure 1: (A) Schematic illustration of the design of a combinatorial library of *de novo* proteins containing six β-strands (arrows) punctuated by turns.
(B) Designed binary sequence pattern. Alternating pattern in the β-strands is indicated with polar residues (O) as white font in black background, and nonpolar residues (●) as black font in gray background. Combinatorial diversity is incorporated at positions marked O, ●, and t (turn). Fixed residues are incorporated at the termini and in some of the turns.
(C) Amino acid sequences (single letter code) of 17 *de novo* proteins from the combinatorial library.

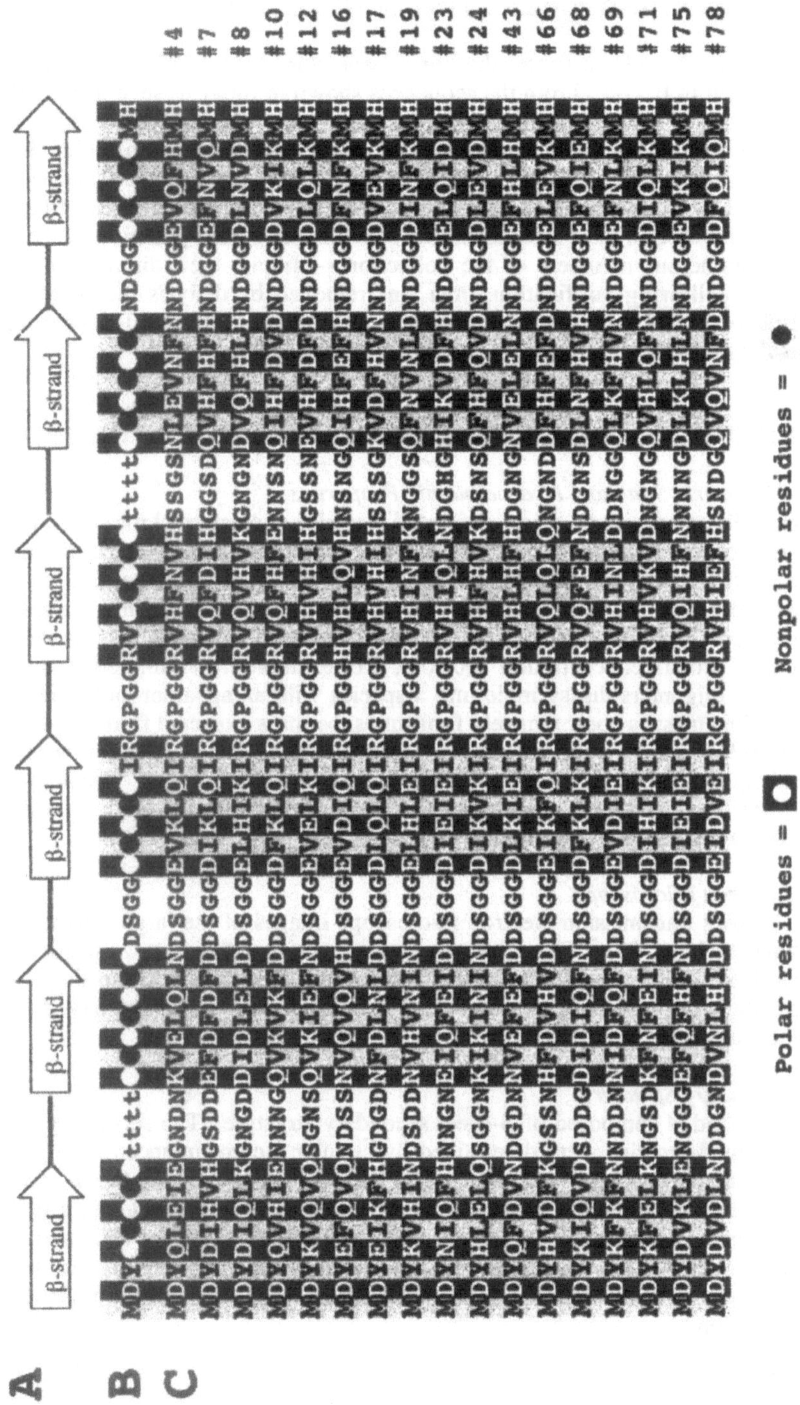
A
B
C
β-strand
β-strand
β-strand
β-strand
β-strand
β-strand
Polar residues =
Nonpolar residues =
#4
#7
#8
#10
#12
#16
#17
#19
#23
#24
#43
#66
#68
#69
#71
#75
#78

3.3. EXPRESSION AND PURIFICATION OF NOVEL PROTEINS

Following assembly of the genes, the library was cloned into a T7 vector and proteins were expressed in E. coli. From the sequences shown in figure 1, several proteins were arbitrarily chosen for expression and purification (14). As expected for sequences designed to form amphiphilic β-strands, all eight expressed as insoluble inclusion bodies. The inclusion bodies were solubilized in 10 M urea, and the *de novo* proteins were purified by anion exchange chromatography in the presence of urea. Because the proteins are purified in their urea-denatured state, the method is generic and was readily applied to numerous members of the collection -- despite their different amino acid sequences. Following purification, urea was removed by dialysis into native buffer. The *de novo* proteins remained in solution. Identities of the purified proteins were confirmed by mass spectrometry.

3.4. BIOPHYSICAL CHARACTERIZATION OF *DE NOVO* AMYLOID PROTEINS

3.4.1. *Oligomeric State and Hydrodynamic Properties*

Size and oligomeric state were estimated by size-exclusion chromatography. At moderate protein concentrations (500 μM), the *de novo* proteins elute in the excluded volume of the sizing column, indicating assembly into an oligomeric state with an apparent molecular mass >1,000,000. For a 63-residue protein, this suggests >140 monomers per oligomer. The large oligomers were dissociated into protomers by lowering the protein concentration 100-fold. For example, when diluted from 500 μM to 5 μM the oligomers disassemble into a species with an apparent molecular mass of ~29,000. This mass is approximately four times the mass expected from the amino acid sequence (6925), suggesting that at low concentrations the protein forms tetramers. [The native, non-amyloid form of transthyretin is also tetrameric (8)]. These protomers, which form upon dilution, are competent to re-assemble into the large oligomers and readily do so upon re-concentration.

3.4.2. *Electron Microscopy*

Figure 2 shows transmission electron microscope images of the large oligomeric form of proteins #17 and #23. These images demonstrate that the *de novo* sequences self-assemble into fibrils. The fibrils have diameters of 25-35 Å, similar to those seen for the protofilaments of natural amyloid (7, 9). Similar images were observed for other sequences shown in figure 1.

3.4.3. *Secondary Structure*

Natural amyloid is composed of β-sheet secondary structure. The secondary structures of the designed proteins were determined by circular dichroism spectroscopy (Fig. 3). The spectra display a single minimum at ~217 nm demonstrating that the *de novo* fibrils are dominated by β-structure (21). Fitting the experimental spectra to the algorithm of Greenfield and Fasman (21, 22) enables an estimation of 76% for #17; and 78% for #23. These estimates are consistent with the design in which 68% of the residues (43 of 63) were designed to be in β-strands. Similar spectra were observed for other sequences in the collection.

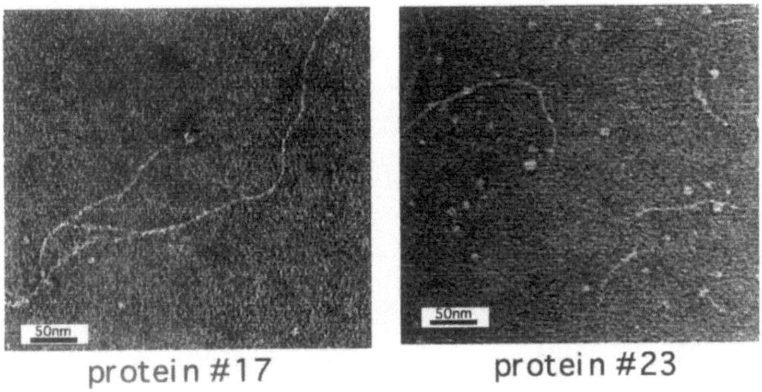

Figure 2: Transmission electron micrographs of *de novo* amyloid-like fibrils.

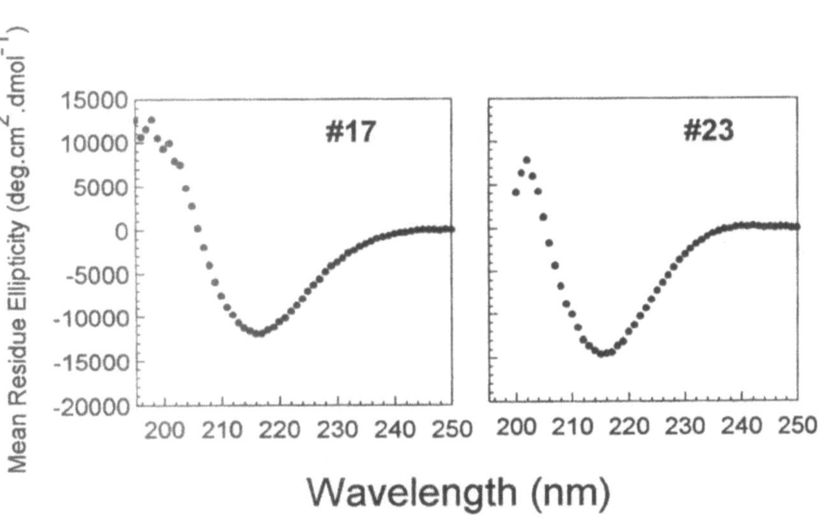

Figure 3: Circular dichroism spectra. A minimum at ~217 nm indicates the presence β-structure

3.4.4. *Binding to Congo Red*

Amyloid structures are routinely identified by their binding to the dye Congo Red. Binding appears to be specific for amyloid proteins with a cross-β structure (23). For our designed proteins, Congo Red binding was demonstrated spectroscopically as shown in figure 4. Visually, this corresponds to a shift from pink to bright red. Binding was observed for the high-order oligomeric form of the *de novo* proteins, but not for monomeric or tetrameric forms. The results shown in figure 4 for the *de novo* proteins mimic those reported for naturally occurring amyloid fibrils (23).

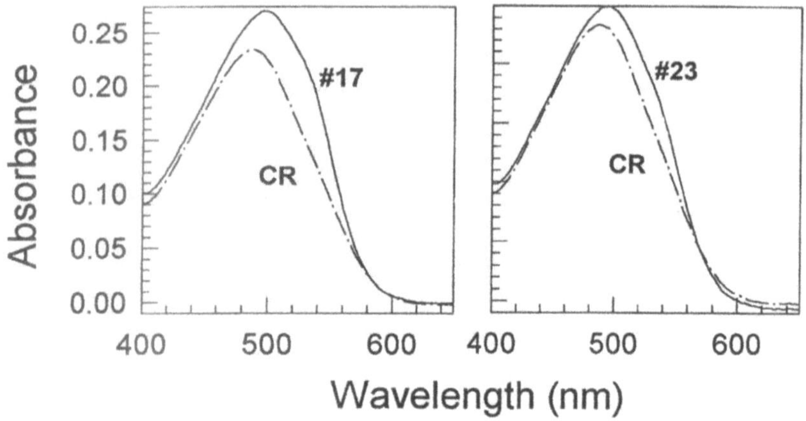

Figure 4: Binding to Congo Red. Each panel compares the spectrum of Congo Red (CR) alone to that of CR in the presence of amyloid protein. The fibrillar form of the Alzheimer's peptide, Aβ (1-40) yielded a similar spectrum. Similar spectra were observed for other sequences shown in figure 1.

3.4.5 *Assembly and Disassembly of Fibrils*

At room temperature (or physiological temperature), the *de novo* proteins form β-strands that assemble into long fibrils capable of binding Congo Red (Figs. 2-4). However, as the temperature is raised, the structure responsible for these properties disassembles (Fig. 5) Upon cooling back to room temperature, β-structure re-forms and the protein reassembles into large oligomers that elute in the excluded volume of a sizing column. Thus the disassembly and re-assembly process is reversible.

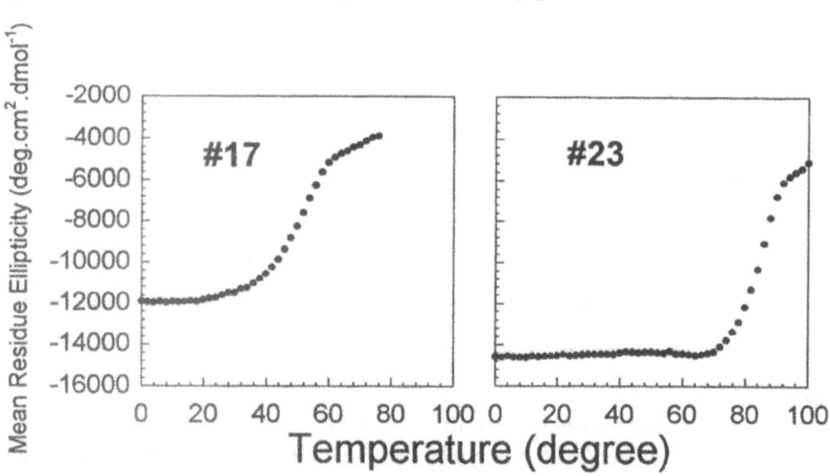

Figure 5: Thermally-induced disassembly. Loss of structure was followed by monitoring ellipticity at 217 nm. >90% of the initial signal is restored upon cooling back to room temperature.

Herein lies a key advantage of *de novo* sequences explicitly designed to contain alternating polar/nonpolar patterns. The designed sequences fold into fibrils quite readily. Moreover, the reversible assembly and disassembly of these *de novo* fibrils suggests their structures are dictated by thermodynamic stability, not kinetic trapping. For these reasons, this library of model amyloid structures may provide important advantages as a tractable system both for basic studies into the mechanisms of fibril assembly, and for the development of molecular therapies that interfere with this assembly.

3.5. ALTERNATING POLAR/NONPOLAR SEQUENCES ARE DISFAVORED BY NATURE.

The results described above demonstrate that for libraries of sequences designed *de novo,* alternating patterns of polar and nonpolar amino acids predispose a sequence to form amyloid-like structures. If this is also true for sequences that evolved in nature, one might expect alternating patterns to be toxic, and therefore disfavored by natural selection. To test this possibility, we analyzed a database (24) of 250,514 natural protein sequences comprising 79,708,024 residues, and calculated the frequencies of alternating patterns relative to other patterns with similar compositions (25).

The expected probability for any binary pattern of polar and nonpolar amino acids can be calculated from the amino acid composition of the database as a whole. (Polar residues were defined as Arg, Lys, Asp, Glu, Asn, Gln, and His; Nonpolar residues were defined as Leu, Ile, Val, Phe, and Met (13)). All cassettes containing the same number of polar and nonpolar residues (e.g. 4 polar and 3 nonpolar) would be expected to occur with the same frequency. In contrast with this expectation, however, our search demonstrated that equal representation is not observed: Some patterns occur far less frequently than others. As shown in figure 6, the alternating binary patterns occur significantly less often than other patterns with the same polar/nonpolar composition. For cassettes between 5 and 9 residues, the alternating patterns are the absolute rarest. For example, there are 35 possible ways of arranging 4 polar and 3 nonpolar residues in a string of 7 residues. The alternating pattern used in our design (O●O●O●O) ranks 35th (Fig. 6). Likewise, there are 70 ways of arranging 4 polar and 4 nonpolar residues in a string of 8 residues. Among these, the two alternating patterns O●O●O●O● and ●O●O●O●O rank as 69th and 70th. (25). For all windows ranging from 5 to 10 residues, alternating patterns invariably occur at least one standard deviation below the mean (25).

Based on these considerations one might expect the sequences of natural amyloidogenic proteins to be rich in alternating polar/nonpolar patterns. We found, however, that the patterns in these sequences are similar to those in the database as a whole. Although initially surprising, this similarity is exactly what should be expected. The amyloid found in diseased tissues is an off-pathway misfolded structure (6, 8). The amino acid sequences of amyloidogenic proteins did not evolve to form amyloid. They evolved to fold into globular structures capable of performing functions beneficial to the organism. As expected, their amino acid sequences reflect this evolutionary history.

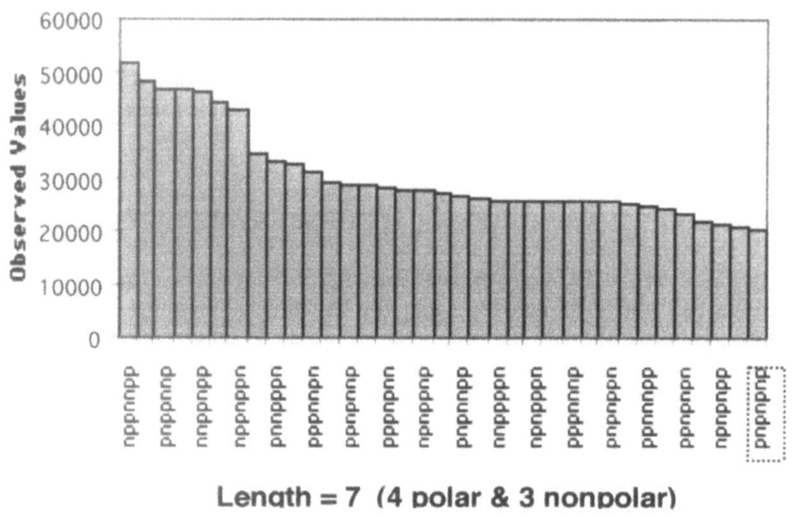

Figure 6: Number of occurrences of various binary patterns of polar ('p') and nonpolar ('n') amino acids. The figure shows the tabulation for a 'window' of 7 residues with a composition containing 4 polar and 3 nonpolar residues. The alternating pattern (dotted rectangle) occurs less frequently than other patters with the same composition of polar and nonpolar residues. Similar results were observed for windows ranging from 5 to 10 residues (25). In all cases, the alternating patterns occur less frequently than other patters with the same composition.

---

4. Discussion

The designed sequences described above self-assemble into amyloid-like structures. Which features in their design predispose these *de novo* proteins to assemble into fibrils? To address this question, we contrast the sequences described in the current study with those described in our earlier work designing α-helical proteins (26). In both cases, combinatorial libraries were designed by constraining the binary pattern of polar and nonpolar residues. Yet the resulting proteins display dramatically different properties: In the previous work, the sequences formed α-helices that folded *intra*molecularly into small globular domains (26-28). In contrast, the sequences described in the current work form β-strands and self-assemble *inter*molecularly into high-order oligomers that assume fibrillar structures.

What causes these dramatically different structures? The lengths of the sequences are not dramatically different (74 versus 63), nor are their overall compositions. We propose that the determining difference is the binary patterning itself. In the earlier work, the library was constrained by the pattern O●OO●●OO●OO●●O, consistent with the periodicity of α-helical structure (26). In contrast, the current library is constrained by the pattern O●O●O●O, consistent with the periodicity of amphiphilic β-strands.

Our findings are consistent with previous observations on several synthetic peptides. In pioneering work, Brack and Orgel demonstrated that polymers composed of alternating valine and lysine residues form β-structures that aggregate into large assemblies (29). More recently, Janek *et al.* showed that alternating valine and lysine residues can serve as model systems for studies of fibril self-assembly (30). Zhang *et al.* found that 16-residue peptides composed of sequences that alternate between alanine and a charged residue spontaneously assemble into large oligomeric structures dominated by β-sheet (31). Finally, Lim *et al.* described a *de novo* protein, Betabellin-15D, formed by the disulfide-mediated dimerization of a 32-residue peptide. The 32-mer was designed to form 4 β-strands, each containing 6 residues of alternating polar and nonpolar residues. The resulting proteins formed structures visible in electron microscopy images as long narrow multimeric fibrils (32).

The amyloidogenic propensity of alternating polar/nonpolar patterns is further supported by our survey of the sequences of natural proteins (25). In that search, we found that nature disfavors alternating patterns. We propose that evolution has selected against such sequences because they have an inherent propensity to aggregate into deleterious fibrillar plaque.

References

1. Alzheimer, A. (1907) *Alg. Z. Psychiatry* 64, 146-148.
2. Lansbury Jr., P.T. (1996) A reductionist's view of Alzheimer's disease. *Acc. Chem. Res.* 29, 317-321.
3. Kelly, J.W. (1996) Alternative conformations of amyloidogenic proteins govern their behavior. *Curr. Opin. Struct. Biol.* 6, 11-17.
4. Kisilevsky, R. & Fraser, P.E. (1997) Aβ Amyloidogenesis: Unique or variation on a systemic theme. *Crit. Revs. Biochem. & Molec. Biol.* 32, 361-404.
5. Pruisner, S.B. (1997) Prion diseases and the BSE Crises. *Science* 278, 245-250.
6. Wetzel, R. (1997) Domain stability in immunoglobulin light chain deposition disorders. *Advances in Protein Chemistry* 50, 183-242.
7. Sunde, M. & Blake, C. (1997) The structure of amyloid fibrils by electron microscopy and X-ray diffraction. *Advances in Protein Chemistry* 50, 123-159.
8. Kelly, J.W., Colon, W., Lai, Z., Lashuel, H.A., McCulloch, J., McCutchen, S.L., Miroy, G.J., Peterson, S.A. (1997) Transthyretin quaternary and tertiary structural changes facilitate misassembly into amyloid. *Advances in Protein Chemistry* 50, 161-181.
9. Harper, J.D., Wong, S.S., Lieber, C.M., & Lansbury Jr., P.T. (1997) Observation of metastable Aβ amyloid protofibrils by atomic force microscopy *Chemistry & Biology* 4, 119-125.
10. Lansbury Jr., P.T., Costa, P.R., Griffiths, J.M., Simon, E.J., Auger, M., Halverson, K.J., Kocisko, D.A., Hendsch, Z.S., Ashburn, T.T., Spencer, R.G.S., Tidor, B. & Griffin, R.G. (1995) Structural model for the β-amyloid fibril based on interstrand alignment of an antiparallel-sheet comprising a C-terminal peptide. *Nature Structural Biology*. 2, 990-997.
11. Xiong, H., Buckwalter, B.L., Shieh, H.M. & Hecht, M.H. (1995) Periodicity of polar and nonpolar amino acids is the major determinant of secondary structure in self-assembling oligomeric peptides. *Proc. Natl. Acad. Sci. USA* 92, 6349-6353.
12. Xiong, H. (1995). The importance of hydrophobic/hydrophilic patterns in the amino acid sequence of peptides and proteins. Ph.D. Dissertation. Department of Chemistry, Princeton University.
13. West, M.W. & Hecht, M.H. (1995) Binary patterning of polar and nonpolar amino acids in the sequences and structures of native proteins. *Protein Science* 4, 2032-2039.
14. West MW, Wang W, Patterson J, Mancias JD, Beasley JR & Hecht MH (1999) De novo amyloid proteins from designed combinatorial libraries. *Proc.Natl Acad. Sci.(USA)* 96, 11211-11216.
15. Creighton, T.E. (1993) *Proteins: Structures and molecular properties (2nd edition).* WH Freeman & Company, New York.
16. Benzinger, T. M., Gregory, D. M., Burkoth, T. S., Miller-Auer, H., Lynn, D. G., Botto, R. E. & Meredith, S. C. (1998) Propagating structure of Alzheimer's β-Amyloid(10-35) is parallel β-sheet with residues in exact register *Proc. Natl. Acad. Sci. USA* 95, 13407-13412.
17. Richardson, J.S. (1981). Anatomy and taxonomy of protein structure. *Advances in Protein Chemistry* 34, 167-339.
18. Zhu, H. Y. & Braun, W. (1999) Sequence specificity, statistical potentials, and three-dimensional structure prediction with self-correcting distance geometry calculations of β-sheet formation in proteins. *Protein Science* 8, 326-342.
19. Hecht, M.H., Richardson, J.S., Richardson, D.C. & Ogden, R.C. (1990) *De novo* design, expression, and characterization of felix: A four-helix bundle protein of native-like sequence.

Science 249, 884-891.
20. Hutchinson, E.G. & Thornton, J.M. (1994) A revised set of potentials for β-turn formation in proteins. *Protein Science* 3, 2207-2216.
21. Greenfield, N. & Fasman, G.D. (1969) Computed circular dichroism spectra for the evaluation of protein conformation. *Biochemistry* 8, 4108-4116.
22. Greenfield, N.J. http://www2.umdnj.edu/cdrwjweb/index.htm#software. The program we used is called G&F, which uses poly-lysine as model structures.
23. Klunk, W.E., Pettegrew, J.W., & Abraham, D.J. (1989) Quantitative evaluation of Congo red binding to amyloid-like proteins with beta pleated sheet conformation *J. Histochem. & Cytochem*. 37, 1273-1281.
24. Bleasby, A.J., Akrigg, D., Attwood, T.K. (1994) OWL - A non-redundant, composite protein sequence database. *Nucleic Acids Research* 22, 3574-3577.
25. Broome BM & Hecht MH (2000) Nature Disfavors Sequences of Alternating Polar and Nonpolar Amino Acids: Implications for Amyloidogenesis. *J. Molec. Biol.* 296, 961-968.
26. Kamtekar, S., Schiffer, J.M., Xiong, H., Babik, J.M. & Hecht, M.H. (1993) Protein design by binary patterning of polar and nonpolar amino acids.*Science* 262, 1680-1685.
27. Roy, S., Ratnaswamy, G., Boice, J.A., Fairman, R., McLendon, G. & Hecht, M.H. (1997) A protein designed by binary patterning of polar and nonpolar amino acids displays native-like properties. *J. Am. Chem. Soc.* 119, 5302-5306.
28. Roy, S. & Hecht M.H. (2000) Cooperative thermal denaturation of proteins designed by binary patterning of polar and nonpolar amino acids. *Biochemistry (in press)*
29. Brack, A. & Orgel, L. E. (1975) β structures of alternating polypeptides and their possible prebiotic significance. *Nature* 256, 383-387.
30. Janek, K, Behlke, J., Zipper, J., Fabian H., Georgalis, Y., Beyermann, M., Bienert, M., & Krause, E. (1999) Water soluble β-sheet models which self-assemble into fibrillar structures. *Biochemistry* 38, 8246-8252.
31. Zhang, S., Holmes, T. Lockshin, C. & Rich, A. (1993) Spontaneous assembly of a self-complementary oligopeptide to form a stable macroscopic membrane. *Proc. Natl. Acad. Sci. USA* 90, 3334-3338.
32. Lim, A., Saderholm, M.J., Makhov, A.M., Kroll, M., Yan, Y., Perera, L., Griffith, J,D,, & Erickson, B.W. (1998) Engineering of betabellin-15D: A 64 residue beta sheet protein that forms long narrow multimeric fibrils. *Protein Science* 7, 1545-1554.

DESIGN OF SYNTHETIC BRANCHED CHAIN POLYMERIC POLYPEPTIDES FOR TARGETING/DELIVERING BIOACTIVE MOLECULES

F. HUDECZ
Research Group of Peptide Chemistry, Hungarian Academy of Science, Eötvös L.University, P.O. Box 32, Budapest 112, Hungary, H-1518

1. Introduction

Soluble synthetic polymers represent a new group of compounds considered as novel class of therapeutics. Recent results proved that such polymeric macromolecules could exhibit promising biological activity attractive for biomedical research and application. *N*-(2-hydroxypropyl)methacrylamide copolymer (HPMA) has been used with success in experimental cancer research [1], while Copaxone® a random copolymer of Lys, Glu, Tyr and Ala residues has been proved by FDA against multiple sclerosis in human [2]. Polyamino acids containing negatively charged and aromatic residues at specific ratios appear to bind HIV type 1 V3 loop and neutralise diverse laboratory isolates [3]. Artificial immunogens as vaccine candidates have been produced by free radical induced polymerization of vinyl monomers such as the acryloyl peptides representing B- and or T-cell epitopes of relevant proteins [4]. Peptide oligomers containing an (Aib-Lys-Gly)$_n$ backbone and epitope sequence at the ε-amino group of Lys residue were built up by co-linear solid phase synthesis [5].

Polymers are also developed for modification of proteins to suppress their undesired properties. Therefore protein/peptide conjugates with a polymeric component

[†] Abbreviations for amino acids and their derivatives follow the revised recommendation of the IUPAC-IUB Committee on Biochemical Nomenclature, entitled "Nomenclature and Symbolism for Amino Acids and Peptides" (recommendations of 1983). Nomenclature of branched polypeptides is used in accordance with the recommended nomenclature of graft polymers (IUPAC-IUB recommendations, 1984,.) For the sake of brevity codes of branched polypeptides were constructed by us using the one-letter symbols of amino acids. The abbreviations used in this paper are the following. ADA, adenosine deaminase; AK, poly[Lys-(DL-Ala$_m$)]; ANS, sodium anilino naphthalene sulfonate; AXK, poly[Lys-(DL-Ala$_m$-X$_i$)]; X = Leu (ALK), Phe (AFK), Ser (ASK), Glu (AEK); CD circular dichroism; PG, phosphatidyl glycerol; DP$_n$, average degree of polymerisation; DPH, 1,6-diphenyl-1,3,5-hexatriene; DPPC, dipalmitoyl phosphatidyl choline; DTPAA, diethylenetriamine pentaacetic acid anhydride; FDA, Food and drug administration; GPC, gel permeation chromatography; GnRH, gonadotropin releasing hormone; HPMA, *N*-(2-hydroxypropyl)methacrylamide copolymer; Pcp, pentachlorophenol; PEG, polyethylene glycol; SK, poly[Lys(DL-Ser$_m$)]; TFA, trifluoroacetic acid; XAK, poly[Lys(X$_i$-DL-Ala$_m$)]; X = Leu (LAK), Phe (FAK), Ser (SAK), Thr (TAK), Glu (EAK), Ac-Glu (Ac-EAK), Suc-Glu (Suc-EAK), Lys (KAK), Orn (OAK), His (HAK) or Pro (PAK); XSK, poly[Lys(X$_i$-DL-Ser$_m$)]; X = Leu (LSK), Glu (ESK), Ac-Glu (Ac-ESK); X$_i$K, poly[Lys(X$_i$)]; Z, benzyloxycarbonyl;. All amino acids are of L-configuration unless otherwise stated.

A. Aggeli et al. (eds.), Self-assembling Peptide Systems in Biology, Medicine and Engineering, 139–160.

could be considered as an other class of compounds with high potential for rational design of therapeutics [6,7]. The coupling of polyethylene glycol (PEG)-polymers to proteins has yielded derivatives with intact protein-related biological activity and reduced immunogenicity. PEG modified adenosine deaminase (ADA) conjugate preparations are in use as replacement therapy for ADA deficiency in patients with severe combined immunodeficiency, while PEG-asparaginase (pegaspargase) has shown promise in patients with acute lymphocytic leukemia [6]. Further examples are discussed in a recent comprehensive review on degradable polymeric delivery systems available for protein pharmaceuticals in relation to both protein stability and protein release kinetics [7].

In a third group of compounds polymers are utilised as macromolecular carriers for target specific delivery and/or for sustained release of drugs, radionuclides. These conjugates exhibit improved solubility, beneficial pharmacological properties (extended blood survival, specific tissue accumulation), diminished cytotoxicity and increased therapeutic efficacy as compared to the free drugs, radionuclides [8-10]. In the field of cancer chemotherapy clinical studies with two conjugates have shown promising results. Neocarcinostatin covalently attached to copolymer of styrene and maleic acid anhydride introduced by Maeda showed increased accumulation in solid tumours and led to the discovery of the "enhanced penetration and retention" [11]. HPMA copolymer bearing doxorubicin represents another highly efficient polymer therapeutics in phase I/II clinical trial [12,13]. A recent extension of this line of research is the use of polymeric macromolecules for construction of synthetic antigens. In these conjugates the polymer component serves also as carrier, but instead of drugs the attached bioactive entities are peptide epitope(s) of viral or bacterial origin [14, 15, 16]. This approach could led to the development of the new generation of fully synthetic subunit vaccines [17, 18].

Despite numerous papers in the literature which include many encouraging results even in experimental and/or preclinical models, very few systematic studies were reported on structural and functional factors required for an optimal polymer carrier. To this end we have initiated investigation for establishing structure - function type correlation to facilitate rational design and/or selection of polymers as synthetic macromolecular carrier.

In order to contribute to this line of research we have prepared new groups of branched chain polymeric polypeptides with the general formula poly[Lys(X_i-DL-Ala_m)] (XAK), poly[Lys(X_i-DL-Ser_m), (XSK), poly[Lys(X_i)] (X_iK) or poly[Lys(DL-Ala_m-X_i)] (AXK) where $i<1$, $m \sim 3$, and X represent an additional optically active amino acid residue [19-22]. These polymeric polypeptides were characterised by their size, chemical (primary structure, solution conformation) and biological (*in vitro* cytotoxicity, pyrogenicity, biodegradation, immunoreactivity and biodistribution) properties [19, 21-24]. In this communication recent progress in the synthesis, structural and biological evaluation of a new set of polymeric polypeptides will be reviewed and discussed in the context of polymers belonging to the same family, but reported earlier. Finally brief section will deal with the preparation and biological activity of branched

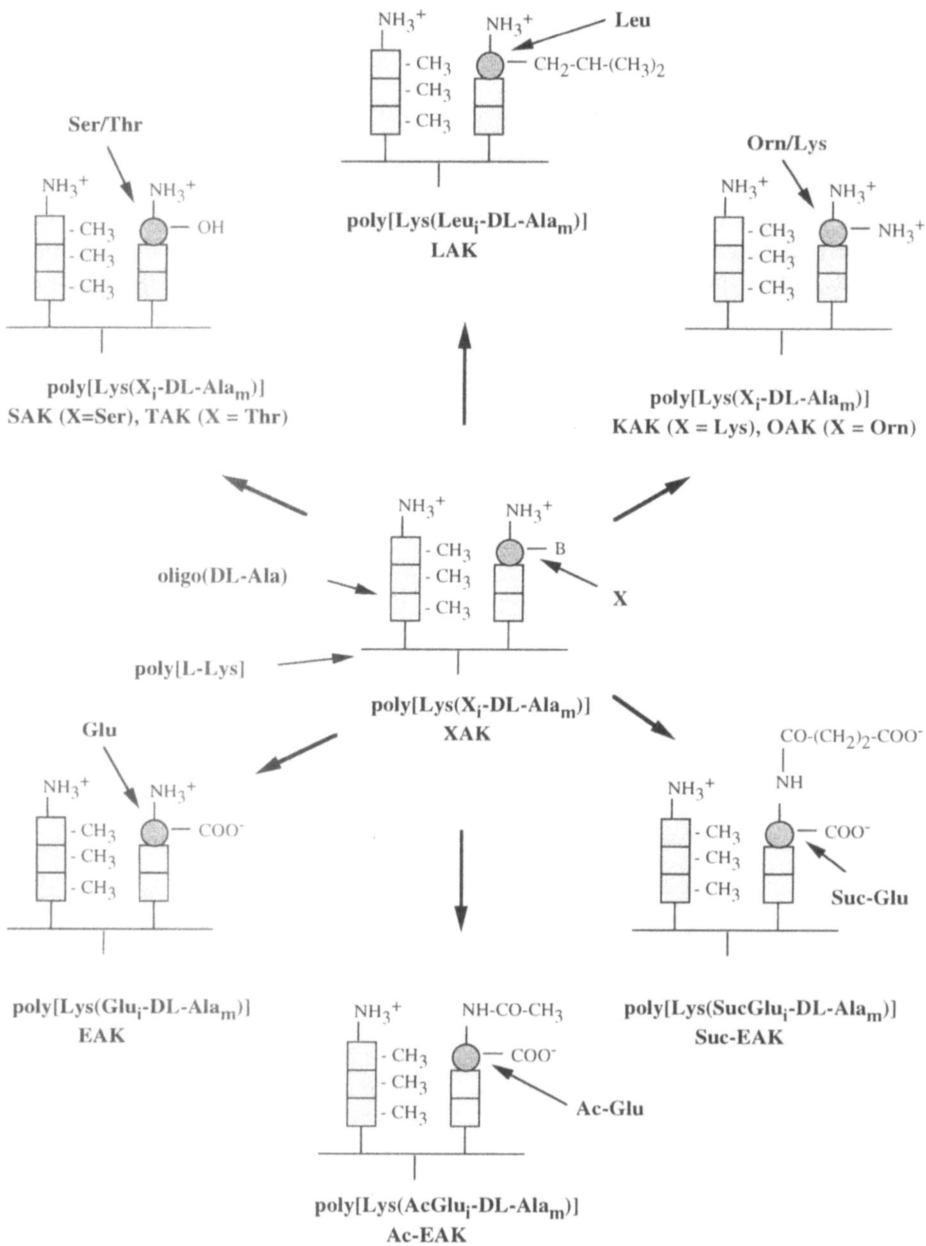

Figure 1A..Simplified structure of the poly[Lys(X$_i$-DL-Ala$_m$)], XAK type branched polymeric polypeptides

polypeptide based conjugates with anticancer drugs (daunomycin, GnRH analogue), radionuclides and epitope peptides.

2. Synthesis and Chemical Structure

Branched chain polymeric polypeptides developed in our laboratory are poly[L-Lys] derivatives substituted by short (3-6 amino acid residue) branches at the ε-amino groups. The side chains are composed of about three DL-Ala [20, 25] (Figure 1A) or DL-Ser [22] (Figure 1B) residues and one other amino acid residue (X) at the N-terminal end of the branches (XAK or XSK) (Figures 1) or next to the poly[L-Lys] backbone (AXK) (Figure 2).

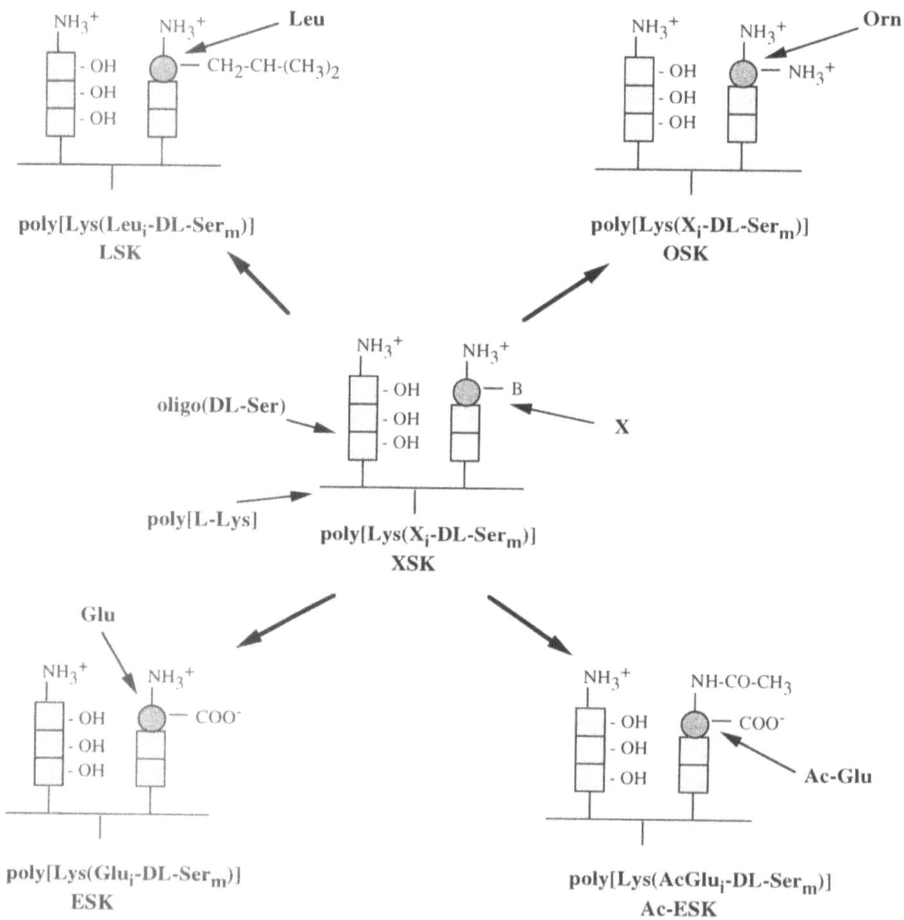

Figure 1B. Simplified structure of the poly[Lys(X$_i$-DL-Ser$_m$)], XSK type branched polymeric polypeptides

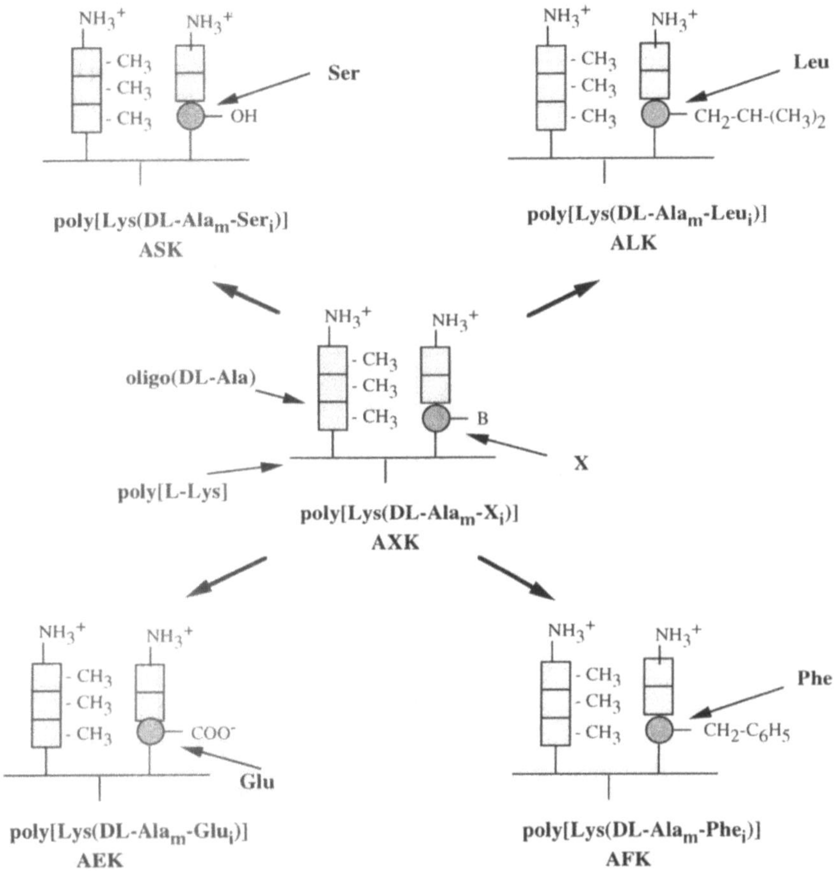

Figure 2. Simplified structure of the poly[Lys(DL-Ala$_m$-X$_i$)], AXK type branched polymeric polypeptides

The strategies for the syntheses of polypeptides designed by us are outlined in Figure 3 for XAK and XSK. It should be noted that somewhat similar XAK type polymeric polypeptides with much longer DL-Ala branches (20-30 residues) and copolymeric N-terminal portions were described by Sela et al. [26]. Experimental details of the procedures for preparation of the XAK [22, 25, 27] and XSK [22] polymers were presented previously. Briefly, poly[L-Lys] was synthesised by the polymerisation of N^{α}-carboxy-N^{ε}-benzyloxycarbonyl-lysine anhydride under conditions that allowed a number average degree of polymerisation [DP_n] approximately 60 or 120. Protecting groups were cleaved and DL-alanine or DL-serine oligomers were grafted to the ε-amino groups of polylysine by polymerisation of N-carboxy-DL-alanine anhydride or of N-carboxy-DL-serine anhydride to produce poly[Lys(DL-Ala$_m$)] (AK) and

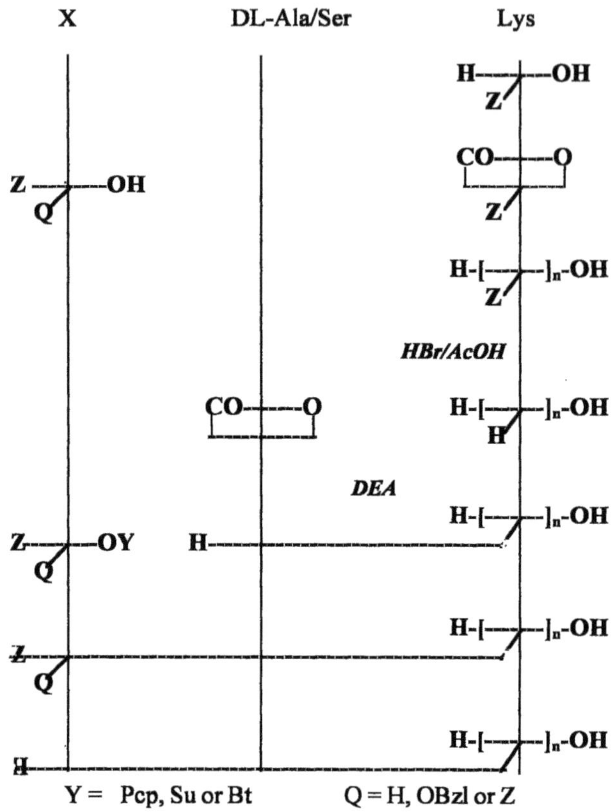

Figure 3. Outline of the synthesis of poly[Lys(X_i-DL-Ala_m)], XAK or poly[Lys(X_i-DL-Ser_m)], XSK type branched polymeric polypeptides.

poly[Lys(DL-Ser_m)] (SK), respectively. The suitably protected and activated amino acid X was coupled to the end of branches of AK or SK by amide bond in the form of Z-X(U)-OPcp residue, where X = Glu (U=OBzl), Orn (U=Z), Lys (U=Z), Ser, Thr (U=O). After the removal of protecting groups samples were dialysed against distilled water, freeze-dried. From sedimentation equilibrium measurements of polylysine, the number average degree of polymerisation the average molar mass, and the relative molar mass distribution were calculated [28]. The average molar mass of the branched polypeptides was estimated from DP_n of polylysine and from the amino acid composition of the side chains determined by quantitative amino acid analysis.

These polypeptides can be classified according to their charge properties. Under physiological conditions (pH 7.3 in 0.15 M NaCl) polycationic polymers like poly[Lys(Leu_i-DL-Ala_m)], (LAK), poly[Lys(Phe_i-DL-Ala_m)], (FAK), poly[Lys(Ser_i-DL-Ala_m)], (SAK) and poly[Lys(Thr_i-DL-Ala_m)], (TAK) possess single positive charge (protonated α-amino group) per side chains. Within this group we found that the water solubility of these compounds differs significantly. The presence of hydrophobic amino acids (e.g. Leu, Ile, Phe) at the branch terminal position resulted in compounds with pronounced tendency to form aggregates after storage for a longer period of time. In contrast polycationic polymers with hydrophilic amino acid residues (Ser, Thr) at the same position possess long shelf-life stability and they are highly soluble in aqueous solution even after extended storage. Polypeptides containing Orn in poly[Lys(Orn$_i$-DL-Ala$_m$)], (OAK) or Lys in poly[Lys(Lys$_i$-DL-Ala$_m$)], (KAK) at the end of the branches are also polycationic. These compounds bear two positive charges per side chain having both α- and ω-amino groups. The side chains of amphoteric poly[Lys(Glu$_i$-DL-Ala$_m$)], (EAK) contains glutamic acid at the end of the branches. Therefore this polymer has not only free α-amino, but also free γ-carboxyl group in the side chain, consequently this compound has balanced charge character. Acetylation of EAK resulted in a polyanionic derivative poly[Lys(Ac-Glu$_i$-DL-Ala$_m$)], (Ac-EAK) [29]. In this polypeptide no positive charge can be observed, which is due to the covalent modification of free α-amino group of Glu. The γ-carboxyl group of the terminal amino acid remained intact providing a polymer with one negative charge per side chain. By acylation the α-amino group of Glu in EAK with dicarboxylic anhydride we have produced a polymer in which the positively charged α-amino group of Glu is replaced by a negative charge. Therefore succinylated EAK (Suc-EAK) possesses not only free γ-carboxyl group, but also a second negative charge per side chain which originates from the dicarboxylic acid used for acylation [29]. Suc-EAK has two negative charges per side chain and represents a polyanionic polypeptide with high negative charge density.

Recently polypeptides with multiple hydroxyl groups in the side chains were prepared [22, 34]. In these compounds oligo(DL-alanine) was replaced by oligo(DL-serine) as shown in Figure 1B. High water solubility and long term shelf stability were achieved even in case of polymer with hydrophobic Leu residue (poly[Lys(Leu$_i$-DL-Ser$_m$)] (LSK). Within this group we have also polycationic (LSK), amphoteric (ESK) and polyanionic (Ac-ESK) polypeptides.

In XAK and XSK type of polypeptides amino acid residue X is situated at the end of the branches. For the analysis of the role of the position of X in conformation and biological properties of the polymers we have invented compounds with reversed sequences in the side chains. In these compounds amino acid X is located next to the polylysine backbone and elongated by short oligo(DL-alanine). The outline of the synthetic route is given in Figure 4, while schematic structure of these polymers is depicted in Figure 2. Detailed description of the synthesis has been published earlier. Briefly, ε-amino groups of poly[L-Lys] was reacted with suitably protected and activated amino acid X to form amide bond. For this Z-X(U)-OPcp amino acid derivatives were used where X = Glu (U=OBzl), Ser, Leu, Pro, His (U=O). Protecting groups were cleaved and DL-alanine oligomers were grafted to the α–amino groups of poly[Lys(X$_i$)] (X$_i$K) by polymerisation of N-carboxy-DL-alanine anhydride to produce poly[Lys(DL-Ala$_m$-X$_i$)] (AXK). Samples were dialysed against distilled water and freeze-dried. Three polycationic variants (poly[Lys(DL-Ala$_m$-Ser$_i$)] (ASK), poly[Lys(DL-Ala$_m$-Leu$_i$)] (ALK), poly[Lys(DL-Ala$_m$-Phe$_i$)] AFK) and an amphoteric polymer, poly[Lys(DL-Ala$_m$-X$_i$)] (AEK) are shown in Figure 2.

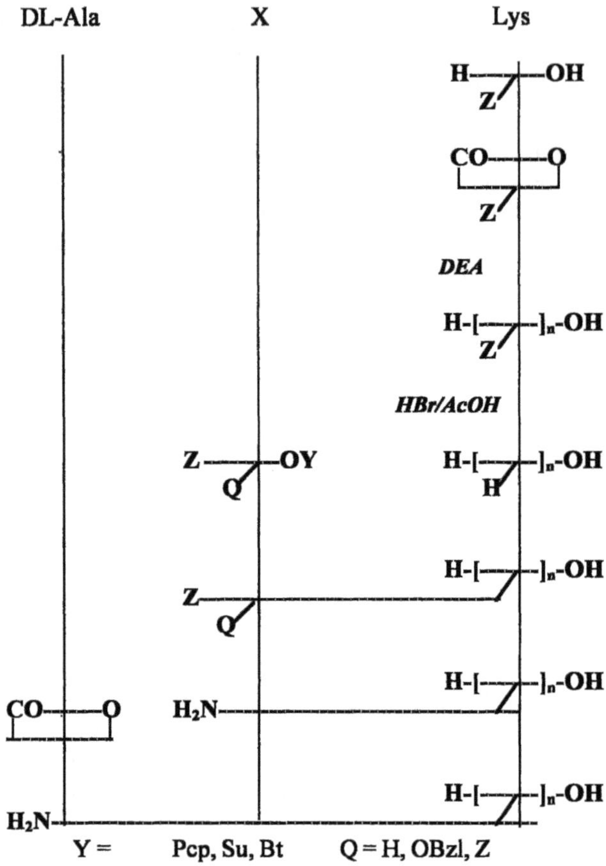

Figure 4. Outline of the synthesis of poly[Lys(DL-Ala_m-X_i)], AXK type branched polymeric polypeptides.

The common feature of AXK polypeptide is that the branch terminating positively charged α-amino group is separated in space from the side chain group of amino acid X. Comparison the structure of ASK (Figure 2) and of SAK (Figure 1A) could point out differences in position of hydroxyl group of the Ser residue relative to the α-amino group of the branch terminal. In SAK both functions are belonging to the same amino acid residue (e.g. Ser) and they are close in space. However, in ASK the terminal alanine possesses α-amino group, while OH group is part of the Ser residue. These two functions are segregated by several Ala residues and are not close in space.

The size and architecture of all these polymeric polypeptides are rather similar, but differences in the identity and position of amino acid X in their side chains resulted in altered relative hydrophobicity and electrical charge. These features provide a logical basis for comparative studies of their conformational properties, membrane interaction and biological features relevant for carrier function.

3. Chemical Characterisation of Branched Polypeptides

We have characterised the structural properties of branched chain polymeric polypeptides as outlined in Figure 5 by their size, primary structure, conformation in solution and their surface activity at the air/water interface.

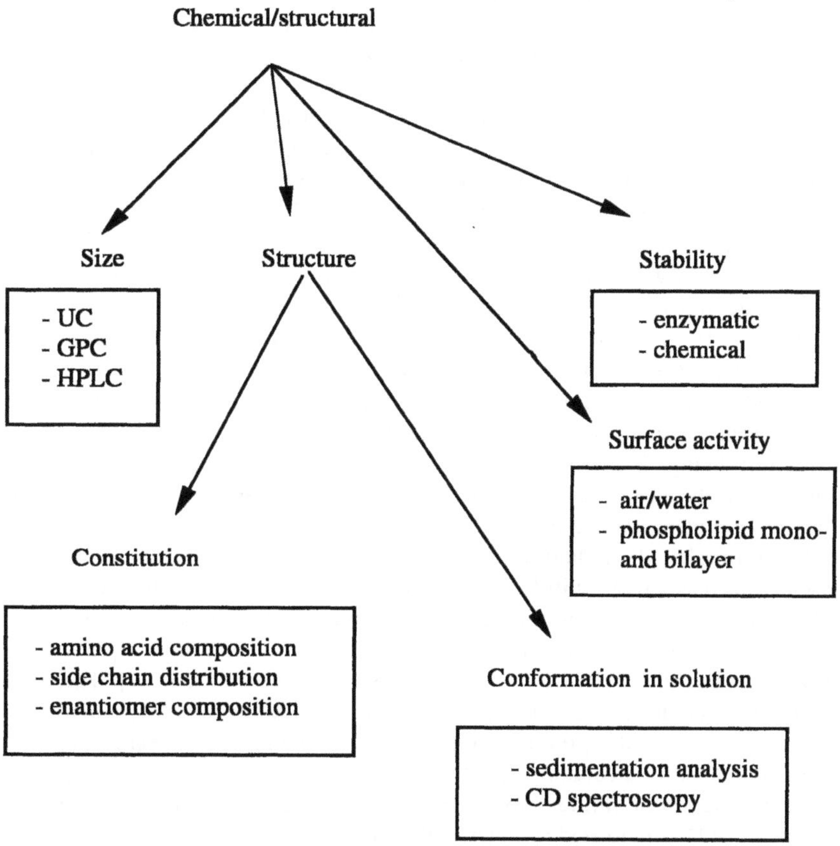

Figure 5. Outline of chemical characterization of branched polymeric polypeptides.

3.1. SIZE, PRIMARY STRUCTURE

The average relative molar mass of XAK, XSK and AXK type polymers were estimated from DP_n of the polylysine backbone obtained from sedimentation equilibrium measurements [28] and the equivalent relative molar mass of the side chains per one lysine residue obtained by quantitative amino acid analysis. Gel permeation column chromatography (GPC) using Sephadex G-100, G-150, G-150 superfine [28] or Sephacryl S-300 [30] gels were also applied. It has been found that the elution profile from GPC columns can be effectively used for a quick characterisation of the relative molar mass distribution of branched polypeptides, but it is not suitable for the determination of average relative molar mass with conventional protein standards. Preliminary data have been reported on the heterogeneity of drug-conjugates studied by capillary electrophoresis [31]. Recently we have developed a reversed phase HPLC based analytical approach for the highly charged branched polymeric polypeptides [32,33]. Polymer samples were studied on Delta Pak C_{18} column with 15 mm silica (300 Å pore size) as a stationary phase, and 0.1 % TFA in water was used as eluent A and 0.1% TFA in acetonitrile-water (80:20, v/v %) as eluent B. Using linear gradient elution and UV detection at λ = 208 and 214 nm we have provided a new set of data for the characterisation of polycationic (SAK, OAK, ALK), amphoteric (EAK) and polyanionic (Ac-EAK, Suc-EAK) polymers. Our data obtained from sedimentation analysis, GPC and RP-HPLC experiments indicate that the branched polypeptides produced in our laboratory possess a fairly narrow distribution of relative molar mass

The amino acid composition and sequence of the side chains proved to be important structural features, which can influence *in vitro* cytotoxicity [23, 34], blood clearance [24, 34] or immunoreactivity [21, 34]. Consequently much emphasis was put on the characterisation of the primary structure of branched polypeptides. The amino acid composition was determined by amino acid analysis and the presence of N-terminal amino acid X in the side chains (in XAK and XSK type compounds) was verified by its identification using 1-dimethylamino-5-naphtalenesulfonyl chloride using HPLC analysis of hydrolyzates of dansylated polypeptides. The enantiomer composition of the branches in polymers was determined by the aid of a chiral derivatizing agent, *N*-(5-fluoro-2,4-dinitrophenyl)-L-alanine amide (Marfey's reagent) [35, 36] and RP-HPLC analysis of the respective hydrolysates for diastereomers. The results from this study show that i) the polymerisation initiated by the polylysine backbone is not stereospecific or stereoselective; ii) no detectable racemization occurs during the coupling of amino acid X active ester to poly[L-Lys] [20], the (AK) or to SK [22, 25, 27]. These studies proved that the synthetic methods applied for the preparation of XAK, XSK and AXK type branched polypeptides produce compounds whose enantiomer compositions are in good agreement with calculated expectations. These findings could be useful for the reliable interpretation of various biological phenomena observed in connection with the presence of L- or D- amino acid residues in the side chains.

3.2. CONFORMATION IN SOLUTION

XAK and AXK types of polymeric branched polypeptides developed in our laboratories provide a versatile model system for the assessment of correlation between primary structure and solution conformation. Therefore conformation of polypeptides was studied by circular dichroism (CD) spectroscopy in water solutions of various pH values (2-12) and ionic strength (0-2.0 M NaCl) and in MeOH or TFE - water mixture or in aqueos SDS solution [37-39]. CD spectra used for the detection of elements of secondary structure and monitor changes induced by alteration of pH, ionic strength or dielectric constant of the solution. Considering that no ordered conformation of the short branches can be expected, the term "ordered conformation" we mean mainly the conformation of the polylysine backbone. Data obtained indicated marked dependence of the polymeric polypeptide conformation on the identity, hydrophobic/hydrophilic nature, charge, configuration and sequential position of amino acid X in the side chain.

In the group of XAK type polypeptides it has been found that the presence of hydrophobic amino aci residues (Leu, Phe, Ile, Nle) at position X induced the formation of α-helix under physiological conditions. In contrast polymers with certain hydrophilic X residues (His, Glu, Lys) at the side chain termini displayed unordered conformation. Interestingly partially ordered secondary structure was detected when Ser or Thr occupied the same side chain terminal position [21, 22, 34, 40].

The comparative conformational analysis of XAK and AXK type polypeptides suggests that the tendency to form an ordered structure is determined by the identity and the position of the chiral amino acid X in the sequence of the branches. The presence of certain hydrophilic residues (Glu, His) at the N-terminal of the branches (in XAK polypeptides) or next to the poly[L-Lys] backbone (in AXK polypeptides) results in an unordered steric arrangement under nearly physiological conditions. Due to the presence of charged α-amino groups at the end of the side chains in polycationic XAK or AXK polypeptides (X = Leu, Phe or Ser) no ordered structure formation could be expected. In contrast, it has been found that both XAK and AXK polypeptides adopt partially ordered conformation, when Leu, Phe or Ser amino acid residue is present. When these amino acids are directly joined to the poly[L-Lys] backbone, slightly more pronounced formation of ordered conformation could be detected [21, 22, 34, 40].

Taken together it can be concluded that the ordered structure of a branched polypeptide might be stabilised by hydrophobic interaction between apolar side chains (in case of XAK or AXK polypeptides X = Leu, Phe), by ionic attraction (EAK vs AEK) or by H-bond formation (SAK vs. ASK). It should be stressed the importance of the appropriate distance between positively charged N-terminal amino groups and the negatively charged carboxyl groups of the glutamic acid residues. For example these groups could produce salt bridge in AEK and contribute to the formation of partly ordered conformation, while in EAK these two groups are well-separated by space resulting in fully disordered steric arrangement. Similarly, the appearance of H-bond between N-terminal α-amino group and OH group of serine residue depends on the relative position of these two functions. SAK has a higher capability to form regular conformation under comparable conditions, while the replacement of Ser by Thr was accompanied by the presence of more helical structure [22, 34]. Results obtained with these three pairs of polymeric polypeptides give rather conclusive evidence that the interaction between the side chains of amino acid X can dominate over the repulsive forces of the charges of the N-terminal amino groups [22, 34].

Based on the correlation between the primary structure (side chain composition, sequence) and solution conformation outlined here it is feasible to design branched polypeptides with predicted secondary structure. By proper selection of the identity (hydrophobic vs. hydrophilic), configuratiom (L- or D-) and sequential position of amino acid X in the branches one can synthesise polycationic or amphoteric polypeptides with ordered (α-helical) or disordered (random) solution conformation.

3.3. SURFACE ACTIVITY OF POLYMERS

The surface activity of polypeptides with polycationic (SAK, OAK), amphoteric (EAK) and polyanionic (Ac-EAK) character at the air/water interface was analysed [32, 41]. Surface pressure values recorded as a function of concentration indicate that all branched polypeptides studied are able to form monolayers at the air/water interface. SAK showed the highest surface activity (π_{max}=11.6 mN/m), while π_{max} was between 8.5 and 6.3 mN/m for other polypeptides. It is interesting to note that π_{max} values for linear polymers with similar molecular mass [poly[Glu(OMe)] [42] and polyethylene glycol [43] were higher. These findings clearly indicate that the surface activity at the air/water interface depends on the identity of the side chain terminal amino acid residue and can be described by the following order: SAK>AK>EAK>Ac-EAK>OAK>>poly[L-Lys].

4. Interaction of Polymers with Phospholipid Mono- and Bilayers

The penetration of branched polypeptides into phospholipid monolayers composed of DPPC, DPPC/PG (95/5 mol/mol) and of DPPC/PG (80/20 mol/mol) was studied by comparing the compression isotherms as well as by monitoring changes in surface pressure at different initial surface pressure (5, 10, 20 and 32 mN/m) and constant area [32, 41]. These data revealed that the presence of SAK resulted in the highest surface pressure increase (6.8 mN/m at 5 mN/m initial pressure) of DPPC monolayer. This value was somewhat lower when other polypeptides were present (2.5 mN/m for OAK, 4.7 mN/m for EAK, 4.4 mN/m for Ac-EAK). Essentially the same tendency was observed at higher initial pressures (at 10, 20 or 32 mN/m). Addition of 5 or 20 % PG to DPPC resulted in less pronounced surface activity for polycationic SAK However, no significant changes were observed with EAK and Ac-EAK. Based on surface pressure increase values observed, penetration of polypeptides into DPPC, DPPC/PG (95/5 mol/mol) and DPPC/PG (80/20 mol/mol) monolayers can be described with SAK>AK>EAK≅Ac-EAK>OAK order. The maximum initial surface pressure we have used in these experiments was 32 mN/m and represents a value described for cellular membranes of biological systems [44]. Therefore differences in surface pressure increases observed with branched polypeptides might be indicative for their potential interaction with plasma membranes. Data obtained by branched polypeptides demonstrate that the side composition of these compounds could markedly influence their surface as well as membrane activity by altering the relative hydrophobicity and the charge properties of the polymer.

The effect of branched polypeptides on phospholipid membranes was further investigated using lipid bilayers with DPPC/PG (95/5, 80/20 mol/mol). Two fluorescent probes of different character were used to analyse the effect of polymers on the outer surface (negatively charged, sodium anilino naphthalene sulfonate, ANS) and on hydrophobic core (hydrophobic, 1,6-diphenyl-1,3,5-hexatriene DPH) of bilayers. For these studies small unilamellar vesicles were used [45]. The effect of branched polypeptides on phospholipid bilayers saturated with fluorescent probes located either at the polar surface (ANS) or within the hydrophobic part (DPH) of the liposome indicate that only polymers with high positive charge density (OAK, polylysine) are capable to initiate moderate influence on the fluidity of hydrophobic alkyl chain region of bilayers.

Mono- or bilayer experiments suggest that the interaction between branched polymers and phospholipid model membranes is highly dependent on the charge properties (SAK, EAK, Ac-EAK) and - for polycationic polypeptides - on the type of amino group (α/δ) belonging to the side chain terminating amino acid (Ser, Orn). Under nearly physiological conditions membrane activity of the polymer can be controlled by the proper selection of the side chain terminating amino acid. Positioning amino acid residues with α-amino group (with pK_a 7.5 - 9) could result in polymers with less pronounced positive charge, while the presence of ε-amino group (with pK_a 10-11) at the same position create highly positive charged polymers.

5. Biological Properties of Branched Polypeptides

We have also investigated the biological characteristics of branched chain polymeric polypeptides by studying their biodegradation, pharmacological properties, cytotoxicity, immunological properties as outlined in Figure 6. Here we have highlighted only a few, but perhaps very interesting findings.

5. 1. CYTOTOXICITY

Cytotoxicity of branched polymers and polylysines (poly-α-lysine, {poly[Lys]} and poly-ε-lysine {poly[Lys(Lysn)]}) were analysed in four *in vitro* systems using rat liver, mouse spleen, HeLa cells and C26 mouse colorectal carcinoma cells. Viability of isolated cells as well as the growth of HeLa or C26 cells in the presence of the polypeptides at various concentration was investigated [21, 34].

5.1.1. Cytotoxicity against Isolated Rat Liver and Mouse Spleen Cells in Vitro

Polypeptides at various concentrations (1.5 - 50.0 μg/ml) were incubated with rat liver cells or with mouse spleen cells and the percentage of cell viability as compared to untreated controls was determined. Poly[Lys] as described previously was rather toxic to human lymphocytes [46], to rat liver cells [47] and even more toxic to mouse spleen cells [47]. In contrast, we found that poly[Lys(Lysn)], containing α-amide bonds, decreased only slightly the viability of liver cells (5 %) and of spleen cells (15 %).

Alteration of the ε-amino groups of poly[Lys] by the introduction of oligopeptide side chains composed of DL-Ala or of DL-Ser and amino acid residue X (XAK, XSK and AXK polypeptides) significantly reduced (> 50 %) or fully eliminated cytotoxicity. The extent of the reduction in both test systems differed according to the structure of the branches.

The presence of Phe, D-Phe, Pro, Leu, D-Leu, His, D-His, Ser at position X in XAK polypeptides [47] or Ala, Leu at position X in XSK polymers [34] resulted in only marginal (< 10 %) loss of viable liver cells, but only at the highest concentration. No

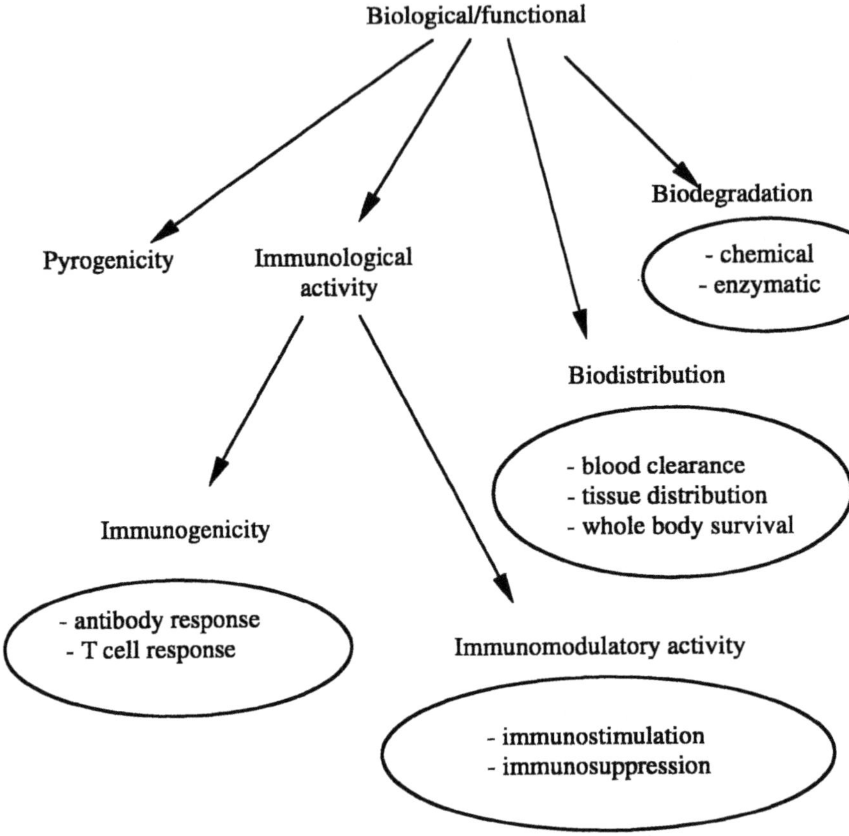

Figure 6. Outline of biological characterisation of branched polymeric polypeptides.

toxic effect of these polypeptides was found on mouse spleen cells at high polypeptide concentration, neither 1 h nor 24 h incubation periods [34,47]. In the case of XAK polypeptides containing L- or D-Lys at the end of the side chains (KAK and D-KAK), a more pronounced cytotoxic effect was observed in both assays. In sharp contrast, no toxic effect could be demonstrated with polypeptides containing one L- or D-Glu residues attached to AK or SK.

The importance of the sequence of the amino acids in the side chain was also investigated, using Leu, Ser or Glu containing AXK and XAK polypeptides. The cytotoxicity of the polycationic ALK and ASK was not considerably different from those of LAK and SAK. In case of amphoteric pair of polypeptides there was only a small decrease in viability, when glutamic acid is linked directly onto the poly[Lys] backbone (AEK) compared to substitution in the terminal side chain position (EAK) [23].

5.1.2. Effect on Growth of Tumour Cells

Cytotoxicity studies were extended to evaluate of growth of HeLa as well as of C26 human carcinoma cells in the presence of various amounts of polymers. It has been demonstrated earlier that poly[Lys] is highly toxic to HeLa cells [48, 49] and also to C26 cells [34]. Cytostatic activity of branched polypeptides could be categorised according to their side chain structure. No cytotoxic effect of amphoteric polypeptides (ESK, EAK vs. D-EAK) was observed. Among polycationic polymers polypeptides containing L- or D-Lys at the end of the side chains (KAK vs. D-KAK) were as toxic as poly[Lys] in HeLa cells. Most of the polycationic compounds studied showed an upper (PAK, D-HAK and LAK) or a lower (HAK, ALK, D-LAK, FAK, D-FAK) medium level toxicity. However, no cytotoxicity was observed with Ser containing polymers on C26 carcinoma cells.

Our data clearly suggest that there is a strong correlation between charge and cell killing activity of branched polypeptides. Amphoteric compounds with balanced charge distribution (α-amino and γ-carboxyl groups) have no cytotoxic or cytostatic activities in the concentration range studied. Among polycations, two groups could be identified. Macromolecules with free α-amino groups with relatively low pK_a=8.95-9.7 at the end of the branches - almost regardless of their sequence, composition and configuration - exhibit no (SAK) or moderate cytotoxicity. In the second subset, polycationic polymers with both α- and ε-amino groups of higher pK_a=10.53 at the branch terminal position are toxic to spleen or liver cells and inhibit HeLa cell growth. These findings could be related to a possible perturbation of cell surface by the charged ε-amino groups of this poly-α-amino acid resulting in an elevated release of ions [49, 50]. Results summarized briefly suggest that the appropriate substitution at the side chain terminal provides a feasible tool for the reduction of the cytotoxicity.

5. 2. BIODISTRIBUTION

Considering *in vivo* application of branched polymeric polypeptides and their conjugates, studies were performed to establish correlations between the structural features and biodistribution profile in normal and tumour bearing mice. For this methods were developed for labelling of branched polypeptides with appropriate radionuclides. By the aid of these derivatives we have examined the blood clearance, whole body survival and tissue distribution profile of polymers with different size [24]. In view of the importance of the side-chain structure in α-helix formation [19, 21, 34], in cytotoxicity [23, 34] and in immunological properties [34, 51, 52] we have studied the effect of amino acid X with different i) identity (e.g., X = Leu, Pro, Ser, or Glu), ii) configuration (X = L- or D-Glu, L- or D-Leu), and iii) position on biodistribution of the branched polypeptides. Here we summarise our data on blood clearance, but further details are presented in the original articles [34, 51, 52].

5.2.1. Labelling of Branched Polypeptides

Due to the high number of free amino group at the side chain terminal position of the branches polypeptides were reacted with N-succinimidyl 3-(4-hydroxyphenyl)-propionate (Bolton and Hunter reagent) prelabelled with ^{123}I, ^{125}I, ^{131}I [53, 54] or radiometals such as ^{111}In [30, 54-56], ^{51}Cr [29] and $^{99}Tc^{m}$ [57-59] using indirect (^{111}In, ^{51}Cr or direct $^{99}Tc^{m}$ labelling procedure. During indirect labelling the amino groups on the terminal amino acid residues were modified first with a chelating agent, diethylene-triamine pentaacetic acid anhydride (DTPAA) [60] followed by addition of ^{111}In [30] or ^{51}Cr [29] using the strong interaction between the negatively charged carboxyl groups of DTPA and the positively charged metal ions. Using the same procedures we have also shown that it is possible to label branched polypeptides conjugated to anti-cancer drugs such as methotrexate, daunomycin, GnRH analogs with similar efficacy [61-63]. Our results suggest that gamma-emitters can be incorporated into polymeric branched polypeptides as well as their antitumor drug conjugates with high specific activity offering potential for the use of scintigraphy in pharmacokinetic studies and in clinical imaging.

5.2.2. Blood Clearance

There was no significant difference between the blood clearance profile of the large (157-213 kDa) and small (34-46 kDa) relative molecular mass versions of ^{125}I labelled polycationic compounds (eg. Leu derivatives). However, for the amphoteric glutamic acid substituted polypeptide pair the blood survival of the small polypeptide was longer than that for EAK with large molecular mass.

Identity of the side chain terminal amino acid in XAK as well as in XSK branched polypeptides determine the overall charge of the compounds. Therefore it has a pronounced effect on blood survival. Among polycationic polypeptides in the XAK group we found that certain compound (eg. X = Leu, Phe) were cleared rapidly from the circulation. In contrast a 10-fold increase in $AUC_{0\text{-}6h}$ was observed when leucine was replaced by glutamic acid in the terminal position (EAK). This conferred an almost neutral charge to the polypeptide. 24 hours after iv injection there was over 70 times more EAK remaining in the circulation compared to LAK. Further alteration of the side chain terminal amino acid residue by incorporation serine resulted in a markedly changed blood clearance even in the group of polycationic polypeptides. The profile of SAK was almost identical with that of EAK. This is the first polycationic polymer reported in the literature, which is capable to circulate in the blood for an extended period of time. Comparison of XAK and XSK polymers showed that replacement of alanine by serine residues in the side chain resulted in significantly increase (eg. LAK vs. LSK) or no change (eg. EAK vs. ESK) in blood circulation time.The incorporation of the D-amino acid into the polypeptide (EAK vs. D-EAK) had no significant effect on blood survival. Similar results were obtained with the polycationic LAK/D-LAK pair.

The effect of altering the position of amino acid X on the blood survival of the polypeptides was also examined. For the amphoteric EAK/AEK as well as for the polycationic SAK/ASK pair of polypeptides there was a 2-2.5 fold reduction in $AUC_{0\text{-}6h}$ and $AUC_{0\text{-}4h}$ when Glu/Ser is linked directly onto the polylysine backbone. In contrast, for the polycationic, but relatively hydrophobic polypeptide pair LAK/ALK a 2-fold increase in $AUC_{0\text{-}6h}$ was observed. Twenty-four hours following injection there was a 4-fold difference in the amount of polypeptide surviving in the circulation.

These data show that the blood survival for branched polypeptides was not only dependent on charge, side chain amino acid sequence, but also on the relative hydrophobicity of amino acid X (eg. Leu vs. Ser). The glutamic acid containing amphoteric and the serine containing polycationic polypeptides showed the longest blood survival [24, 34, 64].

In conclusion, the branched polypeptides with a poly(L-lysine) backbone provide a relatively simple system with which to identify factors (molecular size, ionic charge, hydrophobicity, primary structure) influencing the biodistribution of macromolecules and would allow suitable carriers to be selected according to their intended use. The amphoteric Glu containing as well as the polycationic polypeptide with Ser residue are good candidates for conjugation to cytotoxic drugs with potential use for site-specific drug delivery, either simply as drug-polypeptide conjugates or linked to monoclonal antibodies. The polycationic polymers, exhibiting rapid blood clearance and high spleen uptake, have potential use as carriers for epitopes in the construction of synthetic antigens.

6. Conjugates with Branched Polypeptides

The aim of the synthesis and perform structural (conformation analysis, phospholipid membrane interaction) and biological (cytotoxicity, biodegradation, biodistribution, immunoreactivity) studies briefly outlined above was to provide a rational approach for selection of synthetic branched polypeptide as carriers for the construction of drug-macromolecule conjugates or of epitope-polymer conjugates.

Recently we have prepared several conjugates in which various antitumor agents have been covalently attached to selected structurally related branched polymeric polypeptides. Polymeric polypeptide derivatives of daunomycin [61], methotrexate [62], GnRH antagonist [63, 65, 66], amiloride [67], borono compounds [68] and several radionuclides [54, 59] have been synthesised and their solution conformation [61-63, 67], in vitro cytotoxicity [61-68], biodistribution [59, 61-63] and antitumour properties were analysed.

Antitumor activity of Ac-[D-$Trp^{1\text{-}3}$, D-Cpa^2, D-Lys^6, D-Ala^{10}]-GnRH - poly[Lys(Ac-Glu_i-DL-Ala_m)] (Ac-EAK) conjugate has been clearly documented [63, 65]. The coupling of acid labile derivative of daunomycin to poly[Lys(Glu_i-DL-Ala_m)] (EAK) resulted in compensation for the immunosuppressive effect of the drug and this polymeric conjugate *in vivo* was very effective against L1210 leukaemia producing 66-100 % long-term survivors (> 60 days) in mice [69].

Several groups of branched polypeptide based synthetic compounds were also prepared by introduction of B-cell or T-cell epitope peptides onto the side end of the branches [70]. It has been shown that the composition of the polymeric component has a marked influence on the interaction between phospholipid mono- or bilayers and epitope conjugates [71-73] We have demonstrated that the efficacy of antibody response specific for a *Herpes Simplex virus* glycoprotein D epitope is highly dependent on the chemical structure (sequence and conformation) of branched polypeptide carrier [74,75]. Recently a fully synthetic prototype conjugate of EAK, Ac-EAK and Suc-EAK with one or two independent T cell epitope peptides of *M.tuberculosis* proteins (16 kDa and 38 kDa) were prepared [76-78]. *In vivo* data clearly suggest that the selection of the carrier highly influence T-cell epitope peptide specific immune response [76, 77]. In conjugate with dula specificity both mycobacterial epitopes preserved their capability to induce specific T-cell proliferation [78].

7. Acknowledgement

Experimental work summarised in this paper was supported by grants from the Hungarian-Spanish Intergovernmental Programme (5/1998), from the Hungarian Research Fund (OTKA No. T-3024, T-4217, T-014964, T-03838) and from the Hungarian Ministry of Welfare (ETT No. T405, 017/1993, 115/1996).

8. References

1. Seymour, L. W.: Passive tumour targeting of soluble macromolecules and drug conjugates. *Crit. Rev the Drug. Syst.* **9** (1992), 135-162.
2. Johnson, K. P., Brooks, B. R., Cohen, J. A., Ford, C. C., Goldstein, J., Lisak, R. P., Myers, L. W., Panitch, H. S., Rose, J. W., Schiffer, R. B., Vollmer, T., Weiner, L. P. and Wolinsky, J. S.: Extended use of glatiramer acetate (Copaxone) is well tolerated and maintains its clinical effect on multiple sclerosis relapse rate and degreee of disability. Copolymer 1. Multiple Sclerosis Discussion Group. *Neurology* **50** (1998), 701-708.
3. Javaherian, K., and McDanal, C.: Random copolymers containing specific ratios of negatively charged and aromatic amino acids bind V3 disulfide loop and neutralize diverse HIV type 1 isolates. *AIDS Res. Hum. Retroviruses* **10** (1995) 1163-1168.
4. Jackson, D.C., O'Brien-Simpson, N., Ede, N.J., and Brown, L.E.: Free radical induced polymerization of synthetic peptides into polymeric immunogens.*Vaccine* **15** (1997) 1697-1705.
5. Tsikaris, V., Sakarellos, C., Cung, M.T., Marraud, M., and Sakarellos-Daitsiotis, M. Concept and design of a new class of sequential oligopeptide carriers (SOC) for covalent attachment of multiple antigenic peptides. *Biopolymers*, **381** (1996) 291-293.
6. Burnham, N.L.: Polymers for delivering peptides and proteins. *Am. J. Hosp. Pharm.* **51** (1994) 210-218.
7. Gombotz, W.R., and Pettit, D.K.: Biodegradable polymers for protein and peptide drug delivery. *Bioconjugate Chem.* **6** (1995) 332-351.
8. Friend, D. R., and Pangburn, S.: Site-specific drug delivery. *Medicinal Research Reviews* **7** (1987) 53-106.
9. Duncan, R.: Drug-polymer conjugates; potential for improved chemotherepy. *Anti-Cancer Drugs*, **3** (1992) 153-156.
10. Takakura, Y., and Hashida, M., Macromolecular drug carrier systems in cancer chemotherapy: macromolecular prodrugs. *Critical Reviews in Oncology/Haematology* **18** (1995) 207-231.
11. Maeda, H., Seymour, L.W., and Miyamoto, Y.: Conjugates of anticancer agents and polymers: advantages of macromolecular therapeutics in vivo. *Bioconjugate Chem.*, **3** (1992) 351-362.
12. Vasey, P.A., Kaye, S.B., Morrison, R., Twelves, C., Wilson, P., Duncan, R., Thomson, A.H., Murray, L.S., Hilditch, T.E., Murray, T., Burtles, S., Fraier, D., Frigerio, E., and Cassidy, J.: Phase I clinical and pharmacokinetic study of PK1 [*N*-(2-hydroxypropyl)methacrylamide copolymer doxorubicin]: first

member of a new class of chemotherapeutic agents - drug-polymer conjugates. *Clinical Cancer Research* **5** (1999) 83-94.

13. Julyan, P.J., Seymour, L.W., Ferry, D.R., Daryani, S., Boivin, C.M., Doran, J., David, M., Anderson, D., Christodoulou, C., Young, A.M., Hesslewood, S., and Kerr, D.J.: Preliminary clinical study of the distribution of HPMA copolymers bearing doxorubicin and galactosamine*J. Controlled Release* **5** 7 (1999) 281-290.
14. Del Giudice, G.: New carriers and adjuvants in the development of vaccines.*Current Opinion in Immunology*. **4** (1992) 454-459.
15. Hudecz, F., and Tóth, G.K. (1994) Synthetic peptide constructs to increase the immunogenicity of B-cell epitopes, in É. Rajnavölgyi (ed.), *Synthetic peptides in the search for B-and T-cell epitopes* R.G.Landes Company, Austin, pp. 97-119.
16. Gokulan, K., Khare, S., and Rao, D.N.: Increase in the immunogenicity of HIV peptide antigens by chemical linkage to polytuftsin (TKPR40). *DNA Cell Biol.* **18** (1999) 623-630.
17. Arnon, R.: Synthetic vaccines based on peptides, oligonucleotides and conjugate anigens. *BEHRING Inst. Mitt.* **98** (1997) 184-190.
18. Ben-Yedidia, T., and Arnon, R.: Design of peptide and polypeptide vaccines. *Current Opinion in Biotechnology* **8** (1997) 442-448.
19. Hudecz, F., Votavova, H., Gaál, D., Sponar, J., Kajtár, J., Blaha, K., and Szekerke, M.(1985) Branched polypeptides with a poly(L-lysine) backbone: synthesis, conformation and immunomodulation. in Ch.G. Gebelein, and Ch. E. Carraher (eds.) *Polymeric Materials in Medication* Plenum Press, New York, pp. 265-289.
20. Mező, G., Kajtár, J., Hudecz, F. and Szekerke, M.: Carrier design: conformational studies of amino acid (X) and oligopeptide (X-DL-Ala_m) substituted poly[L-lysine]. *Biopolymers*, **33** (1993) 873-883.
21. Hudecz, F.: Design of synthetic branched-chain polypeptides as carriers for bioactive molecules.*Anti-Cancer Drugs* **6** (1995) 171-193.
22. Mező, G., Kajtár, J., Nagy, I., Majer, Zs., Szekerke, M., and Hudecz, F.: Carrier design: Synthesis and conformational studies of poly[L-lysine] based branched polypeptides with hydroxyl groups. *Biopolymers* **42** (1997) 719-730.
23. Hudecz, F., Gaá, D., Kurucz, I., Lányi, S., Kovács, A.L., Mező, G., Rajnavélgyi, É., and Szekerke, M.: Carrier design: cytotoxicity and immunogenicity of synthetic branched polypeptides with poly(L-lysine) backbone. *J. Controlled Release* **19** (1992) 231-243.
24. Clegg, J.A., Hudecz, F., Mező, G., Pimm, M.V., Szekerke, M., and Baldwin, R.W.: Carrier design: biodistribution of branched polypeptides with poly-(L-lysine) backbone. *Bioconjugate Chem.*, **2** (1990) 425-430.
25. Hudecz, F. and Szekerke, M.: Investigation of drug-protein interactions and the drug-carrier concept by the use of branched polypeptides as model systems. Synthesis and characterisation of the model peptides. *Coll. Czech. Chem. Commun.* **45** (1980) 933-940.
26. Sela, M.: (1980) Chemical synthesis for the understanding of immune response phenomena and for their medical application. in C. Larralde, K. Wills, L. Ortiz-Ortiz, and M.Sela (eds.) *Molecules, cells, and parasites in immunology* Acad Press, New York, pp. 215-228.
27. Hudecz, F., and Szekerke, M.: Synthesis of new branched polypeptides with poly-lysine backbone. *Coll. Czech. Chem. Commun.*, **50** (1985) 103-113.
28. Hudecz, F., Kovács, P., Kutassi-Kovács, S., and Kajtár, J.: GPC, CD and sedimentation analysis of poly-Lys and branched chain poly-Lys-poly-DL-Ala polypeptides. *Colloid and Polymer Sci.* **262** (1984) 208-212.
29. Hudecz, F., Kojima, Y., Miyamoto, Y., Kajt‡r, J., and Maeda, H.: The effect of charge on the biodistribution of synthetic branched polypeptides in tumour bearing mice. *J. Controlled Release* **28** (1994) 301-302.
30. Pimm, M.V., Clegg, J.A., Hudecz, F., and Baldwin, R.W.: In 111 labelling of branched polypeptide drug carrier with poly[L-lysine] backbone. *Int. J. Pharmaceutics* **79** (1992) 77-80.
31. Idei, M., Dibó, G., Bogd‡n, K., Mező, G., Horváth, A., Érchegyi, J., Mészáros, Gy., Teplán, I., Kéri, Gy., and Hudecz, F.: Analysis of macromolecular branched chain polypeptides by capillary electrophoresis and micellar electrokinetic chromatography. *Electrophoresis* **17** (1996) 1357-1360.

32. Hudecz, F., Nagy, I.B., Kóczán, G., Alsina, M.A., and Reig, F. (2000) Carrier design: influence of charge on interaction of branched polymeric polypeptides with phospholipid model membranes. in E. Chiellini (ed.) *Frontiers in Biomedical Polymers* , Plenum Press, New York, (in press)

33. Kóczán, Gy., Reményi, J., Hudecz, F.: Novel RP-HPLC analysis and purification of methotrexate-polymeric branched chain polypeptide conjugates. *J. Chromatography* (submitted)

34. Hudecz, F., Pimm, M.V., Rajnavölgyi, E., Mező, G., Fabra, A., Gaá, D., Kovács, A.L., Horváth, A., and Szekerke, M.: Carrier design: New generation of polycationic branched polypeptides containing OH groups with prolonged blood survival and diminished *in vitro* cytotoxicity. *Bioconjugate Chem.* **10** (1999) 781-790.

35. Marfey, P.: Determination of D-amino acids. II. Use of a bifunctional reagent, 1,5-difluoro-2,4-dinitrobenzene. *Carlsberg Res.Commun.* **49** (1984) 591-596.

36. Szókán, Gy., Mező, G., and Hudecz, F.: Application of Marfey's reagent in racemization studies of amino acids and peptides. *J. Chromatography* **444** (1988) 115-122.

37. Votavova, H., Hudecz, F., Kajtár, J., Sponar, J., Blaha, K., and Szekerke, M.*f* Conformation of branched polypeptides based on poly(L-lysine). Effect of the ionic strength and of the presence of alcohols. *Coll. Czech. Chem. Commun.* **47** (1982) 3437-3446.

38. Votavova, H., Hudecz, F., Kajtár, J., Sponar, J., Blaha, K., and Szekerke, M.: Conformation of branched polypeptides based on poly(L-lysine): the effect of terminal amino acids in the branches. *Collect. Czech. Chem. Commun.* **50** (1985) 228-244.

39. Mező, G., Votavova, H., Hudecz, F., Kajtár, J., Sponar, J., and Szekerke, M.: Conformation of branched polypeptides: The influence of DL-alanine oligomer spacers in the side chains. *Coll. Czech. Chem. Commun.*, **53** (1988) 2843-2853.

40. Mező, G., Kajtár, J., Hudecz, F., and Szekerke, M.: Carrier design: conformational studies of amino acid (X) and oligopeptide (X-DL-Alam) substituted poly[L-lysine] *Biopolymers* **33** (1993) 873-885.

41. Nagy, I.B., Haro, I., Alsina, A., Reig, F., and Hudecz, F.: Interaction of branched chain polymeric polypeptides with phospholipid model membranes. *Biopolymers*, **46** (1998) 169-179.

42. Goupil, D. W., and Goodrich, F. C.: Interfacial conformational modifictions in a polypeptide. *J. Coll.Interf. Sci.* **62** (1977) 142-148.

43. Winterhalter, M., Byrner, H., Marzinka, S., Benz, R., and Kasianowicz, J. J.: 1995, Interaction of poly(ethylene-glycols) with air-water interfaces and lipid monolayers: investigation on surface pressure and surface potential. *Biophys. J.* **69** (1995) 1372-1381.

44. Van Deenen, L. L. M., Demel, R. A., Geurtz van Kessel, W. S. H., Kamp, H. H., Roelfsen, B., Veerkelij, A. J., Wirtz, K. W. A., and Zwall, R. F. A. (1976) in *The structural basis of membrane function* Y. Hatefi and L. Djavadi-Ohaniance (eds.) Academic Press, New York, pp. 21-38.

45. Cajal, Y., Alsina, M. A., Rabanal, F., and Reig, F.: A fluorescence and CD study on the interaction of synthetic lipophilic Hepatitis B virus preS(120-145) peptide analogues with phospholipid vesicles. *Biopolymers* **38** (1996) 607-618.

46. Chu, B.C.F., and Howell, S.B.: Differential toxicity of carrier bound methotrexate toward human lymphocytes,marrow and tumor cells. *Biochem.Pharmacol.* **30** (1981) 2545-2552.

47. Gaál, D., Hudecz, F., Kov‡cs, A.L., and Szekerke, M.: Immunomodulatory effect of synthetic branched polypeptides. II. *J.Biol.Response Modifiers* **5** (1986) 148-159.

48. Hudecz, F., Kajtár, J., Kurucz, I., Kovács, A.L., and Szekerke, M.: Investigation of the carrier function of polypeptides. Synthesis, conformation and cytotoxicity of oxazolone conjugates of poly(Lys) and poly[Lys-(DL-Ala$_m$)]. *Makromol. Chem., Macromol. Symp.* **19** (1988) 107-124.

49. Arnold, L.J. Jr, Gutheil, I., and Kaplan, N.O.: Antineoplastic activity of poly(L-lysine) with some ascites tumor cells. *Proc.Natl.Acad. Sci USA* **76** (1979) 3246-3250.

50. Chu, B.C.F., and Howell, S.B.: Differential toxicity of carrier bound methotrexate toward human lymphocytes,marrow and tumor cells. *Biochem.Pharmacol.* **30** (1981) 2545-2552.

51. Rajnavölgyi, É., Hudecz, F., Mező, G., Szekerke, M., and Gergely, J.: Isotype distribution and fine specificity of the antibody response of inbred mouse strains to four compounds belonging to a new group of synthetic branched polypeptides. *Mol.Immunol.* **23** (1986) 27-37.

52. Gaál, D., Hudecz, F., and Szekerke, M.: Immunomodulatory effect of synthetic branched polypeptides. I. *J. Biol. Response Modifiers* **3** (1984)174-184.

53. Bolton, A.E., and Hunter, W.M.: The labelling of proteins to high specific radioactivities by conjugation to a ^{125}I-containing acylating agent. *BiochemJ.* **133** (1973) 529-538.
54. Pimm, M.V., Gribben, S.J., Mező, G., and Hudecz, F.: Strategies for labelling branched polypeptides with a poly(L-lysine) backbone with radioiodines (^{123}I, ^{125}I, ^{131}I) and radiometals (^{111}In, ^{51}Cr) for biodistribution studies and radiopharmaceutical development. *J. Labelled Compounds* **36** (1995) 157-172.
55. Pimm, M.V., Perkins, A.C., and Hudecz, F.: Scintigraphic evaluation of the pharmacokinetics of a soluble polymeric drug carrier. *Eur J. Nuclear Medicine* **19** (1992) 449-452.
56. Pimm, M.V., Perkins, A.C., and Hudecz, F.: Gamma scintigraphy for evaluating the biodistribution of synthetic polymeric drug carriers. *Nuclear Medicine Communications* **13** (1992) 230-231.
57. Frier, M., Perkins, A.C., Pimm, M.V., and Hudecz, F.: Genetically engineered and synthetic macromolecules as radiopharmaceuticals with potential for liver, lung, blood pool and tumour imaging. *Nuclear Medicine Communications* **16** (1995) 240-241.
58. Perkins, A.C., Frier, M., Pimm, M.V., Hudecz, F., and Duncan, R.: Synthetic polymers as potential designer. *Eur. J. Nucl. Med.* **23** (1996) 1134.
59. Perkins, A.C., Frier, M., Pimm, M.V., and Hudecz, F.: Tc^{99m}-branched chain polypeptide (BCP): a potential synthetic radiopharmaceutical. *J. Labelled Compounds* **41** (1998) 631-638.
60. Hnatowich, D.J., Layne, W.W., Childs, R.L., Lanteigne, D., Davis, M.A., Griffin, T.W., and Doherty, P.W.: Radioactive labeling of antibody: a simple and efficient method. *Science* 220 (1983) 613-615.
61. Hudecz, F., Clegg, J.A., Kajtár, J., Embleton, M.J., Szekerke, M., and Baldwin, R.W.: Synthesis, conformation, biodistribution and in vitro cytotoxicity of daunomycin-branched polypeptide conjugates. *Bioconjugate Chem.* **3** (1992) 49-57.
62. Hudecz, F., Clegg, J.A., Kajtár, J., Embleton, M.J., Pimm, M.V., Szekerke, M., and Baldwin, R.W.: Influence of carrier on biodistribution and in vitro cytotoxicity of methotrexate-branched polypeptide conjugates. *Bioconjugate Chem.* **4** (1993) 25-33.
63. Mező, G., Mező, I., Seprődi, A., Teplán, I., Kovács, M., Vincze, B., Pályi, I., Kajtár, J., Szekerke, M., and Hudecz, F.: Synthesis, conformation, biodistribution and hormon related in vitro anti-tumor effect of a GnRH antagonist-branched polypeptide conjugate. *Bioconjugate Chem.* **7** (1996) 642-650.
64. Pimm, M.V., Gribben, S.J., Bogd‡n, K., and Hudecz, F.: The effect of charge on the biodistribution in mice of branched polypeptides with poly(L-lysine) backbone labeled with ^{125}I, ^{111}In or ^{51}Cr in mice. *J. Controlled Release* **37** (1995) 161-172.
65. Vincze, B., Pályi, I., Daubner, D., Kálnay, A., Mező, G., Hudecz, F., Szekerke, M., Teplán, I., and Mező, I.: Antitumor effect of a GnRH antagonist and its conjugate on human breast cancer cells and their xenografts. *J. Cancer Res. Clin. Oncol.* **120** (1994) 578-584.
66. Mező, G., Mező, I., Vincze, B., Kálnay, A., and Hudecz, F.: Synthesis and antitumour activity of new GnRH analogue-branched polypeptide conjugates. *Cancer Detect.Prev.* **22(S)** (1998) 239-240
67. Pató, J., Ulbrich, K., Baker, P., Mező, G., and Hudecz, F.: Synthesis of macromolecular conjugates of a urokinase inhibitor. *J. Bioactive and Compatible Polymers* **14** (1999) 99-121.
68. Mező, G., Sármay, G., Hudecz, F., Kajtár, J., Nagy, Zs., Gergely, J., and Szekerke, M.: Synthesis and characterization of p-borono-Phe - branched polypeptide - monoclonal antibody ternary systems for potential use in boron neutron capture therapy (BNCT) *J. Bioactive and Compatible Polymers* **11** (1996) 263-285.
69. Gaál, D., and Hudecz, F.: Low toxicity and high antitumour activity of daunomycin by conjugation to immunpotential amphoteric branched polypeptide. *Eur J.Cancer* **34** (1998) 155-161.
70. Hudecz, F.: Alteration of immunogenicity and antibody recognition of B-cell epitopes by synthetic branched polypeptide carriers with poly[L-lysine Backbone. *Biomed. Peptides, Proteins and Nucleic Acids* **1** (1995) 213-220.
71. Garcia, M., Nagy, I.B., Alsina, A., Mező, G., Reig, F., Hudecz, F., and Haro, I.: Analysis of the interaction with biomembrane models of HAV-VP3(101-121) sequence conjugated to synthetic branched polypeptide carriers with poly[L-lysine] backbone. *Langmuir* **14** (1998) 1861-1869.
72. Sospedra, P., Nagy, I.B., Haro, I., Mestres, C., Hudecz, F., and Reig, F.: Physicochemical behaviour of polylysine[HAV-VP3 peptide] constructs at the air-water interface. *Langmuir* **15** (1999) 5111-5117.
73. Nagy, I.B., Alsina, M.A., Haro, I., Reig, F., and Hudecz, F.: Phospholipid - model membrane interactions with branched polypeptide conjugates of a hepatitis a virus peptide epitope. *Bioconjugate Chemistry*, **11** (2000) 30-38.

74. Hudecz, F., Hilbert, Á., Mező, G., Mucsi, I., Kajtár, J. Bősze, Sz., Kurucz, I., and Rajnavölgyi, É. (1994) The use of branched polypeptide carrier based conjugates for the design of synthetic vaccine against HSV infection. in R. Epton (ed.) *Innovation and Perspectives in Solid Phase Synthesis - Peptides, Polypeptides and Oligonucleotides - 1994* Intercept, Andover, pp. 315-320.
75. Hudecz, F., Hilbert, Á., Mező, G., Kajtár, J. and Rajnavölgyi, É.: B-cell epitopes in Herpes simplex Virus-(HSV-1) glycoprotein D (gD). In *Synthetic peptides in the search for B-and T-cell epitopes.* (Rajnavölgyi, É. ed.) R.G.Landes Company, Austin, pp. 157-169., 1994.
76. Venkataprasad, N., Coombes, A.G.A., Singh, M., Rhode, M., Wilkinson, K.A., Davies, S.S., Hudecz, F., and Vordermeier, M.H.: Resorbable lamellar subtrates of lactide polymers induce potent cellular immunity to an adsorbed mycobacterial antigen. *Vaccine*, **17** (1999) 1814-1819.
77. Wilkinson, K.A., Hudecz, F., Vordermeier, H.M., Ivanyi, J., and Wilkinson R.J.: Enhancement of the T cell response to a mycobacterial peptide by conjugation to synthetic branched polypeptide *Eur. J.Immunol.* **29** (1999) 2788-2796.
78. Wilkinson, K.A., Vordermeier, M.H., Wilkinson, R., Ivanyi, J. and Hudecz, F.: Synthesis and in vitro T cell immunogenicity of conjugates with dual specificities: attachment of epitope peptides of 16 kDa and 38 kDa proteins from *M.tuberculosis* to branched polypeptide. *Bioconjugate Chem.*, **9** (1998) 539-547.

Amyloid-like fibrils from a peptide-analogue of the central domain of silkmoth chorion proteins

VASSILIKI A. ICONOMIDOU and STAVROS J. HAMODRAKAS
Faculty of Biology, Department of Cell Biology and Biophysics, University of Athens, Athens 157 01, Greece

The structure and self-assembly properties of a 51-residue peptide-analogue of the central conservative domain of silkmoth chorion (eggshell) proteins were studied in detail. This peptide (cA peptide) forms amyloid-like fibrils, under a variety of conditions.
The fibrils bind Congo red and thioflavin T. Negative staining and shadowing showed that the fibrils are twisted, forming double helices. The average width of the basic double helical unit is ~90 Å and the helical pitch of the double helix ~920 Å. Individual fibrils (protofilaments) which super-coil to form the double helix have a diameter of ~30-40 Å. FT-IR and FT-Raman spectroscopy indicated an antiparallel β-sheet type of structure. X-ray diffraction patterns from fibres of the cA-peptide, taken with a double-mirror camera, clearly indicate a "rich", oriented cross-β fibre pattern characterized by a meridional reflection at ~4.66 Å and an equatorial reflection at ~10.12 Å.
Modeling studies suggest that a twisted β-sheet of 4-residue β-strands alternating with β-turns is the basic structural motif of the fibrils. The models are similar to the cross-β structure proposed a decade and a half ago for silkmoth chorion proteins to dictate the helicoidal architecture of intact, native chorions. Thus, it appears that silkmoth chorion is the first well documented case, where amyloid plaques formed from self-assembly of chorion proteins, *in vivo* protect the oocyte and the developing embryo from a wide range of environmental hazards.

Silkmoth chorion is the major component of the eggshell covering the oocyte and the developing embryo. It is a proteinaceous protective and functional layer with extraordinary mechanical properties and an interesting model system in several current areas of biological research: cellular differentiation, molecular evolution and fibrous protein folding and self assembly (for reviews see Regier & Kafatos, 1985; Goldsmith & Kafatos, 1984; Hamodrakas, 1992). Understanding the relationship between structure, function and assembly of its component proteins might be fruitful into the design of novel biomaterials (Hamodrakas, 1992).

A. Aggeli et al. (eds.), Self-assembling Peptide Systems in Biology, Medicine and Engineering, 161–169.

Biochemically, silkmoth chorion is surprisingly complex: about 200 proteins have been resolved by two-dimensional gel electrophoresis (Regier & Kafatos, 1985). They were classified into a small number of families or classes: the most abundant are the A's and the B's which together account for almost 90% of the total chorion mass. The gene families that encode these proteins are themselves related and constitute a superfamily with two branches, the α-branch and the β-branch (Lecanidou et al., 1986).
Aminoacid sequence comparisons and predictions of protein secondary structure revealed that chorion proteins exhibit a tripartite structure: a central domain, evolutionarily conserved in each protein class (recognizably homologous among the two main A and B classes) and two flanking N- and C- terminal domains or "arms" (Figure 1). These arms are more variable but they are marked by the presence of characteristic tandemly repetitive short peptides that do not appear in the central domain (Hamodrakas et al., 1982; Hamodrakas et al., 1985).

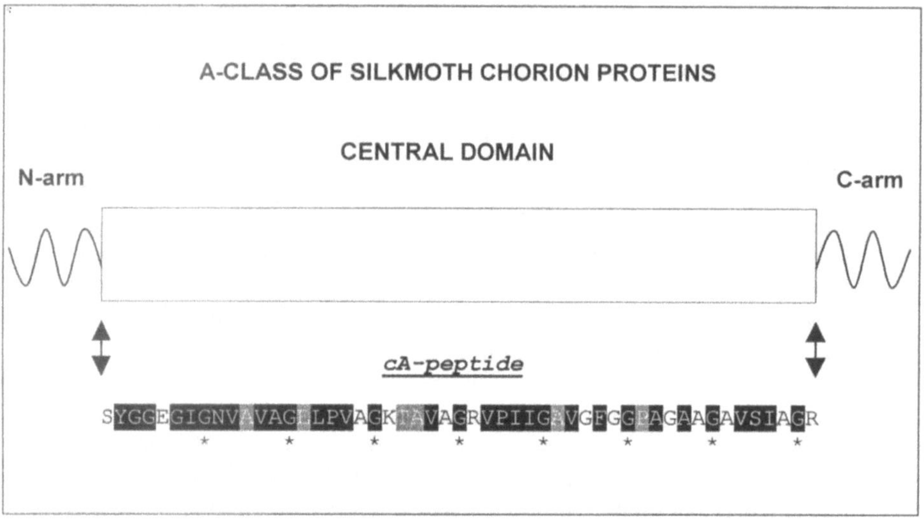

Figure 1. Schematically, the tripartite structure of silkmoth chorion proteins of the A class. A central domain highly conservative and of invariant length, and two more variable flanking "arms" constitute each protein. Characteristic, tandemly repeating peptides are present both in the central domain and in the "arms". (Hamodrakas, 1992 and references therein). The B-class of proteins exhibits a similar tripartite structure.
The aminoacid sequence of the synthetic cA peptide (one letter code), which was designed to be an analogue of the central A-domain is also shown. Invariant glycines repeating every six-residues are marked with an asterisk below the sequence. Black-boxed residues are conservd 100% and gray-boxed residues represent 80% conservative substitutions.

A structural model has been proposed for the central domain and the flanking arms combining data from amino acid sequence comparisons, secondary structure prediction, analysis of amino acid periodicities and modelling (Hamodrakas et al., 1985; Hamodrakas et al., 1988). According to this model chorion proteins adopt a characteristic antiparallel, most probably twisted, β-pleated sheet structure of four residue β-strands connected by β-turns (Hamodrakas et al., 1985; Hamodrakas et al., 1988), in a cross-β conformation. Evidence from X-ray diffraction, laser-Raman, infrared and CD spectroscopy supports this model (Hamodrakas, 1992 and references therein).

However, because of chorion complexity, it is very difficult to isolate and purify individual chorion proteins suitable for structural studies. Therefore, an alternative approach was chosen in attempts to elucidate principles that govern chorion protein folding and assembly: peptides thought to be representative of certain, structurally important, parts of chorion proteins were synthesized and an effort is currently being made to study their structure both in solution and in the solid state. It is hoped that this analysis will reveal folding and packing modes significant in chorion protein structure and assembly.

One such peptide, a 51-residue peptide, hereafter called cA, is an analogue of the entire central domain of the A class of silkmoth chorion proteins (Figure 1). This peptide is representative for about 30% of all the proteinaceous material in the eggshell. We designed this peptide since the central domains of the class A chorion proteins are highly conserved in both sequence and length and this conservation indicates an important functional role for this structural entity in chorion structure formation. Preliminary laser-Raman and FT-IR experiments (Benaki et al., 1998) showed that the structure of this peptide is predominantly anti-parallel β-pleated sheet both in solution and in the solid state.

Results

The cA peptide was synthesized as described by Benaki et al. 1998. It forms, uniform in structure, amyloid-like fibrils by self-assembly in various solvents, pH's, ionic strengths and temperatures (to be published).

The fibrils were judged to be amyloid-like from their tinctorial and structural characteristics: They bind Congo-red and Thioflavin-T (data not shown). They are straight, unbranched uniform in diameter (~90 Å) double helices of indeterminate length as seen in electron micrographs (Figures 2A, 2B). Each double-helical fibril consists of two protofilaments wound around each other. The protofilaments have a uniform diameter of approximately 30-40 Å. The pitch of the double helix (Figure 2A, arrows) is approximately 920 Å.

Suspensions of these fibrils form oriented fibres, which give characteristic "cross-β" X-ray diffraction patterns (Figure 3). In these fibres the long axes of the amyloid-like fibrils seen in the electron micrographs of Figure 2 are most probably oriented more or less parallel to the fibre axis. This is a generally accepted assumption in such cases (Fraser & McRae, 1973).

The oriented X-ray pattern (Figure 3) taken from these fibres indicates the presence of oriented β-sheets in the amyloid-like fibrils of peptide cA. The presence of reflections corresponding to periodicities of 4.66 and 10.12 Å indicates the existence of β-sheets. The strong meridional reflection at 4.66 Å suggests that the β-sheets are oriented so that their β-strands are perpendicular to the fibre axis (consequently, also to the long axis of

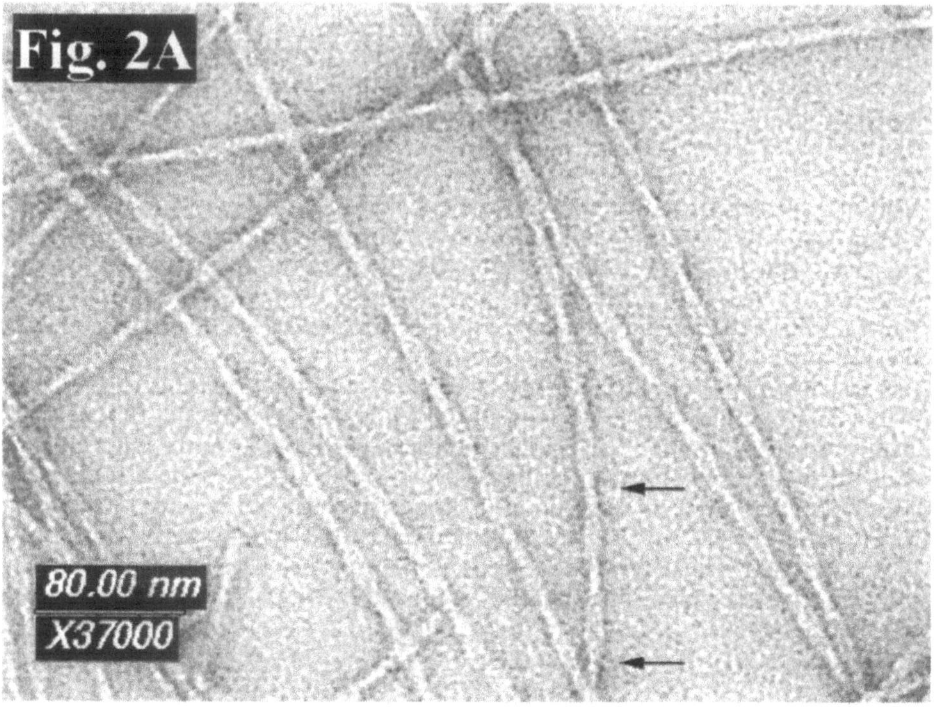

Figure 2A. Electron micrograph of amyloid-like fibrils derived by self-assembly, from a solution 9mg/ml of the cA peptide, in a sodium acetate 50mM buffer, pH 5. Fibrils are negativly stained with 1% uranyl acetate. They are of indeterminate length (several microns), unbranched, approximately 90 Å in diameter and have a double helical structure. The pitch of the double helix is ~ 920 Å (marked with arrows). A pair of protofilaments each 30-40 Å in diameter are wound around each other, forming the double-helical fibrils.

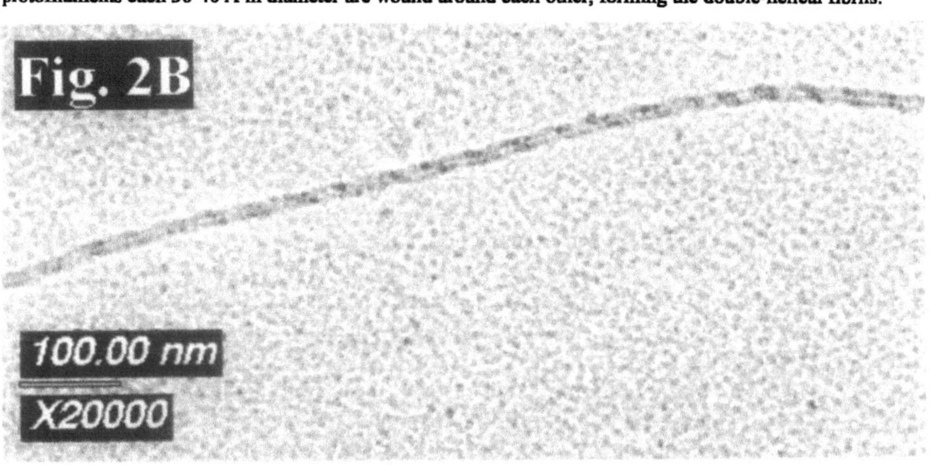

Figure 2B. Electron micrograph of amyloid-like fibrils derived from a solution of the cA peptide (conditions as in Fig. 2A). Fibrils are rotary shadowed with Pt/Pd at an angle of 7 degrees under high vacuum.

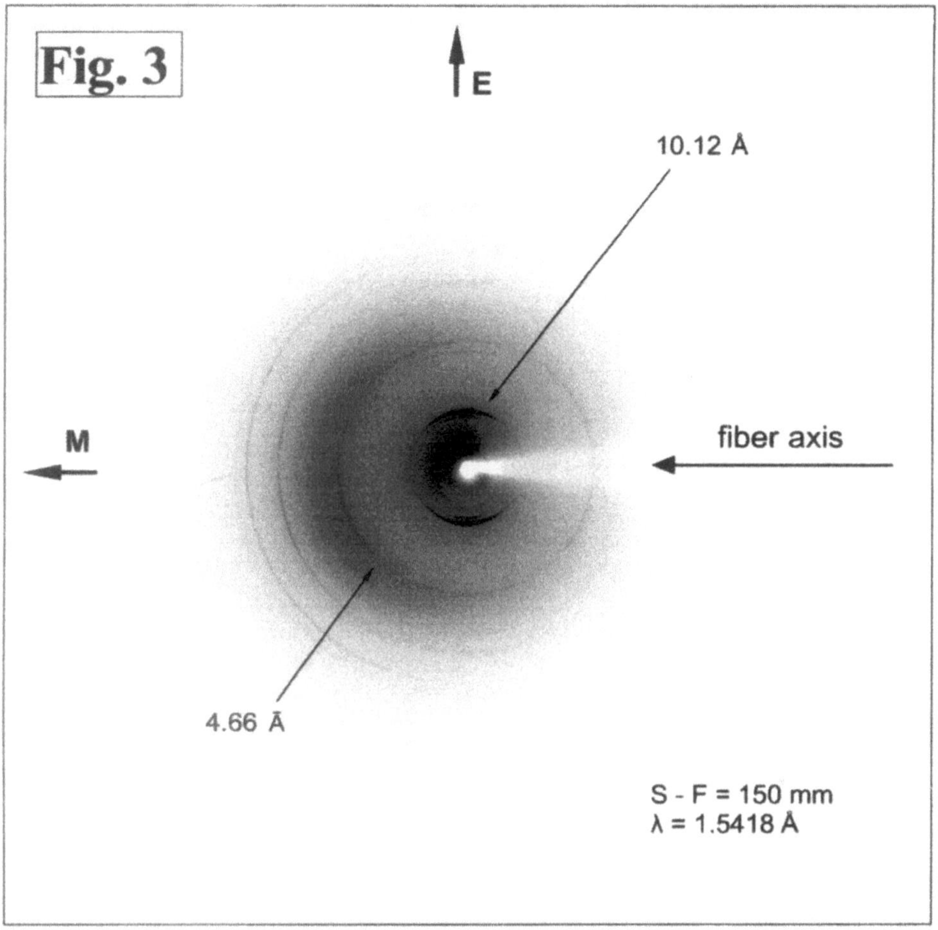

Figure 3. X-ray diffraction pattern from a fibre of cA peptide amyloid-like fibrils. cA peptide was dissolved in a 50 mM sodium acetate buffer pH 5, at a concentration of 9 mg/ml to produce amyloid-like fibrils, after 3-4 weeks incubation. A droplet of fibril suspension was placed between two siliconized glass rods and allowed to dry at room temperature for 1 hr, to form a fibre suitable for X-ray diffraction.
The meridian, M (direction parallel to the fibre axis) is horizontal and the equator, E, is vertical in this display. X-ray diffraction patterns were recorded on a MAR Research 345 image plate utilizing double-mirror focused CuK_α radiation (λ=1.5418 Å), obtained from a GX-21 rotating anode generator operated at 40kV, 75 mA. The specimen to film distance was set at 150 mm and the exposure time was 1 hr.
The X-ray diffraction pattern is a typical "cross-β" pattern showing a 4.66 Å reflection on the meridian and a 10.12 Å reflection on the equator. This indicates a regular structural repeat of 4.66 Å along the fibre axis (meridian) and a structural spacing of 10.12 Å perpendicular to the fibre axis. The structural repeat of 4.66 Å along the fibre axis corresponds to the spacing of adjacent β-strands (which should be perpendicular to the fibre axis, a cross-β structure) and the 10.12 Å spacing perpendicular to the fibre axis corresponds to the face-to-face seperation (packing distance) of the β-sheets.

the amyloid-like fibrils). The strong equatorial reflection at 10.12 Å, which corresponds to the inter-sheet distance, suggests that the packing of the β-sheets is done in a manner parallel to the fibre axis. This X-ray pattern closely resembles typical cross-β patterns taken from amyloid fibres (Sunde & Blake, 1997 and references therein). Therefore, it should arise from a cross-β structure.

The most plausible model that can be proposed for the structure of the cA peptide, to account for all the evidence gathered from this study and from data collected previously on the structure of silkmoth chorion proteins (Hamodrakas, 1992 and references therein), is a model structure shown in Fig. 4.

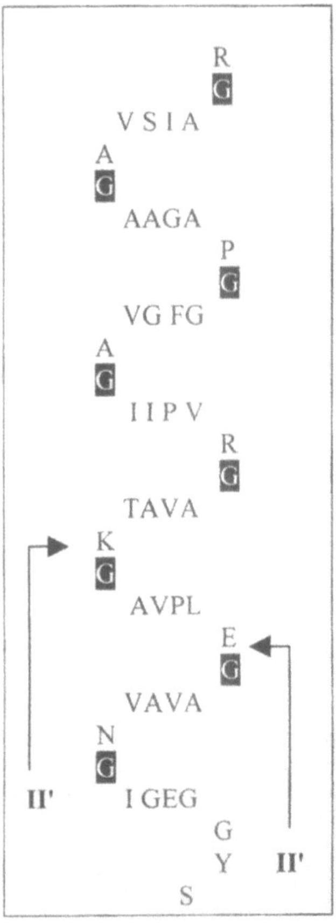

Fig. 4A

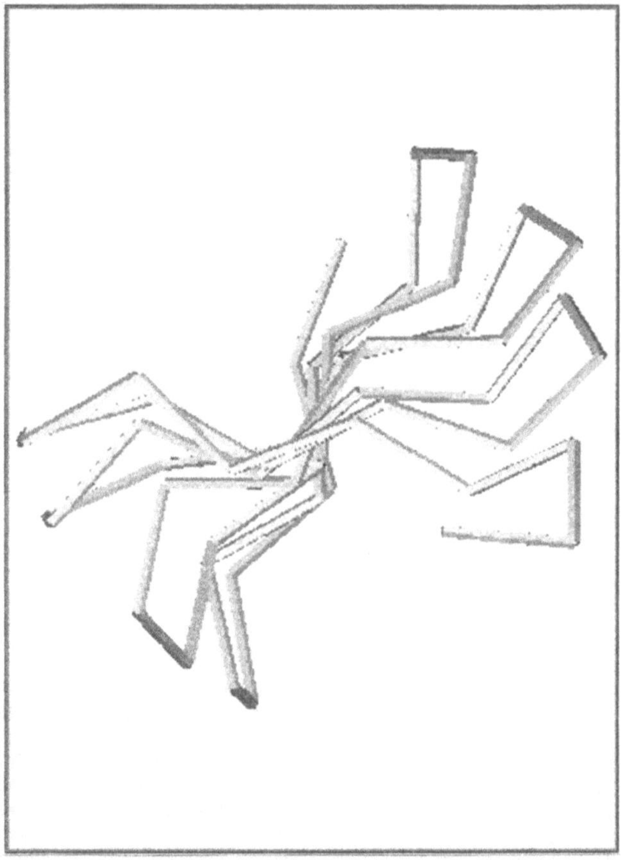

Fig. 4B

Figure 4A. Antiparallel twisted β-sheet model proposed for the cA peptide. Sequence should be read continuously, beginning at the bottom. Invariant glycines (G) occupying the 2nd position in the β-turns are black boxed. Tentative II' β-turns alternate with four-residue β-strands.

Figure 4B. A skeletal C_α model of the cA peptide showing the characteristic antiparallel β-pleated sheet fold. View perpendicular to the β-strands, parallel to the face of the twisted (helical) sheet.

This model is a slightly modified version of the model proposed previously for the central domain of silkmoth chorion proteins (Hamodrakas et al., 1985; Hamodrakas et al., 1988): An antiparallel twisted β-pleated sheet structure of four-residue β-strands alternating with II' β-turns is proposed to be the structural fold of cA.

The main difference between this model structure and the structure proposed previously (Hamodrakas, 1992 and references therein) is that in the current model, all β-turns are proposed to be II', whereas in the previous model, II' and I' turns alternate along the sequence. The modification of the model was judged to be necessary from the presence of invariant Gly residues in the second position of the β-turns. This is a location especially favourable for Gly in II' turns of twisted β-sheets of globular proteins (Sibanda & Thorton, 1985; Hamodrakas, 1992).

Some 25 years ago, it was proposed that because of the inherent twist of the β-sheets in the monomer, the polymeric protofilaments might form long spacing helical structures in which the protofilaments are intertwined to produce 100 Å doubly helical amyloid fibrils (Cooper, 1976). The verbal description of this model (Cooper, 1976) which is reminiscent of the structure of the β-helix of transthyretin amyloid protofilament produced 20 years later by high-resolution X-ray studies (Blake and Serpell, 1996) fits well to our data.

We propose that successive cA units are forming continuous twisted antiparallel β-sheets (β-sheet helices) along the protofilaments of Fig. 2A, with their β-strands perpendicular to the long axis of the protofilaments (cross-β structures).

The thickness of each individual cA unit is of the order of 30 Å (one has also to consider the effect of the negative stain to fully account for protofilament thickness).

Suspiciously, the pitch of the double helix which is formed by the two intertwined protofilaments is 920 Å (Figure 2A), a multiple of the 115 Å spacing of the β-helix in the transthyretin amyloid protofilament (Blake and Serpell, 1996). Furthermore, the β-sheet model of the cA peptide (Figure 4) is an eight-stranded β-sheet, in contrast to the six-stranded β-sheet of transthyretin. A detailed account of our data analysis will be given elsewhere (Iconomidou & Hamodrakas, in preparation).

Discussion

To our knowledge, this is the first well documented case whereby amyloid-like fibrils are formed from a peptide which has a sequence so clearly folded in an antiparallel β-pleated sheet type of structure, which should be of the cross-β type (the β-strands perpendicular to the long axis of the fibrils). Nature, after millions of years of molecular evolution, has designed these peptides to play an important functional role: to protect the oocyte and the developing embryo from a wide range of environmental hazards (Hamodrakas, 1992). Clearly, amyloids were designed by nature to play a protective role in this case. The fact that the cA peptide, which corresponds to about 30% of the total chorion mass, forming the central domain of silkmoth chorion proteins, produces by self-assembly mechanisms amyloid-like fibrils under a great variety of conditions (Iconomidou & Hamodrakas, In preparation), suggests that it should be folded in an amyloid fashion even in the physiological state. Chorion proteins self assemble to form the chorion of silkmoths far away from the follicle cells that synthesize and secrete them (Hamodrakas, 1992 and references therein). Experimental and theoretical data support this assumption (Hamodrakas, 1992). However, it might be argued that this is not the only case where amyloids appear *in vivo*. The *Chrysopa flava* silk should be another such case (Geddes et al., 1968).

Another plausible molecular model for the structure of the cA peptide might be that of the left handed parallel β-helix present in the structure of UDP-N-acetylglucosamine acyltransferase (Raetz & Roderick, 1995). The left-handed parallel β-helix is a uniform in cross section structure with a hydrophobic core and a polar coat, composed of a set of three parallel β-sheets forming the three faces of a prism with a cross section that of an equilateral triangle. Each turn of the left handed β-helix consists of a β-strand of four residues followed by two residues in a β-turn (six residues in total), repeated three times, thus forming an equilateral triangle. This is obvious in the aminoacid sequence of the UDP-N-acetylglucosamine acyltransferase, which shows hexapeptide sequence motifs (Raetz & Roderick, 1995). It should perhaps be mentioned that, models similar to the right-handed parallel β-helix found in the pectate lyases were proposed to be the main molecular components of amyloid protofibrils (Lazo & Downing, 1998). Despite the obvious regularities in sequence of the cA peptide, which also shows hexapeptide periodicities both in Gly and in hydrophobic residues (Hamodrakas, 1992 and references therein) and could well fold into a left-handed β-helix (data not shown), there are certain characteristics in the X-ray diffraction patterns and the spectroscopic data (Benaki et al, 1998; Iconomidou & Hamodrakas, In preparation) in clear favour of the antiparallel β-sheet model shown in Figure 4.

A detailed analysis of our model in comparison to the left-handed β-helix model of transthyretin (Blake & Serpell, 1996) which is very similar to ours, will be given elsewhere (Iconomidou & Hamodrakas , In preparation).

Acknowledgements

S.J.H finds this opportunity (after ~30 years!) to express his gratitude to Dr. A.J. Geddes in public, for teaching him the secrets of molecular architecture. V.A.I. gratefully acknowledges the help of a short term EMBO fellowship to her work. We thank Dr. L. Serrano, Dr. B. Agianian, Dr. K. Leonard, Dr. A. Hoenger and Prof. F.C. Kafatos for their help with the experiments. Special thanks are due to Prof. G. Vriend for his unfailing help and for many useful and stimulating discussions. S.J.H thanks the Greek Ministry of Research and Technology for financial support.

REFERENCES

Benaki, D.C., Aggeli, A., Chryssikos, G.D., Yiannopoulos, Y.D., Kamitsos, E.I., Brumley, E., Case, S.T., Boden, N., Hamodrakas, S.J. (1998) Laser-Raman and FT-IR spectroscopic studies of peptide-analogues of silkmoth chorion protein segments, *Int. J. Biol. Macromol.* **23**, 49-59

Blake, C.C.F., and Serpell, L.C. (1996) Synchrotron X-ray studies suggest that the core of the transthyretin amyloid fibril is a continuous β-sheet helix, *Structure* **4**, 989-998

Cooper, J.H. (1976) In "Amyloidosis" (Wegelius, O. and Pasternak, A., eds.), pp. 61-68. Academic Press, London, New York.

Fraser, R.D.B. and MacRae T.P. (1973) Conformation in fibrous proteins and related synthetic polypeptides, Academic Press, New York and London

Geddes, A.J., Parker, K.D., Atkins, E.D.T., and Beighton, E. (1968) Cross-β conformation in proteins, *J. Mol. Biol.* **32**, 343-358

Goldsmith, M. R. and Kafatos, F.C. (1984) Developmentally regulated genes in silkmoths, *Annu. Rev. Genet.* **18**, 443-487

Hamodrakas, S.J. (1992) Molecular architecture of helicoidal proteinaceous eggshells in "Results and Problems in Cell Differentiation" (S.T. Case, ed.), Vol. **19** (Ch. 6), pp. 115-186, Springer-Verlag, Berlin and Heidelberg

Hamodrakas, S.J., Jones, C.W. and Kafatos, F.C. (1982) Secondary structure predictions for silkmoth chorion proteins, *Biochim. Biophys. Acta* **700**, 42-51

Hamodrakas, S.J., Etmektzoglou, T. and Kafatos, F.C. (1985) Amino acid periodicities and their structural implications for the evolutionarily conservative central domain of some silkmoth chorion proteins, *J. Mol. Biol.* **186**, 583-589

Hamodrakas, S.J, Bosshard, H.E. and Carlson, C.N. (1988) Structural models of the evolutionarily conservative central domain of silkmoth chorion proteins, *Prot. Eng.* **2(3)**, 201-207

Lazo, N.D. and Downing, D.T. (1998) Amyloid fibrils may be assembled from β-helical protofibrils, *Biochem.* **37(7)**, 1731-1735

Lekanidou, R., Rodakis, G.C., Eickbush, T.H. and Kafatos, F.C. (1986) Evolution of the silkmoth chorion gene superfamily: gene families CA and CB, *Proc. Natl. Acad. Sci. USA* **83**, 6514-6518

Raetz, C.R.H. and Roderick, S.R. (1995) A left-handed parallel β-helix in the structure of UDP-N-Acetylglucosamine acyltransferase, *Science* **270**, 997-1000

Regier, J.C. and Kafatos, F.C. (1985) Molecular aspects of chorion formation In "Comprehensive Insect Biochemistry, Physiology and Pharmacology" (Gilbert, L.I. and Kerkut, G.A., eds.), Vol I., pp.113-151, Pergamon Press, Oxford and New York

Sibanda, B.L. and Thornton, J.M. (1985) β-hairpin families in globular proteins, *Nature* **316**, 170-174

Sunde, M. and Blake, C. (1997) The structure of amyloid fibrils by electron microscopy and X-ray diffraction In "Advances in Protein Chemistry", **50**, 123-159

AMYLOIDOGENESIS OF ISLET AMYLOID POLYPEPTIDE (IAPP)

A. KAPURNIOTU
Physiological-chemical Institute
University of Tübingen
Hoppe-Seylerstr. 4
72076 Tübingen
Germany

Introduction

Amyloid formation and tissue deposition is a common pathological symptom of more than 15 known diseases including Alzheimer's disease (AD), prion protein (PrP)-related encephalopathies, familial amyloid polyneuropathy (FAP), and type II diabetes mellitus (Sipe, 1994). The various amyloid fibrils originate from several distinct polypeptides and proteins but share similar physicochemical, ultrastructural, and cytotoxic properties, although there is no primary structure homology between the various amyloidogenic peptide chains (Lansbury, 1992; Sipe, 1994). The reported cytotoxicity of amyloid is indicative of at least a strong association of amyloid with the pathological sequelae of these diseases. Due to the non-crystallinity of amyloid structure and its strong insolubility, physical methods that are usually applied for structure elucidation such as NMR and X-ray crystallography, cannot give adequate information about the exact arrangement of the peptide chains in an amyloid fibril. Accordingly, the precise structural arrangement within the amyloid fibril remains to be elucidated (Lansbury, 1992). However, both earlier and recent models of the amyloid fibril are consistent with the notion that it consists of stacked β-sheets (Jimenez *et al.*, 1999; Lansbury, 1992).

In the past few years it has been suggested that not only the amyloid structure but also the conversion of a peptide chain from its soluble state into insoluble fibrils may proceed *via* specific steps at the molecular level that may be common for the different amyloidogenic polypeptides (Lansbury, 1999; Wetzel, 1996). It has been proposed that soluble, non-aggregated peptide transforms into insoluble amyloid *via* a conformational transition of unordered or α-helical regions into β-sheets (Shen & Murphy, 1995; Soto & Frangione, 1995; Thomas *et al.*, 1995). Furthermore, it has recently been suggested that amyloid fibril formation may proceed *via* aggregation of a partly unfolded state of the protein and supported the notion that amyloid formation follows specific and most likely general molecular rules (Colon & Kelly, 1992; Lansbury, 1999; Wetzel, 1996).

Pancreatic islet amyloid is found in the majority of type II diabetes patients and forms by the aggregation of islet amyloid polypeptide (IAPP) which is a 37-residue peptide (Cooper *et al.*, 1987; Westermark *et al.*, 1987). While soluble IAPP is thought to play a

A. Aggeli et al. (eds.), Self-assembling Peptide Systems in Biology, Medicine and Engineering, 171–185.

role in carbohydrate metabolism, most likely as an insulin counter-regulator, insoluble IAPP-derived amyloid has been found to be strongly cytotoxic (Lorenzo *et al.*, 1994; Luskey, 1992). Formation of IAPP-derived amyloid has been associated with the pathology of type II diabetes (Lorenzo *et al.*, 1994; Luskey, 1992).

We have studied the molecular mechanism of IAPP amyloidogenesis and here, we present several of our studies concerning the conformational transitions that we have observed to precede and underly amyloid formation by IAPP *in vitro* (Kayed *et al.*, 1999).

Results and Discussion

IAPP AMYLOID FORMATION IS NUCLEATION-DEPENDENT

It has been suggested for several amyloidogenic peptides that amyloid formation proceeds via a nucleation-dependent polymerization mechanism (Jarrett & Lansbury, 1993). According to this mechanism, nucleus formation is the defining step of the rate of amyloid formation that typically exhibits a lag time. Seeding of a supersaturated solution by preformed amyloid eliminates this lag time.

To measure the time dependence of IAPP amyloid formation, a filtration assay was developed that was based on the co-precipitation of ^{125}I-labeled IAPP with non-labeled IAPP and on the retention of insoluble aggregates on a filter membrane (Kapurniotu *et al.*, 1998). Thus in this assay setting, ^{125}I-IAPP which was added at an about 1000-fold lower concentration than non-labeled peptide acted as a tracer of IAPP amyloid formation. Amyloid formation by 1 µM IAPP in 10 mM phosphate, pH 5.0, containing 1% 1,1,1,3,3,3-hexafluoro-2-propanol (HFIP) was found to exhibit a lag time of about 6 h. IAPP insolubilization immediately ensued thereafter as usually observed in nucleation-dependent polymerization processes. IAPP immediately aggregated into amyloid at concentrations higher than 1 µM, indicating that the rate of IAPP aggregation is concentration-dependent. Moreover, the lag time observed at 1 µM was completely eliminated by the addition of traces of preformed fibrils, suggesting that insoluble amyloid formation by IAPP indeed is a nucleation-dependent process. The fact that IAPP amyloid formation can be seeded at concentrations as low as 1 µM was indicative of a very low critical concentration of this peptide and is reminiscent of the reported low critical concentration of Aβ(1-42) (Harper & Lansbury, 1997).

IAPP AMYLOID FORMATION PROCEEDS VIA A NUCLEATION-DEPENDENT CONFORMATIONAL TRANSITION INTO HYDROPHOBIC B-SHEETS

The nucleation dependence of IAPP amyloid formation as found by the filtration assay does not give information about early molecular events, including early self-association events and conformational transitions of soluble IAPP that may precede amyloid formation. This is because the filtration assay can only detect aggregates that are retained on the membrane, whereas smaller and probably still soluble aggregates that may form during the lag time period as defined by the filtration assay might not be detected (Harper & Lansbury, 1997).

We became interested in the early events of IAPP aggregation into fibrils, i.e. molecular steps that precede and accompany the conversion of soluble molecules into insoluble aggregates. To be able to detect possible conformational changes during transition, we chose to apply far-UV circular dichroism spectroscopy (CD). Far-UV CD cannot offer information about local conformational features of a polypeptide of 37 amino acid residues but is a valuable tool for the monitoring of time-dependent changes in the overall conformation of aggregating systems in solution, including amyloidogenic peptides .

First, we studied the pathway of amyloid formation of a supersaturated but kinetically soluble IAPP solution (5 μM in 10 mM phosphate, pH 7.4, containing 1% HFIP) (Figure 1A). For the preparation of this solution, a stock IAPP solution in HFIP was used. HFIP was chosen as a solvent for the IAPP stock solution because we had previously found that IAPP populates a non-aggregation-prone conformeric state, i.e. an α-helix, in this solvent (Kapurniotu *et al.*, 1998; Kayed *et al.*, 1999). Similar disaggregating properties have been reported for HFIP in the case of other amyloidogenic peptides including Aβ sequences (Wood *et al.*, 1996). Using these conditions, we are able to store IAPP for several months at 4°C, obtaining essentially the same starting conformation following dilution into aqueous buffers over the entire storage period (Kapurniotu *et al.*, 1998; Kayed *et al.*, 1999). CD spectra of the aqueous IAPP solution (5 μM in 10 mM phosphate, pH 7.4, containing 1% HFIP) immediately after its preparation indicated that IAPP contained significant amounts of both ordered and unordered structure (Figure 1A). Ordered elements under these conditions were found to comprise 24% β-sheet, 8% β-turn, and 18% α-helix according to secondary structure analyses using the reference spectra of Brahms and Brahms (Brahms & Brahms, 1980). This solution was then seeded with traces of *in vitro*-prepared IAPP fibrils. Immediately thereafter, a steep conformational change towards more ordered structures was observed. The shapes of the CD spectra were indicative of a time-dependent increase of the β-sheet component which reached a maximum at 26 min following the begin of the transition. The increase of β-sheet structure was also confirmed by secondary structure analyses of the spectra (Figure 2). Insoluble IAPP then started to precipitate out of solution. The β-sheet content that corresponded to the spectrum that was obtained just prior insolubilization was nearly twice as high as the β-sheet content before seeding (Figure 2). It should be noted that the amounts and molecular weights of the several different IAPP assemblies that might be present in the supersaturated solutions examined here, are not known. IAPP as used in the above supersaturated solutions was partially retained by gel filtration columns and it also only partially entered SDS-PAGE gels, which hampered studies on its degree of self-association (Kapurniotu *et al.*, 1998). Therefore, estimations of molar peptide concentration and secondary structure analyses were performed by using the molecular weight of monomeric IAPP, assuming that the solution consists of soluble IAPP monomers.

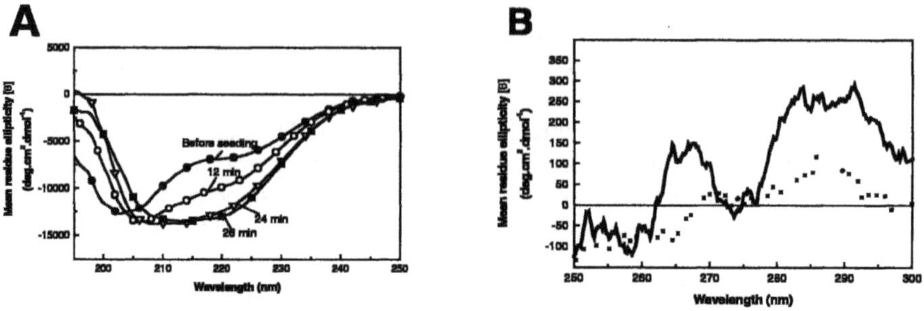

Figure 1. CD spectra of nucleated IAPP amyloid formation. IAPP (5 µM in 10 mM phosphate, pH 7.4, and 1% HFIP) was seeded with 500 nM preformed fibrils. CD spectra at several time points up to insolubilization (28 min) are shown. (A) Far-UV CD spectra before, at 12 min, at 24 min, and at 26 min following seeding. (B) Near-UV CD spectra of IAPP before seeding (broken line) and at 26 min after seeding (continuous line) or just prior to insolubilization.

Insoluble aggregates were examined by electron (EM) and atomic force microscopy (AFM) and found to consist exclusively of amyloid fibrils that exhibited the typical morphology of *in vivo*-formed amyloid (Sipe, 1994) (Figure 3A). These fibrils also bound Congo red (CR) and exhibited the typical amyloid green/gold birefringence under polarized light. The soluble β-sheet-rich IAPP aggregates that were observed just prior insolubilization may already correspond to high molecular weight fibrillar aggregates, i.e. protofibrils, that may still be visibly soluble. Such early and soluble self-assemblies have previously been described to be formed by the Aβ peptide (Harper *et al.*, 1997; Wood *et al.*, 1996). Such a notion would be consistent with the SDS-PAGE profile of glutaraldehyde-crosslinked IAPP samples that was obtained during the above amyloidogenesis process (not shown). The SDS-PAGE data also indicated that there is very little peptide capable of entering the running gel at 30 min following seeding. By contrast, before and several minutes after seeding, monomeric, dimeric, and some multimeric species that could enter the gel were detected.

In contrast to the immediate conformational transition into β-sheets and amyloid formation following seeding, non-seeded IAPP (5 µM in 10 mM phosphate, pH 7.4, and 1% HFIP) remained visibly clear and no changes in overall conformation were usually observed for about 5 to 7 days following preparation of the solution. At that time, however, the same transition towards β-sheet and insoluble amyloid ensued as observed in the seeded solution. Comparison of the profile of the (absolute) increase of the mean residue ellipticity at 218 nm ($[\theta]_{218}$) *versus* time between the seeded and non-seeded IAPP solutions showed that they were virtually identical (Kayed *et al.*, 1999). Moreover, the kinetic profiles of the transition into β-sheets were typical profiles of a nucleation-dependent polymerization mechanism (Jarrett & Lansbury, 1993). These results together suggested that amyloid formation by IAPP proceeds *via* a conformational transition towards β-sheet aggregates and that this transition is nucleation-dependent. While it is likely that the conformational transition observed „accompanies" IAPP aggregation into amyloid, we do not know yet if it is caused by the aggregation event or *vice versa*. Studies towards obtaining a clearer picture of the molecular assembly forms of IAPP in the process of nucleated amyloid formation are now under way in our laboratory.

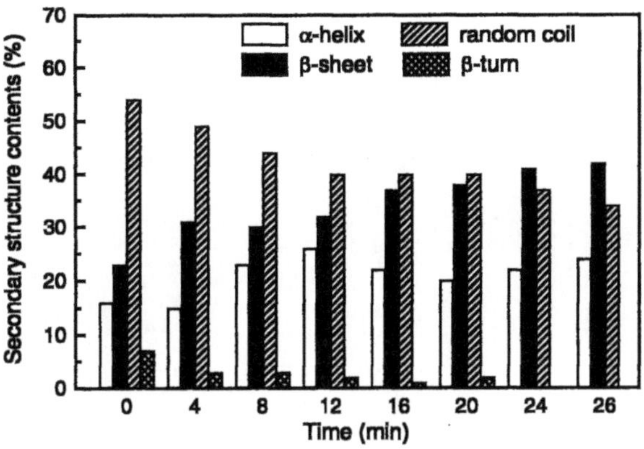

Figure 2. Calculated secondary structure contents of IAPP (5 μM in 10 mM phosphate, pH 7.4, and 1% HFIP) before seeding (0 min) and at several time points following seeding with preformed amyloid (10%) up to appearance of insoluble amyloid (28 min following seeding). Analyses of the CD spectra were performed by the reference spectra set of Brahms and Brahms (Brahms & Brahms, 1980).

Nucleated amyloid formation was also followed by near-UV CD (Figure 1B). No significant changes were observed in the spectra immediately following seeding and up to 24 min thereafter, which was the time necessary for the β-sheet transition to proceed (see above). At 26 min, however, which was the time point just prior to IAPP insolubilization, the spectrum became markedly different from the spectra before and two maxima at 265 and 285 nm appeared (Figure 1B) that are often assigned to aromatic interactions (Woody & Dunker, 1997). The changes observed in the near-UV CD region just before formation of insoluble fibrils or at 26 min following seeding would be consistent with a dramatic conformational rearrangement just before insolubilization ensues. Of note, the changes in the near-UV CD region were accompanied by a red shift of the crossover point of the corresponding far-UV CD spectrum which, however, retained a very similar shape to the spectrum that was obtained 24 min following seeding (Figure 1A). This suggested that the conformational rearrangement that occured just before insoluble amyloid had appeared could be at the level of tertiary or quartenary structure. Previous studies have related a red shift of the crossover point of β-sheets to the formation of stacked β-sheets (Maeda, 1987; Maeda & Ooi, 1981). Such a stacking would be consistent with the invasive IAPP insolubilization that ensued just after the red-shifted spectrum was observed (28 min).

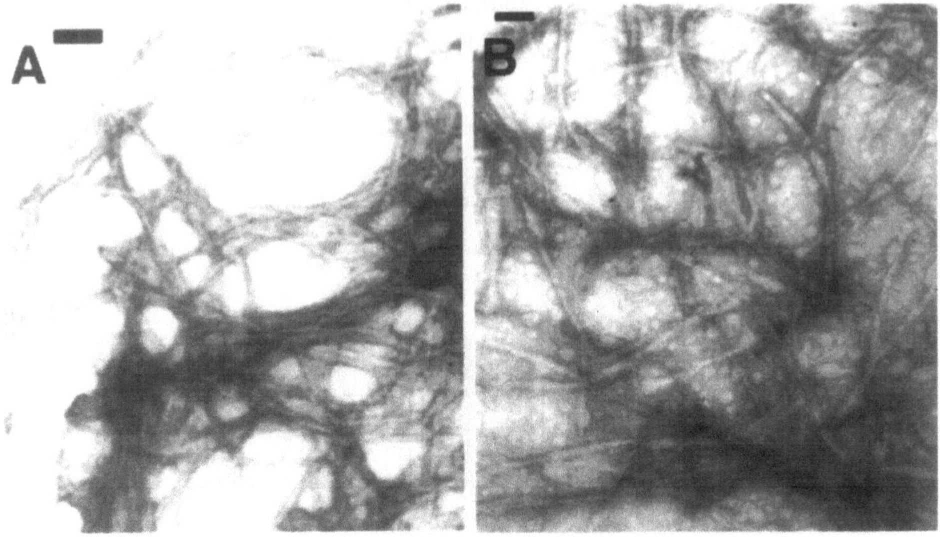

Figure 3. EM examination of IAPP-derived insoluble amyloid formed (A) in a nucleation-dependent CD aggregation experiment 28 min following seeding with IAPP fibrils (10%) and (B) after a 20 min exposure of an unseeded IAPP solution to 45°C (5 μM IAPP in 10 mM phosphate, pH 7.4, and 1% HFIP).

We also applied 1-anilino-naphthalenesulfonic acid (ANS) binding to test for possible exposure of hydrophobic surfaces in the process of IAPP amyloid formation (Semisotnov *et al.*, 1991). Soluble IAPP did not bind ANS. However, following seeding in the presence of ANS, a strong affinity of IAPP for ANS was observed during the transition into β-sheets and amyloid (Figure 4). Both the increased amplitude of the fluorescence emission maxima and the blue shift from 510 to 475 nm were consistent with this notion. The enhanced ANS binding reached its maximum at about 15 to 25 min following seeding. Thereafter, insoluble IAPP precipitated out of solution and the fluorescence intensity decreased. Of importance, we obtained the same enhancement of ANS binding following a lag time of several days, when non-seeded IAPP was allowed to spontaneously aggregate into amyloid under the same conditions as above.

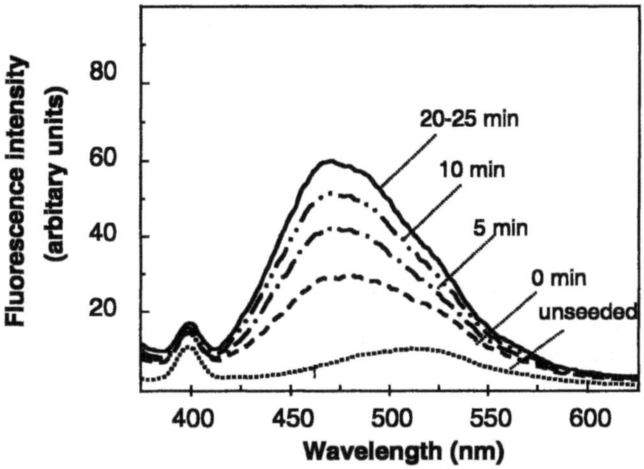

Figure 4. ANS binding of IAPP during amyloid formation. Fluorescence emission spectra of a 5 μM IAPP solution (10 mM phosphate, pH 7.4, containing 1% HFIP and 10 μM ANS) before and after seeding at various time intervals up to insolubilization are shown.

High ANS binding affinity is commonly attributed to the exposure of hydrophobic surfaces of a protein to the solvent (Semisotnov *et al.*, 1991), which is associated with or can cause protein aggregation (Kim & Baldwin, 1990). The enhanced ANS binding properties of IAPP just prior to amyloid formation would suggest that the observed β-sheet formation following seeding is accompanied by a strong exposure of hydrophobic surface. This could be either a driving factor or a result of IAPP self association into soluble precursors of amyloid.

Together, the far- and near-UV CD data and the ANS binding results are consistent with the notion that intermolecular β-sheet formation accompanied by solvent exposure of hydrophobic surfaces and possibly β-sheet stacking are critical steps of IAPP amyloid formation and insolubilization.

AMYLOID FORMATION BY PARTLY UNFOLDED IAPP DURING THE TEMPERATURE-INDUCED DENATURATION PATHWAY

It has been long known that protein folding intermediates may easily self-associate into insoluble, amorphous aggregates (Jaenicke & Rudolph, 1986). More recently, it has also been demonstrated that folding intermediates can aggregate into ordered amyloid fibrils (Wetzel, 1996). To obtain more information about soluble precursors of amyloidogenesis, we next studied thermal denaturation of IAPP by far-UV CD. The denaturation profile that was obtained for IAPP (5 μM in 10 mM phosphate buffer, pH 7.4, and 1% HFIP) was not the typical profile of a two-state denaturation of a folded polypeptide (Figure 5).

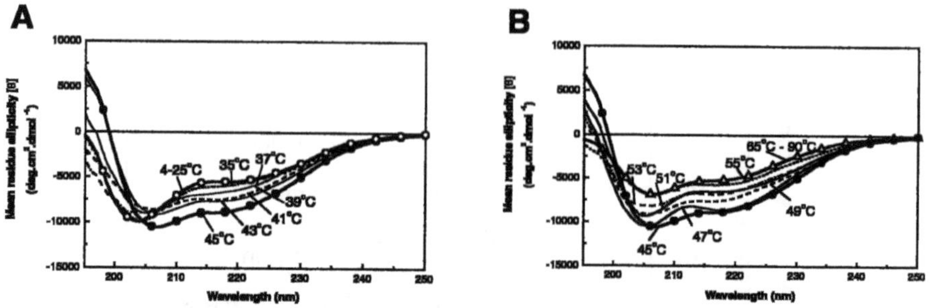

Figure 5. Thermal denaturation of IAPP at 5 μM (in 10 mM phosphate, pH 7.4, 1% HFIP). CD spectra from 4-45°C (A) and 45-90°C (B) recorded at the indicated temperatures are shown.

When the temperature of the IAPP solution was increased stepwise between 4 and 25°C, no changes in overall conformation were seen (Figure 5A). At temperatures higher than 25°C, however, a red shift of the CD minima was observed that was indicative of a gradual increase of ordered conformeric states (Figures 5A and 5B). In the plot of the mean residue ellipticity at 218 nm ($[\theta]_{218}$) *versus* temperature (Figure 6), a steep and cooperative increase of $[\theta]_{218}$ was evident between 40 and 45-47°C.

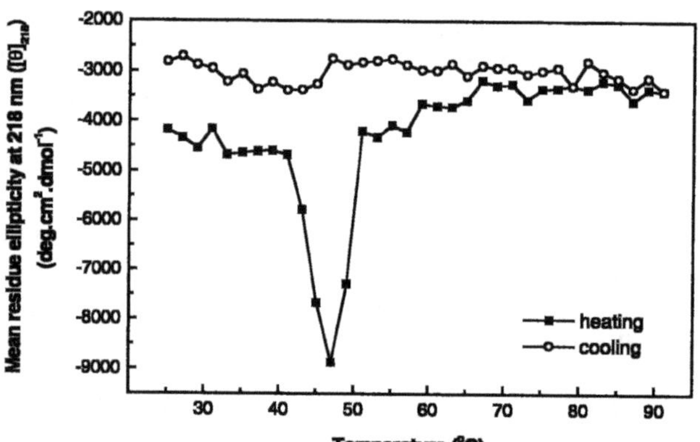

Figure 6. Plot of $[\theta]_{218}$ *versus* temperature increase or decrease (2°C/min) of IAPP at 5 μM (in 10 mM phosphate, pH 7.4, and 1% HFIP) as indicated. No thermodynamic reversibility of the thermal denaturation pathway was observed at this peptide concentration.

At about 45-47°C, a maximum value of $[\theta]_{218}$ was reached (Figure 6). The shape and the secondary structure analysis of the CD spectrum obtained at this temperature suggested increases in both α-helical and β-sheet contents as compared to the contents of the spectrum obtained at room temperature. In detail, the following ordered elements were assigned to the spectrum obtained at 45°C: about 26% of α-helix, 28% of β-sheet, and 5% of turn elements according to the reference spectra of Brahms and Brahms.

When the 45°C state (at 5 μM IAPP) was allowed to stand for several minutes at this temperature, an immediate conformational transition towards β-sheet-containing structures was observed and the ellipticity at 218 nm started to (absolutely) increase (Figure 7A). By contrast, the minimum at about 206 nm that was present in the spectrum at 0 min and 45°C gradually disappeared and CD spectra typical for β-sheet conformations were obtained. This was also confirmed by secondary structure analysis which suggested that after 16 min at 45°C, IAPP would contain about 50% β-sheet structure. After 18 min at 45°C, insoluble aggregates started to precipitate. Insolubilization also occured when attempts were made to reverse formation of the β-sheets by fast cooling of the solution.

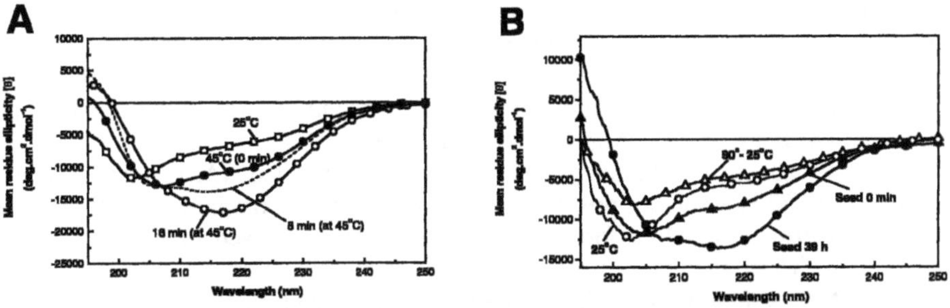

Figure 7. Amyloidogenicity of the 45°C state (A) and the 65°C state (B). (A) IAPP (5 μM in 10 mM phosphate, pH 7.4, and 1% HFIP) was gradually heated to 45°C and allowed to stand for several minutes at 45°C and spectra were recorded at various time intervals up to insoluble amyloid formation 18 min later. (B) IAPP (5 μM in 10 mM phosphate, pH 7.4, and 1% HFIP) (CD spectrum of IAPP at 25°C as indicated) was heat-denaturated up to 90°C and then cooled down stepwisely to 4°C and heated again up to 25°C. No changes were observed in the spectrum obtained at 65 or 90°C during the above procedure (spectrum indicated by: 90-25°C). Thereafter, IAPP was seeded with about 500 nM IAPP fibrils and an immediate conformational change was observed. Insoluble IAPP aggregates precipitated out of solution 39 h later.

Insoluble aggregates were found to consist exclusively of amyloid fibrils of the same morphology as those fibrils formed when native IAPP aggregated at room temperature (Figure 3B). Formation of the 45°C state was only thermodynamically reversible when IAPP concentrations did not exceed 1.25 μM, while invasive insolubilization into amyloid was observed at higher IAPP concentrations when solutions were cooled down. Together, these results suggested that the 45°C state is an aggregation-prone, soluble precursor of IAPP amyloid which is in equilibrium with a, possibly, non-amyloidogenic IAPP conformation that IAPP populates mainly between 4 and 25°C and for which we use the term „native" IAPP.

At higher temperatures, a fast melting of the 45°C state was observed. Heat-denaturated IAPP was obtained at 65°C and its conformation did not change when the temperature was further increased. Heat-denaturated IAPP retained significant amounts of structure (12% α-helix, 34% β-sheet, and 8% turn) and there was no thermodynamic reversibility of its formation at IAPP concentrations higher than 0.1 μM. This indicated that it may consist of soluble, irreversibly self-associated, denaturated IAPP chains, i.e. „wrong" aggregates. Such aggregates have been reported to form during the denaturation processes of several proteins (Jaenicke & Rudolph, 1986). Of note, no amyloid formation was observed for heat-denaturated IAPP upon standing at room temperature for indefinite time, while, as mentioned, native IAPP forms amyloid in about 7 days following solution preparation. Furthermore, seeding of heat-denaturated IAPP with IAPP fibrils resulted in formation of visibly insoluble aggregates only 39 h later instead of 30 min later as observed when native IAPP was seeded (Figure 7B). This data indicated that irreversibly heat-denaturated IAPP is markedly less amyloidogenic than native IAPP which might be related to the presence of kinetic and/or thermodynamic barriers in its conversion into significant amounts of the amyloidogenic conformeric state. High kinetic barriers have been reported to exist in the refolding pathway of transthyretin and have been suggested to play a role in its amyloidogenicity (Lai *et al.*, 1997). Further studies on the nature and molecular assembly form of irreversibly heat-denaturated IAPP are now in progress in our laboratory.

Together, the above results suggested that partly unfolded IAPP that is mainly populated at 45°C is a soluble precursor of amyloid fibrils. Comparison of CD spectra obtained just prior to amyloid formation when IAPP was allowed to stand for several minutes at 45°C to spectra obtained just prior insolubilization of both seeded and unseeded IAPP solutions at 25°C (Figure 8A) suggested that similar β-sheet-rich conformeric states preceded insoluble amyloid formation when unseeded IAPP was heated to 45°C or seeded IAPP aggregated at 25°C. We therefore hypothesized that the amyloidogenic conformeric state that was populated at 45°C (0 min) might also appear in the process of amyloid formation at room temperature. In fact, we found that the CD spectrum of the state populated at 12 min following seeding of IAPP at 25°C (Figure 1A) was very similar to that of the 45°C state (Figures 7A and 8B). Of importance, ANS binding was nearly at its maximum value at 12 min following seeding (Figure 4) which indicated that the state at 12 min had strongly exposed hydrophobic surfaces, a notion that would be consistent with its strong aggregation potential.

The thermal denaturation pathway was also followed by near-UV CD and no differences were observed between the native (at 25°C) and the amyloidogenic or 45°C state of IAPP. Both CD spectra had no distinct maxima suggesting the absence of any constraint at side chains of aromatic residues which might be due to the lack of their persistent quartenary or tertiary structural interactions. Thus, taken together, the CD and the ANS-binding studies suggested that the partly folded state that precedes amyloid formation at room temperature or at 45°C exhibits molten globule-like properties that include a pronounced secondary structure, exposed hydrophobic surface and lack of tertiary structure interactions (Ptitsyn, 1995).

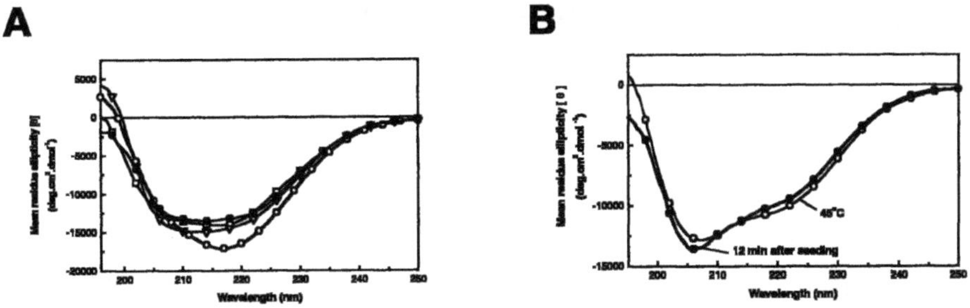

Figure 8. Comparison of CD spectra obtained following nucleated amyloid formation by IAPP at 25°C with spectra obtained following heating of an unseeded solution at 45°C. Experiments were performed with 5 μM IAPP in 10 mM phosphate buffer, pH 7.4, containing 1% HFIP. (A) The spectrum obtained at 26 min following seeding of IAPP or just prior to insolubilization (closed symbols) is compared to several spectra obtained after 10-16 min at 45°C or before insoluble amyloid formation (open symbols). (B) The spectrum of the partly unfolded amyloidogenic state that was populated at ~45°C (0 min) is compared to the spectrum of the state formed at ~12 min following seeding at 25°C.

AMYLOID FORMATION BY PARTLY UNFOLDED IAPP IN THE DENATURANT-INDUCED DENATURATION PATHWAY

Since partially heat-denaturated IAPP was found to aggregate into amyloid, we next studied the denaturant-induced denaturation pathway (5 μM IAPP in 10 mM phosphate buffer, pH 7.4, and 1% HFIP). As also observed in the heat-induced denaturation pathway, a profile that was non-typical for a denaturation process of a polypeptide was obtained when $[\theta]_{218}$ was plotted *versus* concentration of the denaturant. Thus, while increases in denaturant concentration usually result in loss of structure, IAPP was found to populate two structured states during the increase of denaturant concentration. The first state was maximally populated at 0.75 M while the second was at 4.25 M GdnHCl (Figure 9A). Interestingly, IAPP precipitated out of solution between 24 and 48h following formation of both states and EM showed that these aggregates consisted exclusively of fibrils (Figures 10A and 10B). Thus, both partly folded states were soluble precursors of amyloid. In addition, the CD spectrum of the state populated at 4.25 M GdnHCl (Figure 9B) was very similar to the spectrum of the 45°C state (see above), suggesting that IAPP populates the same amyloidogenic state when heated to 45°C or when subjected to 4.25 M GdnHCl. The spectrum of IAPP in 0.75 M GdnHCl, however, had a similar shape but exhibited about half of the signal intensity of the 4.25 M GdnHCl spectrum. It should be noted that, due to the strong absorbance of GdnHCl in the region below 215 nm, it was not possible to obtain CD data for denaturant-induced IAPP below this wavelength.

As shown in Figure 9, the 4.25 M GdnHCl state was populated only in a narrow GdnHCl concentration range and was thereafter converted into GdnHCl-denaturated IAPP at 6 M GdnHCl. Judging from the signal intensities of the CD spectra, IAPP was denaturated to a greater extent in 6M denaturant as compared to the heat-denaturated peptide, indicating that heat-induced denaturation was less effective than denaturant-induced denaturation.

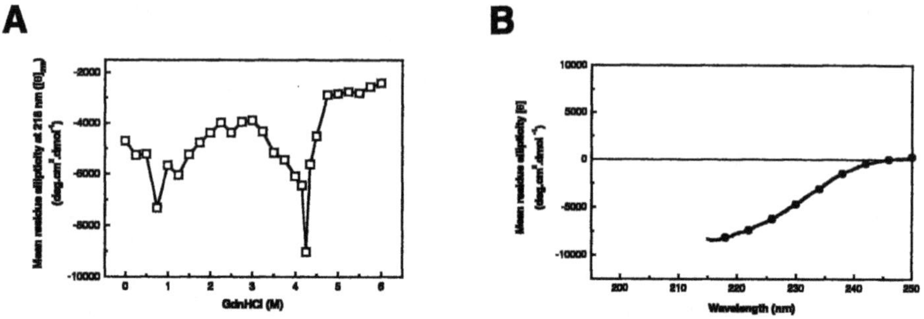

Figure 9. GdnHCl-induced denaturation of IAPP at 25°C (5 μM IAPP in 10 mM phosphate buffer, pH 7.4, 1% HFIP). (A) Plot of $[\theta]_{218}$ *versus* GdnHCl concentration and (B) CD-spectrum of IAPP in 4.25 M GdnHCl.

Contrary to the irreversibility of the heat denaturation process of IAPP at 5 μM, denaturant-induced denaturation was found to be reversible. At GdnHCl concentrations, however, that corresponded to increased populations of the amyloidogenic states, i.e. at 0.75 M and between 3.5 and 4.5 M GdnHCl, insoluble amyloid precipitated immediately (at 4.0 and 3.75 M GdnHCl) or after several minutes following solution preparation. The increased aggregational susceptibility of the amyloidogenic state under refolding conditions may be related to the high amounts of partially unfolded chains causing a kinetic competition between refolding and aggregation into amyloid (Goldberg *et al.*, 1991). Increased fibril-forming potential was also reported for the amyloidogenic intermediate of transthyretin when formed during reconstitution *versus* the acid-induced denaturation (Lai *et al.*, 1996). The results obtained by the denaturant-induced denaturation studies strongly supported the findings of the thermal denaturation studies and suggested that partly unfolded IAPP is strongly amyloidogenic.

When denaturant-induced denaturation was performed at lower peptide concentrations than 5 μM, essentially the same denaturation profiles as at 5 μM were obtained (Figure 11). Thus, in the denaturation pathways of IAPP between 0.1 μM and 5 μM, an amyloidogenic and partly folded state was populated between 3 and 5 M GdnHCl, while the 0.75 M GdnHCl state that was observed at 5 μM IAPP did not appear at peptide concentrations lower than or equal to 0.5 μM. Notably, when 0.05 μM IAPP was applied, the denaturation profile corresponded to an apparent two-state denaturation and no structured state was found to be populated at any GdnHCl concentration. Together, these results suggested that both states, populated at 0.75 and 4.25 M GdnHCl, were self-assembled forms of partly unfolded state(s) of IAPP.

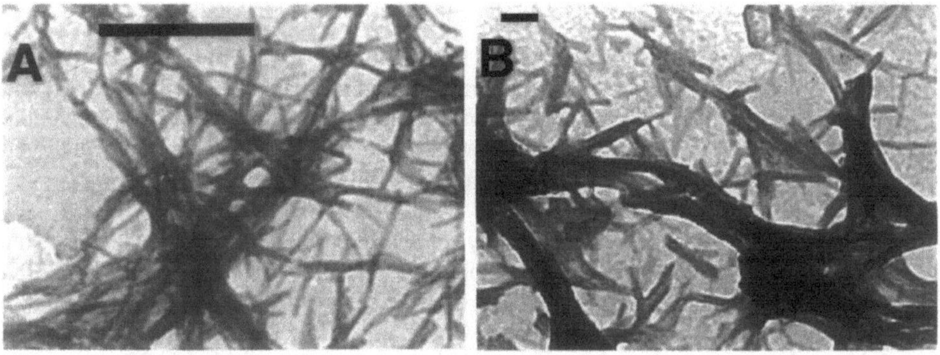

Figure 10. EM of insoluble IAPP aggregates formed (A) in 0.75 M GdnHCl (bar: 500 nm) and (B) in 4.25 M GdnHCl (bar: 100 nm) after standing for 24-48 h at room temperature. IAPP concentration was 5 μM for (A) and 0.25 μM for (B) (in 10 mM phosphate buffer, pH 7.4, 1% HFIP).

It could be possible that both states originate from the same partially unfolded state and correspond to different assembly states. However, due to experimental difficulties that we have encountered with the gel chromatography of IAPP, our current data are not sufficient to exactly characterize the self-association degree of IAPP in the above structured states and work towards this issue is in progress.

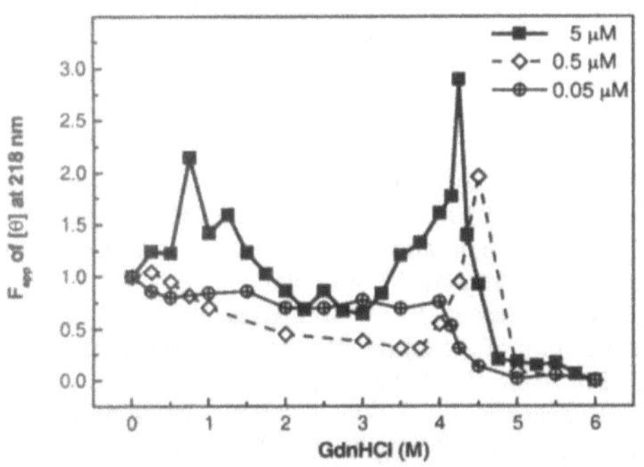

Figure 11. Concentration dependence of GdnHCl-induced denaturation of IAPP at 25°C (in 10 mM phosphate buffer, pH 7.4, 1% HFIP). Plot of the apparent fraction of folded state at 218 nm ($F_{app} = ([\theta]_{218observed} - [\theta]_{218unfolded}) / ([\theta]_{218native} - [\theta]_{218unfolded})$ *versus* GdnHCl concentration for 5, 0.5, and 0.05 μM IAPP is shown as indicated.

Conclusions

Our studies suggest that insoluble amyloid formation by IAPP follows kinetics that are consistent with a nucleation-dependent polymerization mechanism. Thus, IAPP amyloid formation can be accelerated by seeding with preformed IAPP amyloid fibrils. At the molecular level, IAPP amyloid formation was found to proceed *via* a conformational transition into hydrophobic β-sheet containing conformeric states.

The transition into β-sheets could also be seeded by IAPP fibrils and proceeded *via* formation of a structured state with strongly solvent-exposed hydrophobic patches. This amyloidogenic state was found to be populated in both the temperature- and the denaturant-induced denaturation pathways of IAPP and led to formation of insoluble amyloid fibrils. Concentration dependence CD studies suggested that the partly folded amyloidogenic state may form by self-association of partially unfolded IAPP.

Based on these results and on the observed nucleation-dependent protein polymerization mechanism, we propose that partially unfolded IAPP and its self-associated forms may be in equilibrium with native or non-amyloidogenic IAPP conformers and act as early and soluble precursors of β-sheet and amyloid formation.

Acknowledgements

I am grateful to N. Greenfield and J. Bernhagen for useful discussions and their contribution to this work. I thank R. Kayed for performing several of the CD experiments and H. Brunner and W. Voelter for supporting this work. This work was financially supported in part by an institutional grant of The Fraunhofer Institute for Interfacial and Biological Engineering (FhIGB, Stuttgart, Germany).

References

Brahms, S. and Brahms, J. (1980) Determination of Protein Secondary Structure in Solution by Vacuum Ultraviolet Circular Dichroism, *J. Mol. Biol.* **138**, 149-178.

Colon, W. and Kelly, J. W. (1992) Partial denaturation of transthyretin is sufficient for amyloid fibril formation in vitro, *Biochemistry* **31**, 8654-8660.

Cooper, G. J. S., Willis, A. C., Clark, A., Turner, R. C., Sim, R. B. and Reid, K. B. M. (1987) Purification and characterization of a peptide from amyloid-rich pancreases of type 2 diabetic patients, *Proc. Natl. Acad. Sci. U.S.A.* **84**, 8628-8632.

Goldberg, M. E., Rudolph, R. and Jaenicke, R. (1991) A kinetic study of the competition between renaturation and aggregation during the refolding of denaturated-reduced egg white lysozyme, *Biochemistry* **30**, 2790-2797.

Harper, J. D. and Lansbury, P. T., Jr. (1997) Models of amyloid seeding in Alzheimer's disease and scrapie: Mechanistic truths and physiological consequences of the time-dependent solubility of amyloid proteins, *Annu. Rev. Biochem.* **66**, 385-407.

Harper, J. D., Wong, S. S., Lieber, C. M. and Lansbury, P. T., Jr. (1997) Observation of metastable Aβ amyloid protofibrils by atomic force microscopy, *Chem. Biol.* **4**, 119-125.

Jaenicke, R. and Rudolph, R. (1986) Refolding and association of oligomeric proteins, *Methods Enzymol.* **131**, 218-250.

Jarrett, L. L. and Lansbury, P. T. J. (1993) Seeding one-dimensional crystallization of amyloid: a pathogenic mechanism in Alzheimer's disease and scrapie?, *Cell* **73**, 1055-1058.

Jimenez, J. L., Guijarro, J. I., Orlova, E., Zurdo, J., Dobson, C. M., Sunde, M. and Saibil, H. R. (1999) Cryo-electron microscopy structure of an SH3 amyloid fibril and model of the molecular packing, *Embo J.* **18**, 815-821.

Kapurniotu, A., Bernhagen, J., Greenfield, N., Al-Abed, Y., Teichberg, S., Frank, R. W., Voelter, W. and Bucala, R. (1998) Contribution of advanced glycosylation to the amyloidogenicity of islet amyloid polypeptide, *Eur. J. Biochem.* **251**, 208-216.

Kayed, R., Bernhagen, J., Greenfield, N., Sweimeh, K., Brunner, H., Voelter, W. and Kapurniotu, A. (1999) Conformational transitions of islet amyloid polypeptide (IAPP) in amyloid formation *in vitro*, *J. Mol. Biol.* **287**, 781-796.

Kim, P. S. and Baldwin, R. L. (1990) Intermediates in the folding reactions of small proteins, *Annu. Rev. Biochem.* **59**, 631-660.

Lai, Z., Colon, W. and Kelly, J. W. (1996) The acid-mediated denaturation pathway of transthyretin yields a conformational intermediate that can self-assemble into amyloid, *Biochemistry* **35**, 6470-6482.

Lai, Z., McCulloch, J., Lashuel, H. A. and Kelly, J. W. (1997) Guanidine hydrochloride-induced denaturation and refolding of transthyretin exhibits a marked hysteresis: Equilibria with high kinetic barriers, *Biochemistry* **36**, 10230-10239.

Lansbury, P. T., Jr. (1999) Evolution of amyloid: What normal protein folding may tell us about fibrillogenesis and disease, *Proc. Natl. Acad. Sci. USA* **96**, 3342-3344.

Lansbury, P. T. Jr. (1992) In pursuit of the molecular structure of amyloid plaque: New technology provides unexpected and critical information, *Biochemistry* **31**, 6865-6870.

Lorenzo, A., Razzboni, B., Weir, G. C. and Yankner B.A. (1994) Pancreatic islet cell toxicity of amylin associated with type-2 diabetes mellitus, *Nature* **368**, 756-760.

Luskey, K. L. (1992) Possible links between amylin and diabetes, *Diabetes Care* **41**, 297-299.

Maeda, H. (1987) Irreversible nature of the stacked β-pleated sheets of a model polypeptide, *Bull. Chem. Soc. Jpn.* **60**, 3438-3440.

Maeda, H. and Ooi, K. (1981) Isodichroic point and the xβ-random coil transition of poly(S-carboxymethyl-L-cystein) and poly(S-carboxyethyl-L-cysteine) in the absence of added salt, *Biopolymers* **20**, 1549-1563.

Ptitsyn, O. B. (1995) Molten globule and protein folding, *Adv. Proein Chem.* **47**, 83-229.

Semisotnov, G. V., Rodionova, N. A., Razgulyaev, O. I., Uversky, V. N., Gripas, A. F. and Gilmanshin, R. I. (1991) Study of the "molten globule" intermediate state in protein folding by a hydrophobic fluorescent probe, *Biopolymers* **31**, 119-128.

Shen, C.-L. and Murphy, R. M. (1995) Solvent effects on self-assembly of β-amyloid peptide, *Biophys. J.* **69**, 640-651.

Sipe, J. D. (1994) Amyloidosis, *Critical Reviews in Clinical Laboratory Sciences* **31**, 325-354.

Soto, C. and Frangione, B. (1995) Two conformational states of amyloid β-peptide: implications for the pathogenesis of Alzheimer's disease, *Neurosci. Letters* **186**, 115-118.

Thomas, P. J., Qu, B.-H. and Pedersen, P. L. (1995) Defective protein folding as a basis of human disease, *Trends Biochem. Sci.* **20**, 456-459.

Westermark, P., Wernstedt, C., Wilander, E., Hayden, D. W., O'Brien, T. D. and Johnson, K. H. (1987) Amyloid fibrils in human insulinoma and islet of Lagerhans of the diabetic cat are derived from a neuropeptide-like protein also present in normal islet cells, *Proc. Natl. Acad. Sci. USA* **84**, 3881-3885.

Wetzel, R. (1996) For protein misassembly it's the "I" decade, *Cell* **86**, 699-702.

Wood, S. J., Maleeff, B., Hart, T. and Wetzel, R. (1996) Physical, morphological and functional differences between pH 5.8 and 7.4 aggregates of the Alzheimer's amyloid peptide Aβ, *J. Mol. Biol.* **256**, 870-877.

Woody, R. W. and Dunker, K. (1997) Aromatic and cystine side-chain circular dichroism in proteins, in G. D. Fasman (ed.), *Circular Dichroism and the Conformational Analysis of Biomolecules.* Plenum Press, New York and London, pp. 109-158.

ENGINEERING SELF-ASSEMBLY OF PEPTIDES BY AMPHIPHILIC 2D MOTIFS : α-TO-β TRANSITIONS OF PEPTIDES

H. MIHARA,[a,b] Y. TAKAHASHI,[a] I. OBATAYA,[a] S. SAKAMOTO [a]
[a]Department of Bioengineering, Faculty of Bioscience and Biotechnology, Tokyo Institute of Technology, and [b]Form and Function, PRESTO, JST, Nagatsuta, Midori-ku, Yokohama 226-8501, Japan

1. Introduction

A number of studies on disease-related proteins, such as those involved in Alzheimer's or the prion diseases have led to improvements in our understanding of protein misfolding and transformation [1-11]. The studies have pointed out common and possible mechanisms for the formation of protein amyloid fibrils. Environments and mutations which might destabilize the native structures of proteins and facilitate the formation of partially unfolded intermediates have been demonstrated to be possible causes of the fibril formation of proteins [1-5]. Such intermediates undergo peptide conformational transitions, which initiate protein aggregation and then fibril formation. Especially in the prion protein (PrP), a highly α-helical structure in the monomeric PrP^{C} is transformed to a β-sheet conformation in aggregated protease-resistant PrP^{Sc}[6-8]. Similarly, β-amyloid peptides composed of 39–43 amino acid residues are also transformed to the amyloid form with a cross-β-sheet structure [9-11]. Although the mechanisms and intermediates of these amyloid formations have not yet been fully understood, it has been suggested that the aggregation process for the proteins from a less-β-structured monomer plays a key role in the conformational changes and the formation of amyloid with a higher β-sheet content [3,4]. In general, one cause of protein misfolding and transformation is thought to be the exposure of the hydrophobic region of proteins in an unstable form to water environments, and the formation of aggregates that follows.

On the other hand, similar α→β transitions also occur through the correct-folding pathway from intermediates to native forms in proteins with a non-hierarchical folding mechanism such as β-lactoglobulin [12,13]. Moreover, the same peptide sequence named 'chameleon' adapted an α-helix or a β-strand at the different position in the protein, the B1

A. Aggeli et al. (eds.), Self-assembling Peptide Systems in Biology, Medicine and Engineering, 187–205.

domain of protein G (GB1) [14]. The protein named 'Janus' was successfully designed from the β-sheet-dominant protein, GB1, to the α-helical protein, Rop, by changing no more than half the sequence [15]. These studies suggest that the key feature in the α–β transition is the conversion from short-range interactions between nearby amino acid residues stabilizing an α-helix structure to long-range interactions between secondary structures stabilizing a β-sheet structure.

A simplified model peptide designed by the *de novo* strategy can provide useful information for constructing and manipulating peptide conformation, and elucidating complex folding and misfolding mechanisms [16,17]. The design method often utilizes the amphiphilic nature of peptide secondary structures to construct tertiary structures of artificial proteins [18]. Peptides designed for β-sheet folding have been extensively studied [19], and subsequent formation of fibrils has been characterized [20]. Model peptides homologous to parts of native proteins have been characterized as fibrillogenic peptides and their applications to molecular materials have been attempted [21,22]. Furthermore, some model peptides have been designed to allow α–β transitions *via* the replacement of environments from organic solvent to water [23,24], and the alteration of pH [25]. As a consequence, these studies have revealed that control of the periodicity of hydrophobic and hydrophilic residues in a peptide sequence is essential in determining a secondary structure and an α–β transitional properties. In other words, hydrophobic clustering between amphiphilic peptides seems to determine the secondary structures and their transformation.

Using the *de novo* strategy with amphiphilic 2D motifs, we have found peptides that undergo α→β structural transition and amyloid formation, and characterized their transitional properties [26,27]. Moreover, we have pointed out that a protein cofactor such as heme can regulate a peptide folding from β-aggregates to α-helix bundles [28,29]. The strategy employing simplified model peptides will lead to the studies providing insights in α→β structural transition and amyloid formation of peptides, and applying them to new materials.

2. Peptides That Undergo Autocatalytic α → β Transitions and Amyloid Formation

We have found that an exposed hydrophobic nucleation domain caused an α→β structural transition and fibril formation of a peptide (Figure 1) [2,26,27]. The hydrophobic domain was designated as a hydrophobic defect, since it caused the aggregation and fibril formation of the peptide. The first success of this approach for designing α→β transitional peptides was accomplished by using the adamantanecarbonyl (Ad) group as the N-terminal hydrophobic defect of the 2α-helix peptide [26]. The Ad-linked 2α-peptide (Ad-2α) underwent an α→β transition and a β-fibril formation in an autocatalytic manner. Using this strategy of adding hydrophobic defects of various acyl groups (Figure 1) has revealed that the stability of the α-helix conformation in its initial state has a key role in determining whether or not the peptide structure will be transformed [27]. The designed peptides mimicked an α→β transitional property of proteins such as prion proteins, a nucleation-dependent autocatalytic transformation.

2.1. DESIGN OF PEPTIDES

Peptides composed of two amphiphilic α-helices [18] was designed and the 1-adamantane-carbonyl (Ad) group or an acyl chain of various lengths was attached to the N-termini as an exposed hydrophobic domain, *i.e.*, as a hydrophobic defect (Figure 1). The two-α-helix part was constructed from amino acid sequences of coiled-coil proteins [30], which have heptad repeats $(abcdefg)_n$ with hydrophobic residues at the a and d positions. The $\langle P_\alpha \rangle$ value of the core 14-peptide as calculated with Chou-Fasman parameters [31] was 1.34, and the $\langle P_\beta \rangle$ value was 0.94. Therefore, judging only from the amino acid sequence, the peptide was expected to form an α-helix structure. However, when the peptide sequence was drawn as a β-sheet model, the peptide could take a kind of amphiphilic β-sheet structure, in which hydrophobic Leu residues and hydrophilic Glu and Lys residues were separated on the different faces (Figure 1) [18]. The adamantane group was selected as the hydrophobic defect so that β-cyclodextrin could prevent the α→β structural transition by complexation [32]. The 17-peptide was synthesized by the solid-phase method using 9-fluorenylmethyloxycarbonyl (Fmoc) chemistry [33]. The dimeric peptides (Ad-2α and Cn-2α) were synthesized *via* the disulfide linkage between Cys residues at the 17th position.

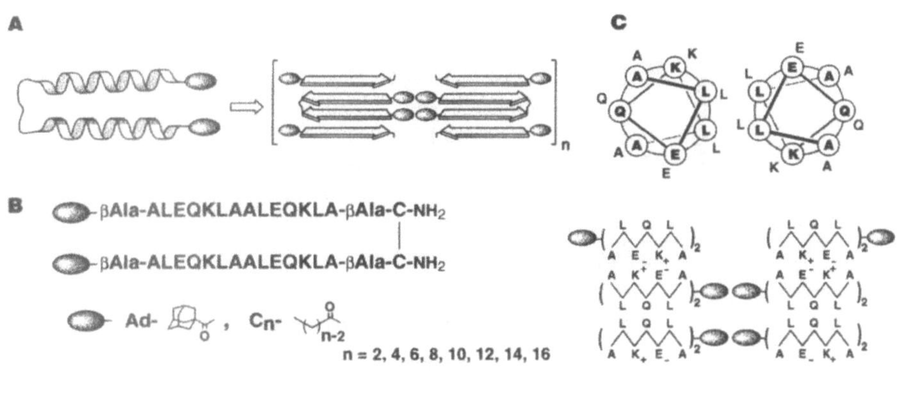

Figure 1. Design of α→β transitional and amyloidal peptides. (A) Illustration of α→β transition. (B) Structure of the peptides. (C) Illustration of amphiphilic α-helices as coiled-coil and β-sheet.

2.2. α→β STRUCTURAL TRANSITION OF PEPTIDES

Adamantane-peptide. The Ad-2α peptide (10 μM) showed a circular dichroism (CD) spectrum typical for an α-helix structure shortly after dilution in 20 mM Tris·HCl buffer (pH 7.4) from the trifluoroethanol (TFE) solution (final TFE was 2.5%) [26]. Without TFE, the α→β transition underwent faster than that in 2.5% TFE. Observation of the CD spectrum revealed a gradual change to a spectrum typical for a β-structure, with a single negative maximum at 218 nm and a positive maximum at 198 nm after 4 h at 25 °C (the half transition time $t_{1/2}$ = 130 min). The time course of the structural transition was quite sigmoidal, that is, after the lag time (ca. 90 min under the conditions) the α-helix conformation was rapidly transformed to the β-structure (Figure 2). The slow and autocatalytic shape of the transition bore a resemblance to the shapes of the fibril formation of β-amyloid peptides and prion peptides [9,34]. The autocatalytic pathway of the transition was also suggested by the observation that the transition was accelerated by seeding of a small amount (5%) of pre-formed β-structural Ad-2α to the α-helical Ad-2α (10 μM), resulting in the elimination of the lag time.

Fourier-transform infrared (FTIR) measurements confirmed the α→β structural transition of peptides. The dried film of peptides from TFE solution showed an absorption spectrum typical for an α-helix structure (amide I peak maximum at 1654 cm^{-1} and amide II peak maximum at 1546 cm^{-1}). The exposure to water in shorter time (<1 h) did not change the spectrum. On the other hand, elongation of the water-exposure period to 20 h changed the spectrum into that typical for β-sheet (amide I peak maximum at 1619 cm^{-1}) [35,36].

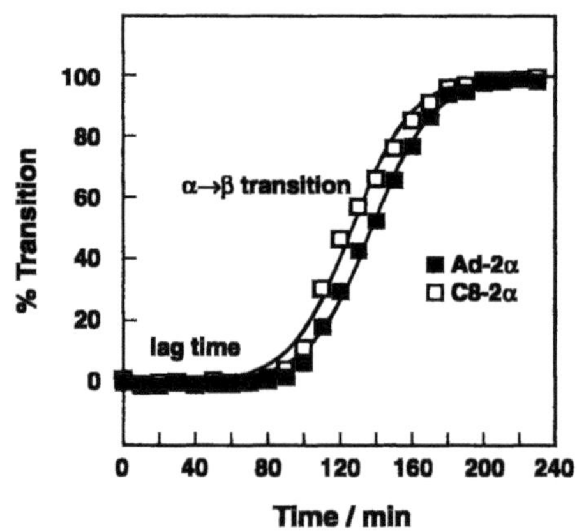

Figure 2. Time courses of α→β transitions of the peptides, Ad-2α and C8-2α, examined by CD measurements.

Acyl-chained peptides. To explore the effect of the hydrophobicity of the conformational defects, we newly synthesized 2α-peptides with a variety of acyl chains and examined the conformational properties [27]. It was interesting that the α→β structural transition of peptides occurred optimally with octanoyl (C8) chain (TABLE 1). That is, the α→β transition of the 2α-peptide with octanoyl groups (C8-2α) occurred at the highest transitional rate among all of the peptides with acyl chains, the rate and the transitional shape being almost identical to that of the adamantanecarbonyl peptide (Ad-2α). On the other hand, the CD spectra of C2-2α did not show a transition to a β-structure, but remained in α-helix. The CD spectra of C4- and C6-2α changed over 14 and 7 h, respectively, but their shapes indicated the presence of a mixture of α-helix and β-sheet, with the content of β-structure of C6-2α being higher than that of C4-2α. As the chain length increased, the α→β transition appeared to take place more rapidly and completely. The peptides with C8 and C10 showed an almost complete α→β transition, though the rate of transition for C10-2α was lower than that of C8-2α. In contrast, the peptides with acyl chains longer than C10 (C12- to C16-2α) formed rather stable α-helix (larger ellipticity at 222 nm), but showed no sign of making an α→β transition. These results indicated that the hydrophobic defect was important for the 2α-helix peptide to transform to β-structure, but there was an optimum in the hydrophobicity of the defect. Whether or not the peptides are able to make the α→β transition appears to be related to the peptide ellipticity in the initial α-helix state (TABLE 1). Peptides with longer acyl chains showed a larger ellipticity at 222 nm. The longer acyl chains provided a higher degree of stabilization for the α-helix structure in its initial state.

TABLE 1. The initial α-helical and α→β transitional properties of the peptides

	Initial α-helical state			
Peptide	$-[\theta]_{222}$ / deg cm^2 dmol^{-1}	$-\Delta G_{H_2O}$ / kcal mol^{-1}	Oligomerization	α→β Transition
C2-2α	15 800	0.5	1 mer	α
C4-2α	15 500	0.9	1 mer	α→β*
C6-2α	16 200	1.1	1 mer	α→β*
C8-2α	21 000	0.8	1 mer	α→β
Ad-2α	23 000	1.0	1 mer	α→β
C10-2α	25 000	1.7	1–3 mer	α→β
C12-2α	25 300	2.9	3 mer	α
C14-2α	25 600	2.9	3 mer	α
C16-2α	26 000	2.4	3 mer	α

* The α→β transition was incomplete.

2.3. AMYLOID FORMATION

Transmission electron microscopy (TEM) studies clearly showed that the β-structural Ad-2α formed a fibrillar structure (Figure 3). The fibrils were ~10 nm in width and of indeterminable length (several hundreds nanometers to several micrometers), and aggregated into dense bundles. The morphology of the fibrils formed by Ad-2α is identical to that of C8-2α, and comparable to that by naturally-occurring amyloid proteins and peptides, such as β-amyloid peptides and prion proteins [37,38].

The amyloid formation of Ad-2α was also examined by amyloid-specific dye binding analyses. It is known that a fluorescent dye, thioflavin T (ThT), associates with aggregated β-sheet fibrils, and that the binding gives rise to a significant enhancement in fluorescence according to the amount of amyloid [39]. The amyloidogenesis of Ad-2α was examined by employing the ThT fluorescence. In the presence of Ad-2α (10 μM) in the initial α-helix structure (shortly after dilution to the buffer from TFE), ThT (6 μM) showed little fluorescence. However, in the presence of Ad-2α (10 μM), which accomplished the transition to the β-structure (8 h after dilution), ThT showed a new excitation maximum at 435 nm and an enormously enhanced emission at 482 nm (52 times greater than in the case of α-helix).

To monitor the amyloid formation process of the peptide, ThT was added to the Ad-2α solution after various incubation periods (Figure 4). The time course of fluorescence intensity at 482 nm was superimposed on that of ellipticity at 205 nm which corresponded to the secondary structure transition, suggesting that the amyloidogenesis occurred coincidentally with the α→β structural transition. After the peptide took the β-sheet, significant enhancements in ThT emission were not observed (1.2-fold at 50 h compared

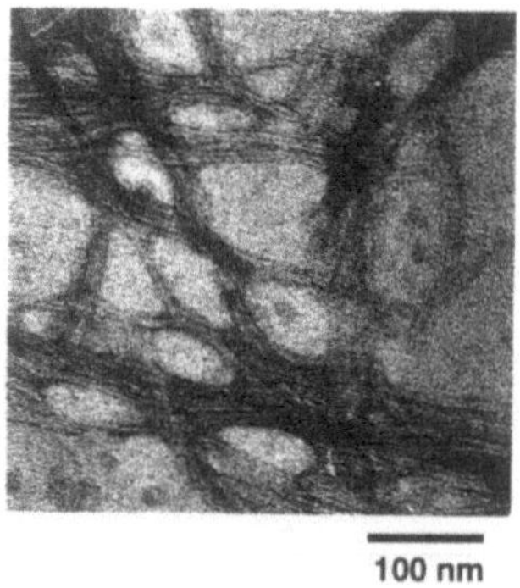

Figure 3. TEM image of the amyloid fibrils of Ad-2α.

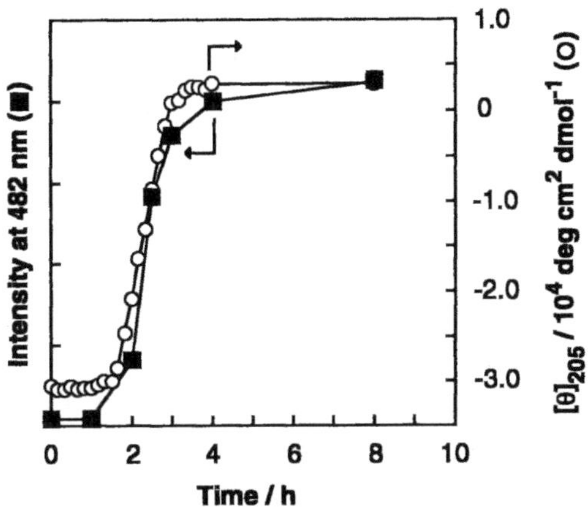

Figure 4. The α→β transitional profile of Ad-2α examined by CD and fluorescence measurements with ThT.

to 8 h). These results indicate that the peptide transformed to a β-structure assembles simultaneously into amyloid fibrils and the amyloid formation completes almost during the secondary structure transitional process.

In order to compare the quaternary structure of the β-structural Ad-2α to the cross-β-structures formed from amyloid proteins, Congo red (CR) binding studies were also carried out [40]. CR is a sulfonated azo dye which binds preferentially to protein adopting a cross-β-structure. The absorption spectrum of CR with Ad-2α in an α-helix structure was not different from that of the dye alone. In contrast, when Ad-2α transformed to a β-sheet was added, CR showed an absorption spectrum characteristic for that of CR bound to a cross-β-sheet amyloid fibril, which exhibits double maxima at approximately 514 nm and 530 nm [20]. This result indicates that β-structural Ad-2α assembles into a regular quaternary structure consistent with a cross-β-sheet structure.

Furthermore the β-fibrils assembled into larger deposits, which were able to be observed by a photomicroscope [26]. During Ad-2α was in an α-helix structure (0–1 h in the buffer at 25 °C), no visible aggregate was observed, as suggested by the fluorescence spectroscopic measurements. However, when the α→β secondary structural transition occurred and the ThT fluorescence enhancement was observed (2 h), visible aggregates appeared. The aggregates formed by the β-structural Ad-2α fibrils propagated along with the process of the α→β structural transition (2–4 h). Further significant growth and propagation of the aggregates were not observed even after 1 day.

Moreover, the peptide converted to a protease-resistant form along with the α→β transition. Ad-2α in an α-helix disappeared by a treatment of trypsin or proteinase K within 1 h at 25 °C. In contrast, the peptide which converted to the β-sheet was hardly degraded by these proteases under the same conditions. A conversion from protease-sensitive to resistant form is one of the most critical characters of prion proteins [6].

2.4. INHIBITION OF TRANSITION AND AMYLOID FORMATION

Inhibition of the transition with β-cyclodextrin. It is well known that β-cyclodextrin (βCDx) strongly captures an adamantane derivative into the hydrophobic cavity at an association constant of ~10^5 M^{-1} [32]. Therefore, βCDx was expected to prevent the peptide aggregation by capturing the exposed Ad group within the cavity. The addition of βCDx to the solution of peptide Ad-2α while it was in the α-helical state retarded the α→β transition with increasing concentration of βCDx [26]. The addition of 10 equiv. of βCDx to the Ad group completely inhibited the transition over the observation period (2 days). Below 5 equiv. of βCDx, CD spectra after the transition seemed to be a mixture of α-helix and β-strand, with the α-helix content being increased depending on the amount of βCDx. The CD spectra resulting from a mixture of α-helix and β-strand appeared to shrink as compared with the spectrum resulting from α-helix (10 equiv. βCDx) or β-sheet (without βCDx). These spectra might be composed of those from intermediate structures as well as from α-helix and β-strand structures.

Size-exclusion chromatography revealed that the association of peptides did not take place in the presence of 10 equiv. of βCDx during the examined time (2 days). The CD spectrum at the initial α-helix state was not changed by the addition of βCDx, indicating that βCDx did not affect the initial α-helix structure. The prevention of the aggregation with βCDx inhibited the structural transition. These results suggest that the Ad groups are defects in the α-helix peptide and nucleate the peptide aggregation, which subsequently causes the structural transition. After the peptide took the β-structure, however, the addition of excess amounts of βCDx could not reverse the conformation from β to α, suggesting that the β-structure formed was packed so tightly that the Ad groups could not be exposed.

Effects of temperature and peptide concentrationn. The temperature dependence of the α→β transition of Ad-2α was examined [26]. The increasing temperature accelerated the structural transition, and chilling decreased the transitional rate. The temperature-dependence of the reaction supports the idea that hydrophobic interactions are responsible for the aggregation and the transition. It was interesting that the lag time (Figure 2) was proportional to $t_{1/2}$ of the α→β transition curve, indicating that the lag time is essential for the transition. The lag times are considered to be necessary for the nucleation of the amyloid formation [9]. The Arrhenius plots of the temperature dependence of the transition were linear as a function of the reciprocal value of lag time, and the analysis revealed that the activation energy, E_a(lag time), was 25 kcal mol^{-1} [26]. On the other hand, the slope, which corresponds to the velocity of the α→β transition step (Figure 2), was inversely proportional to $t_{1/2}$. The Arrhenius analysis revealed that the activation energy, E_a(slope), was 15 kcal mol^{-1} [26]. Therefore, the reaction occurring during the lag time seems to be the rate-determining step. These results suggest that the first step corresponds to the reaction occurring during the lag time, after which the second α→β transitional step follows. During the lag time, the CD spectra indicated that the peptide was predominantly in the α-helix structure. Formation of an α-helical intermediate (an initial nucleus) for the transition reaction might occur during the lag time.

The structural transition of Ad-2α was dependent on the concentration of peptide. The α→β transition occurred at the concentration of 2 μM. Increases of peptide concentration of up to 10 μM linearly enhanced the reaction rate (slope) while shortening the lag time period for β-sheet formation. This concentration dependence supports the above assumptions that the lag time is essential to the structural transition and the aggregation process coincides with the reaction. Further increasing the concentration to more than 20 μM prevented the CD measurements due to the rapid formation of insoluble materials.

Effects of TFE, pH, salt, and detergents. As shown in the illustration of the peptide as a β-form (Figure 1), the hydrophobic Leu residues and hydrophilic Glu and Lys residues are separated to form a kind of amphiphilic β-structure [18]. The hydrophobic interaction between Leu residues and the electrostatic interaction between Glu and Lys residues seem to contribute to the structural transition. Therefore, the addition of an organic solvent TFE and detergents should affect the transition of Ad-2α. The addition of TFE to weaken the hydrophobic interaction between Ad groups or Leu residues decreased the transitional rate linearly with increasing content of TFE [26]. When more than 5% TFE was added, the transition did not occur. After 1 day under this condition, the peptide showed the same retention time as the monomer on the size-exclusion column. The ellipticity at 222 nm was increased to –25200 deg cm^2 $dmol^{-1}$ in 5% TFE from –23000 deg cm^2 $dmol^{-1}$ in 2.5% TFE. TFE of 5% increased the stability of α-helix structure and was sufficient to prevent the peptide aggregation, resulting in no transition.

The highest transition rate took place at neutral pH, but was retarded by changing to acidic or basic pH [26]. At pH 9, the transition rate was one third of that at neutral pH. At pH 3, the structural transition did not take place and the peptide was monomeric on the size-exclusion column. Under the conditions, the ellipticity at 222 nm was increased to –26000 deg cm^2 $dmol^{-1}$ from that of –23000 deg cm^2 $dmol^{-1}$ at pH 7.4, suggesting that change in pH affected the peptide aggregation step as well as the α-helix stability. The addition of NaCl also retarded the structural transition in a concentration-dependent manner, the rate with 1 M NaCl being one third of that without the salt. The ellipticity at 222 nm was increased to -24900 deg cm^2 $dmol^{-1}$ from that without the salt. These pH and salt effects suggested that the electrostatic interactions also contribute, along with the hydrophobic interactions, to the α→β transition. Furthermore, the inhibition experiments in which βCDx or TFE was added or pH was changed showed that prevention of the peptide aggregation inhibits the α→β transition, supporting that the peptide aggregation is an important step for the transition. It is worth noting that the fibril formation of the β-amyloid peptide is pH-dependent [37].

A cationic detergent, cetyltrimethylammonium bromide (CTAB), and an anionic detergent, sodium dodecylsulfate (SDS), similarly retarded or inhibited the structural transition of Ad-2α and C8-2α in a concentration-dependent manner [26,27]. Both detergents completely inhibited the transition at concentrations over 500 μM. The transition was retarded at detergent concentrations under 500 μM. The detergents could be expected to affect the hydrophobic interactions between the hydrophobic groups and between helices, resulting in perturbation of the 3D structure and separation of the two α-helices, allowing each helix to move independently. It is worth noting that some cationic detergents were reported to be inhibitors of the fibril formation of β-amyloid peptide [41].

When the detergents were added, the α-helicity of Ad-2α and C8-2α at the initial state was increased depending on the amount of the detergents, meaning that the increase in α-helix stability of the peptides was also responsible for retardation of the α→β transition. The effect of the detergents seemed to be different from that of βCDx, since the βCDx successfully inhibited the transition of Ad-2α without changing the α-helix structure, but did not inhibit that of C8-2α [27], whereas the detergents inhibited the transitions of both peptides. It is noteworthy that the non-ionic detergent NP-40 (nonylphenoxy polyethoxyethanol) did not influence the structural transition up to 5 mM. Perturbation of electrostatic as well as hydrophobic interactions by the cationic and anionic detergents might be effective at inhibiting the reaction.

2.5. STABILITY OF PEPTIDES

Denaturation experiments. Denaturation experiments using guanidine hydrochloride (GuHCl) were carried out for the α-form and β-form of Ad-2α [26]. Shortly after dilution from the TFE solution, the peptide in the α-form was completely unfolded in the presence of 4 M GuHCl. The concentration of GuHCl for the half-denaturation ($[GuHCl]_{1/2}$) was 1.0 M. On the other hand, the β-form peptide (after incubation at 40 °C for 2 h) was more resistant to the denaturant ($[GuHCl]_{1/2}$ was 1.8 M). From the denaturation curve, the stabilities, ΔG_{H_2O} [42], of the α-form and β-form were estimated as -1.0 and -2.5 kcal mol^{-1}, respectively. Thus the β-form is more stable than the α-form. The reason that the peptide undergoes the α→β transition may be related to the fact that the α-helix form of

Ad-2α is not stable, unlike that of other peptides designed by the *de novo* method, such as DeGrado's peptides (the stability of their 2α-peptide: -13 kcal mol^{-1}) [16,17], and the fact that the difference in stability between α- and β-forms was faint. In other words, this peptide has the property of a 'chameleon' sequence [14], which can form either an amphiphilic α-helix or an amphiphilic β-strand (Figure 1). The slow transition is probably due to the barrier of intermediate formation.

2.6. POSSIBLE MECHANISM OF α→β TRANSITION AND AMYLOYD FORMATION

These studies using the hydrophobic defects in the α-helix peptides [26,27], suggest that unstable α-helix aggregates (oligomer or larger) are an intermediate (the initial nucleus) of the structural transition (Figure 5) [27]. The α→β transition and the amyloid fibril formation of the peptide examined by the CD and fluorescence spectroscopy showed a protracted lag time in α-helix, followed by a sigmoidal (cooperative) structural transition and fibril growth (Figure 2). This type of process is considered to be a nucleation-dependent autocatalytic polymerization pathway, as suggested in, for instance, aggregation of β-amyloid peptides [9] or prion proteins [6], protein crystallization [43], sickle-cell fibril formation [44], vesicle formation of phosphatidyl nucleoside [45], and nanocluster formation of transition metal [46]. The autocatalytic mechanism was also supported by the observation that the lag time was eliminated by the addition of exogenous pre-formed β-sheet aggregates, which seemed to act as templates to induce the α→β transition. Based on this assumption (Figure 5), the conformation of a peptide with moderate hydrophobicity, such as that of Ad-2α or C8-2α, would exist in an equilibrium between a monomer and aggregates in α-helix. When a critical amount of α-helix aggregates accumulate, which give rise to new long-range (intermolecular) interactions, the peptide conformation transforms to β-sheet in an autocatalytic manner. The time necessary for the accumulation of α-helix aggregates may correspond to the lag time for nucleation, although the α-helix aggregates are so unstable that they cannot be detected by the size-exclusion method.

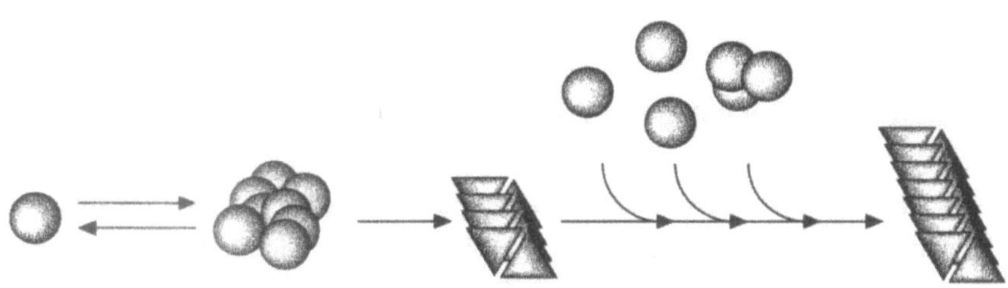

Figure 5. A possible mechanism of the α→β transition and amyloid formation.

Once the β-aggregates appear, monomeric or oligomeric α-helix species could further transform to the β-sheet on the template aggregates. In contrast, when a peptide forms more stable α-helix oligomers such as C12- to C16-2α, the conformation of the peptide is not transformed. A peptide with a less hydrophobic domain, such as C2-2α, remains stable as a monomeric 2α-helix and cannot form an α-helix nucleus, thus cannot acquire the long-range interaction which would stabilize the β-structure. Therefore, there appears to be an optimum length for the hydrophobic defects.

One role of the hydrophobic defects would thus be the formation of an α-helix nucleus. In other words, the hydrophobic clustering by the conformational defects may form unstable α-helix aggregates, which would then initiate the cooperative conversion from short-range interactions to long-range interactions, thereby triggering the autocatalytic transition to β-sheet conformation. The two-segmental peptide is designed to form either an amphiphilic α-helix or an amphiphilic β-sheet using Leu residues for the hydrophobic core (Figure 1), indicating that it may be a chameleon [14]. In fact, there is only a marginal difference (~1 kcal mol^{-1}) between the free energies of stabilization of the α and β conformations of Ad-2α model. The conformation of such a *de novo* designed peptide has been characterized as a molten-globule-like structure [16,17]. Although the mechanism of the α→β transition step is certainly more complex and will require further clarification, it is plausible that the formation of unstable α-helix aggregates like a molten-globule-like structure occurs during the lag time and is the rate-determining step and, therefore, that the initial α-helix structure and its stability are significant determinants for the transition. It has also been suggested that the formation of relatively unstable intermediates of proteins in folding pathways or by unfavorable mutations triggers the aggregation and structural transition [3, 47]. Especially in the prion proteins, it has been proposed that the transformation from an α-helix to a β-sheet structure induces the formation of aggregates as a seed for amyloidogenesis [48,49]. The initiation of α→β transition by introducing hydrophobic domains may provide insight into the conformational changing process that has been widely observed in protein research. The design concept employed here could potentially lead to a model system for clarifying off-pathway aggregations of proteins as well as for controlling self-assembly of polypeptides, which will also lead to the development of peptidyl self-assembling materials.

3. Regulation of α/β-Folding of a Designed Peptide by a Heme Cofactor

Cofactors such as heme in proteins have an important role for not only the function but also the native folding. For example, iron porphyrins occur widely in nature as cofactors of hemeproteins and display diverse functions. In addition to the wide range of iron porphyrin reactivities, iron porphyrins are important factors in defining the 3D structure of hemeproteins, because the removal of a porphyrin cofactor from natural hemeproteins caused perturbation and/or destabilization of protein 3D structures [50-53]. In the field of *de novo* protein design, considerable effort has been devoted to construction of polypeptide 3D structure conjugated with porphyrin molecules *via* chelation or covalent linkage with peptide [54-62]. Although these studies of designed peptides have focused on the functionalization of peptides using heme as a cofactor, we have found that the heme-binding can regulate the α/β folding of designed peptides. A two-segmental peptide, which was designed by an amphiphilic α-helix motif [18] similar to that used in the above section, forms a β-sheet aggregate, but the heme-binding as a cofactor can refold it to an α-helix bundle structure [28,29].

3.1. DESIGN OF PEPTIDES

The design of two-segmental peptides, H2α17-I and H2α17-V, was based on that of the template peptide H2α17-L [57], which took an α-helix structure in a neutral buffer both with and without bound heme. A 17-peptide segment was designed to take an amphiphilic α-helix structure [18], which was stabilized by two sets of E-K salt bridges (Figure 6). The two segments were dimerized by the disulfide linkage of the Cys^{21} residues. As axial ligands of heme, His was introduced at the 9th position to deploy a heme parallel to the helix. Four X residues (X = Leu, Ile, Val) per helix were arranged at the 5th, 6th, 12th and 13th positions around the His to construct a hydrophobic heme-binding site. The amino acid sequences of H2α17-I and H2α17-V were identical to that of H2α17-L except for their hydrophobic residues. In the cases of H2α17-I and H2α17-V, however, it was predicted that the folded-state of amphiphilic α-helical sequences would be significantly destabilized by the introduction of four Ile or Val residues, which prefer a β-strand rather than an α-helix structure [31]. The β-branched amino acids are considered to be α-helix breakers, because their side-chain torsional angle χ_1 is severely restricted when they are forced to reside within a helix. Additionally, when the peptide sequences were drawn as a β-sheet model, the peptide could take a kind of amphiphilic β-sheet structure [18]. However, the heme will bind to the folded-state of an α-helical structure with higher affinity than to that of a β-sheet structure, because the hydrophobic residues, which form the heme-binding site, can be arranged around the His^9, only when the peptides take an α-helix structure. The peptides were synthesized by solid-phase method using Fmoc chemistry [33] and purified with HPLC.

3.2. REGULATION OF α→β TRANSITION

The CD studies revealed that the peptide H2α17-I and H2α17-V showed a typical α-

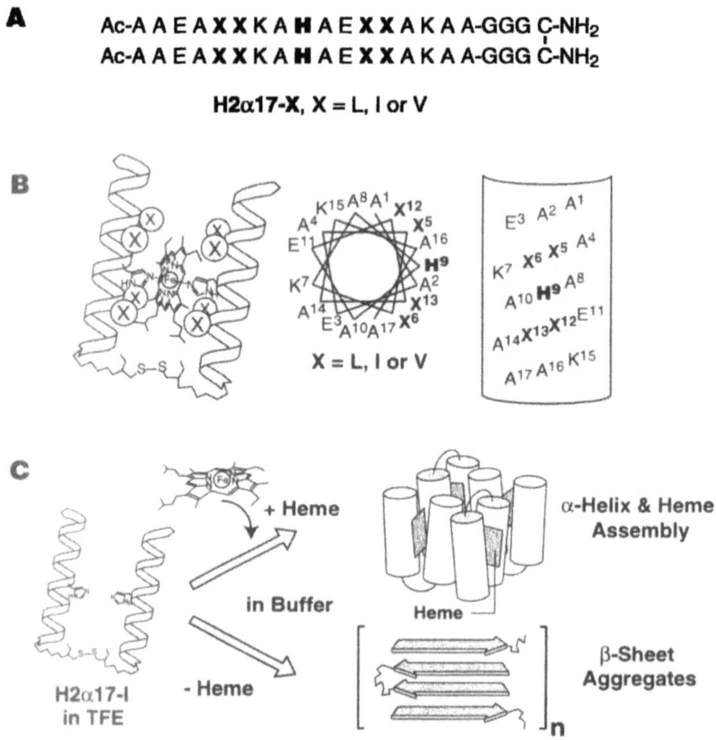

Figure 6. Design of heme-binding peptides and the α/β-folding. (A) Structure of the peptides. (B) Illustration of the amphiphilic α-helix structure. (C) Regulation of the α/β-folding with heme.

helical pattern with double negative maxima at 207 and 222 nm in TFE, which was known to be an α-helix stabilizing solvent [63]. Upon the dilution in 20 mM Tris·HCl buffer (pH 7.4) from TFE stock solution of H2α17-I and H2α17-V (final TFE concentration was 1.0%), the peptides showed CD spectra typical for a β-structure with a single negative maximum at 218 nm and a positive maximum at 198 nm. Additionally, FTIR spectra of the peptides showed the wavenumbers characteristic for a β-structure [28,36]. In the absence of heme, H2α17-I and H2α17-V gave no peak on the size-exclusion column, suggesting that the peptides formed β-sheet aggregates in multiple states. The formation of β-sheet aggregates was also confirmed by the binding of Congo red dye, which stains the β-sheet proteins, such as amyloid and prion proteins [26,40]. Because the template peptide H2α17-L took an α-helix structure and was in a monomeric form under the same conditions, the β-conformation of the peptides in the buffer is responsible for the high β-propensity of Ile and Val residues.

On the contrary to the above results, by the addition of Fe(III)-mesoporphyrin as a heme (1.0 equiv.), the peptide H2α17-I showed a CD typical for α-helix. From the ellipticity at 222 nm ($[\theta]_{222} = -22500$ deg cm^2 $dmol^{-1}$), the α-helicity was estimated as 71% [64]. Size-exclusion chromatography and sedimentation equilibrium study revealed that the peptide H2α17-I was in a tetrameric form ($MW_{obs} = 18000$, 3.9 mer) upon the heme-binding, as was observed in H2α17-L [57]. That is, the heme-binding induced drastic alternation of secondary structure and molecular-association state from β-sheet aggregates to an α-helical tetrameric assembly. The peptide H2α17-I folded into the β-aggregates at acidic pH (4.0–6.0) in the presence and absence of heme. Because the pK_a of His is *ca.* 6.4 [65], the pH effect is attributed to the protonation of the His side chains such that they cannot act as a ligand. Therefore, we concluded that the H2α17-I folded into an α-helix structure *via* the heme-binding through a ligation with His residues. In contrast, the peptide H2α17-V took a β-structure even in the presence of heme at pH 7.4, suggesting that H2α17-V could not bind the heme effectively. CD studies demonstrated that the secondary structure and molecular assembly of the designed peptide was regulated by the heme-binding. Although the α-form (with heme) was completely denatured with 6.0 M urea (urea denaturation midpoint; $[\text{urea}]_{1/2} = 2.7$ M, free energy of unfolding; $\Delta G_{H_2O} = -1.8$ kcal mol^{-1}), the CD spectrum of the β-form (without heme) was not changed even at the 7.0 M concentration of urea, indicating that the β-form is much stabler than the α-form. Thus, the alternation of peptide α/β folding appeared to be regulated by a kinetic mechanism rather than a thermodynamic one. Even in the absence of heme, immediately after dilution from the TFE solution, it is predicted that the peptide takes an α-helix structure, although the α-helix structure is unstable and the rapid transition from α- to β-form follows. In the presence of heme, however, the heme binds the metastable α-form and the binding prevents the peptide to progress into a β-form. Indeed, after formation of the β-sheet structure, the peptide could not bind the heme and α-helix structure was not recovered by the addition of heme.

UV-VIS studies further characterized the heme-binding with the peptides. With increasing concentration of H2α17-I, an increase of the Soret band at 406 nm and a decrease of the band at 355 nm of heme were observed. The UV-VIS spectrum of the heme in the presence of peptide resembles those of natural cytochromes with low-spin 6-coordinate iron [56-58]. The binding constant (K_a) determined from the absorbance change at the Soret band was 1.0×10^7 M^{-1}, comparable to that of H2α17-L ($K_a = 1.1 \times 10^7$ M^{-1}). On the other hand, addition of H2α17-V caused little increase of the Soret band, indicating that the peptide cannot bind the heme effectively in this concentration range. The computer modeling study suggested that the side chain of Val residue was too small to make effective van der Waals contact with the porphyrin plane, when the heme was positioned between the helices with energetically favorable side-chain tortional angles of His ligands [62].

Additionally, the hydrophobicity of Val is lower than that of Ile and Leu [66]. These results revealed that the amino acid composition constructing the heme-binding site is important for the effective heme-binding by the 2α-helix peptides, and that the effective cofactor-binding can correct the folding of peptide from β-aggregates to a globular state. There are enough hydrophobicity and van der Waals volume in Leu and Ile for the effective heme-binding in the 2α-helix form.

In conclusion, we successfully designed and synthesized the peptide H2α17-I with a unique property that the folding state was controlled by the heme-binding as a cofactor. The heme-binding prevented the peptide to form β-sheet aggregates by facilitating the formation of an α-helix tetramer. These findings indicate the importance of heme cofactor as a structural element of the artificial protein. This kind of work will extensively lead to studies to elucidate peptide folding and assembly, and apply their functions.

4. Conclusion and Future Directions

Now that a variety of data on structural transformations including α-to-β transitions have been accumulated, attention is now being focused on the mechanism of polymerization associated with structural transformation. The studies of the non-hierarchical protein folding and the protein fibrillogenesis demonstrate that an intermediate state with a partially unfolded and unstable conformation is a critical crossroad at which the protein will either fold correctly or will misfold. An α-to-β conformational transition is the common determinant for protein folding and misfolding, both in which formation of α-helix may decrease a kinetic barrier of transitions. All amyloid peptides and proteins make fibrils of similar structures through the structural transitions, which are critical to cause of fatal diseases. The continuous efforts at engineering peptides and proteins will eventually reveal the mechanism for each transformation and for fibril formation which will be useful for our understanding of protein folding and for the development of therapeutic drugs. Furthermore, the common structure of polypeptides as an amyloid β-fibril will be potentially useful to develop a new field of polypeptide materials.

5. Acknowledgements

We thank Prof. F. Arisaka, Tokyo Institute of Technology, for the measurements of electron microscopy and ultracentrifugation.

6. References

1. Taubes, G.: Misfolding the way to disease, *Science* **271** (1996), 1492-1492.
2. Mihara, H. and Takahashi, Y.: Engineering peptides and proteins that undergo α-to-β transitions, *Curr. Opin. Struct. Biol.* **7** (1997), 501-508.
3. Kelly, J. W.: The alternative conformations of amyloidogenic proteins and their multi-step assembly pathways, *Curr. Opin. Struct. Biol.* **8** (1998), 101-106.
4. Fink, A. L.: Protein aggregation: folding aggregates, inclusion bodies and amyloid, *Fold. Des.* **3** (1998), R9-R23.
5. Dobson, C. M.: Protein misfolding, evolution and disease, *Trends Biochem. Sci.* **24** (1999), 329-332.
6. Prusiner, S. B.: Prion diseases and the BSE crisis, *Science* **278** (1997), 245-251.
7. Harrison, P. M., Bamborough, P., Daggett, V., Prusiner, S. B., and Cohen, F. E.: The prion folding problem, *Curr. Opin. Struct. Biol.* **7** (1997), 53-59.
8. Riek, R., Hornemann, S., Wider, G., Billeter, M., Glockshuber, R., and Wüthrich, K.: NMR structure of the mouse prion protein domain PrP(121-231), *Nature* **382** (1996), 180-182.
9. Lansbury, P. T., Jr.: A reductionist view of Alzheimer's disease, *Acc. Chem. Res.* **29** (1996), 317-321.
10. Selkoe, D. J.: Amyloid β-protein and the genetics of Alzheimer's disease, *J. Biol. Chem.* **271** (1996), 18295-18298.
11. Forloni, G., Tagliavini, F., Bugiani, O., and Salmona, M.: Amyloid in Alzheimer's disease and prion-related encephalopathies: studies with synthetic peptides, *Prog. Neurobiol.* **49** (1996), 287-315.
12. Kuwata, K., Hoshino, M., Era, S., Batt, C., and Goto, Y.: α→β Transition of β-lactoglobulin as evidenced by hetetronuclear NMR, *J. Mol. Biol.* **283** (1998), 731-739.
13. Kuwajima, K., Yamaya, H., and Sugai, S.: The burst-phase intermediate in the refolding of β-lactoglobulin studied by stopped-flow circular dichroism and absorption spectroscopy, *J. Mol. Biol.* **264** (1996), 806-822.
14. Minor, D. L., Jr. and Kim, P. S.: Context-dependent secondary structure formation of a designed protein sequence, *Nature* **380** (1996), 730-734.
15. Dalal, S., Balasubramanian, and S., Regan, L.: Protein alchemy: changing β-sheet into α-helix, *Nature Struct. Biol.* **4** (1997), 548-552.
16. Betz, S. F., Raleigh, D. P., and DeGrado, W. F.: De novo protein design: from molten globules to native-like states, *Curr. Opin. Struct. Biol.* **3** (1993), 601-610.
17. Betz, S. F., Bryson, J. W., and DeGrado, W. F.: Native-like and structurally characterized designed α-helical bundles, *Curr. Opin. Struct. Biol.* **5** (1995), 457-463.
18. Kaiser E. T. and Kézdy, F. J.: Amphiphilic secondary structure: design of peptide hormones, *Science* **223** (1984), 249-255.
19. Schneider, J. P. and Kelly, J. W.: Templates that induce α-helical, β-sheet, and loop conformations, *Chem. Rev.* **95** (1995), 2169-2187.
20. Choo, D. W., Schneider, J. P., Graciani, N. R., and Kelly, J. W.: Nucleated antiparallel β-sheet that folds and undergoes self-assembly: a template promoted folding strategy toward controlled molecular architectures, *Macromolecules* **29** (1996), 355-366.
21. Aggeli, A., Bell, M., Boden, N., Keen, J. N., Knowles, P. F., McLeish, T. C. B., Pitkeathly, M., and Radford, S. E.: Responsive gels formed by the spontaneous self-assembly of peptides into polymeric β-sheet tapes, *Nature* **386** (1997), 259-262.
22. Zhang, S. and Rich, A.: Direct conversion of an oligopeptide from a β-sheet to an α-helix: a model for amyloid formation, *Proc. Natl. Acad. Sci. USA* **94** (1997), 23-28.
23. Mutter, M. and Hersperger, R.: Peptides as conformational switch: medium-induced conformational transitions of designed peptides, *Angew. Chem. Int. Ed. Engl.* **29** (1990), 185-187.
24. Ono, S., Kameda, N., Yoshimura, T., Shimasaki, C., Tsukuromichi, E., Mihara, H., and Nishino, N.: Super-secondary structure with amphiphilic β-strands probed by pyrenylalanine, *Chem. Lett.* (1995), 965-966.
25. Mutter, M., Gassmann, R., Buttkus, U., and Altmann, K.-H.: Switch peptides: pH-induced α-helix to β-sheet transitions of bis-amphiphilic oligopeptides, *Angew. Chem. Int. Ed. Engl.* **20** (1991), 1514-1516.
26. Takahashi, Y., Ueno, A., and Mihara, H.: Design of a peptide undergoing α–β structural transition and

amyloid fibrillogenesis by the introduction of a hydrophobic defect, *Chem. Eur. J.* **4** (1998), 2475-2484.
27. Takahashi, Y., Ueno, A., and Mihara, H.: Optimization of hydrophobic domains in peptides that undergo transformation from α-Helix to β-Fibril, *Bioorg. Med. Chem.* **7** (1999), 177-185.
28. Sakamoto, S., Obataya, I., Ueno, A., and Mihara, H.: Regulation of α/β-folding of a designed peptide by haem binding, *Chem. Commun.* (1999), 1111-1112.
29. Sakamoto, S., Obataya, I., Ueno, A., and Mihara, H.: Effects of amino acids substitution of hydrophobic residues on haem-binding properties of designed two-α-helix peptides, *J. Chem. Soc., Perkin Trans. 2* (1999), 2059-2070.
30. Kohn, W. D., Kay, C. M., Sykes, B. D., and Hodges, R. S.: Metal ion induced folding of a de novo designed coiled-coil peptide, *J. Am. Chem. Soc.* **120** (1998), 1124-1132.
31. Chou, P. Y. and Fasman, G. D.: Prediction of protein conformation, *Biochemistry* **13** (1974), 222-245.
32. Hamasaki, K., Ikeda, H., Nakamura, A., Ueno, A., Toda, F., Suzuki, I., and Osa, T.: Fluorescent sensors of molecular recognition. Modified cyclodextrins capable of exhibiting guest-responsive twisted intramolecular charge transfer fluorescence, *J. Am. Chem. Soc.* **115** (1993), 5035-5040.
33. Atherton, E. and Sheppard, R. C.: *Solid Phase Peptide Synthesis: A Practical Approach*, IRL Press, Oxford, 1989.
34. Harper, J. D. and Lansbury, P. T., Jr.: Models of amyloid seeding in Alzheimer's disease and scrapie: mechanistic truths and physiological consequences of the time-dependent solubility of amyloid proteins, *Annu. Rev. Biochem.* **66** (1997), 385-407.
35. Zhang, H., Stöckel, J., Mehlhorn, I., Groth, D., Baldwin, M. A., Prusiner, S. B., James, T. L., and Cohen, F. E.: Physical studies of conformational plasticity in a recombinant prion protein, *Biochemistry* **36** (1997), 3543-3553.
36. Surewicz, W. K., Mantsch, H. H., and Chapman, D.: Determination of protein secondary structure by Fourier transform infrared spectroscopy: a critical assessment, *Biochemistry* **32** (1993), 389-394.
37. Wood, S. J., Maleeff, B., Hart, T., and Wetzel, R.: Physical, morphological and functional differences between pH 5.8 and 7.4 aggregates of the Alzheimer's amyloid peptide Aβ, *J. Mol. Biol.* **256** (1996), 870-877.
38. Nguyen, J. T., Inouye, H., Baldwin, M. A., Fletterick, R. J., Cohen, F. E., Prusiner, S. B., and Kirschner, D. A.: X-ray diffraction of scrapie prion rods and PrP peptides, *J. Mol. Biol.* **252** (1995), 412-422.
39. LeVine, H., III: Thioflavine T interaction with synthetic Alzheimer's disease β-amyloid peptides: detection of amyloid aggregation in solution, *Protein Sci.* **2** (1993), 404-410.
40. Klunk, W. E., Pettegrew, J. W., and Abraham, D. J.: Quantitative evaluation of Congo red binding to amyloid-like proteins with a beta-pleated sheet conformation, *J. Histochem. Cytochem.* **37** (1989), 1273-1281.
41. Wood, S. J., MacKenzie, L., Maleeff, B., Hurle, M. R., and Wetzel R.: Selective inhibition of Aβ fibril formation, *J. Biol. Chem.* **271** (1996), 4086-4092.
42. Santoro, M. M. and Bolen, D. W.: Unfolding free energy changes determined by the linear alpha-chymotrypsin using different denaturants, *Biochemistry* **27** (1988), 8063-8088.
43. Blow, D. M., Chayen, N. E., Lloyd, L. F., and Saridakis, E.: Control of nucleation of protein crystals, *Protein Sci.* **3** (1994), 1638-1643.
44. Eaton, W. A. and Hofrichter, J.: The biophysics of sickle cell hydroxyurea therapy, *Science* **268** (1995), 1142-1143.
45. Veronese, A. and Luisi, P. L.: An autocatalytic reaction leading to spontaneously assembled phosphatidyl nucleoside giant vesicles, *J. Am. Chem. Soc.* **120** (1998), 2662-2663.
46. Watzky, M. A. and Finke, R. G.: Transition metal nanocluster formation kinetic and mechanistic studies. A new mechanism when hydrogen is the reductant: slow, continuous nucleation and fast autocatalytic surface growth, *J. Am. Chem. Soc.* **119** (1997), 10382-10400.
47. Booth, D. R., Sunde, M., Bellotti, V., Robinson, C. V., Hutchinson, W. L., Fraser, P. E., Hawkins, P. N., Dobson, C. M., Radford, S. E., Blake, C. C. F., and Pepys, M. B.: Instability, unfolding and aggregation of human lysozyme variants underlying amyloid fibrillogenesis, *Nature* **385** (1997), 787-793.
48. Mestel, R.: Putting prions to the test, *Science* **273** (1996), 184-189.
49. Lundberg, K. M., Stenland, C. J., Cohen, F. E., Prusiner, S. B., and Millhauser, G. L.: Kinetics and mechanism of amyloid formation by the prion protein H1 peptide as determined by time-dependent ESR, *Chem. Biol.*

4 (1997), 345-355.

50. Hargrove, M. S. and Olson, J. S.: Stability of holomyoglobin is determined by heme affinity, *Biochemistry* **35** (1996), 11310-11318.
51. Fisher, W. R., Taniushi H., and Anfinsen, C. J.: On the role of heme in the formation of the structure of cytochrome *c*, *J. Biol. Chem.* **248** (1973), 3188-3195.
52. Moore, C. D., Al-Misky O. N., and Lecomte, J. T.: Similarities in structure between holocytochrome *b*5 and apocytochrome *b*5: NMR studies of the histidine residues, *Biochemistry* **30** (1991), 8357-8365.
53. Feng, Y. Q. and Sligar, S. G.: Effect of heme binding on the structure and stability of Escherichia coli apocytochrome *b*562, *Biochemistry* **30** (1991) 10150-10155.
54. Sharp, R. E., Moser, C. C., Rabanal, F., and Dutton, P. L.: Design, synthesis, and characterization of a photoactivatable flavocytochrome molecular maquette, *Proc. Natl. Acad. Sci. USA* **95** (1998) 10465-10470.
55. Rau, H. K., DeJonge, N., and Haehnel, W.: Modular synthesis of de novo-designed metalloproteins for light-induced electron transfer, *Proc. Natl. Acad. Sci. USA* **95** (1998) 11526-11531.
56. Sakamoto, S., Sakurai, S., Ueno, A., and Mihara, H.: Haem binding and catalytic activity of two-helix peptide annealed by trifluoroethanol, *Chem. Commun.* (1997), 1221-1222.
57. Sakamoto, S., Ueno, A., and Mihara, H.: Molecular assembly of two-α-helix peptide induced by haem binding, *Chem. Commun.* (1998), 1073-1074.
58. Sakamoto, S., Ueno, A., and Mihara, H.: Design and synthesis of haem-binding peptides. Relationship between haem-binding properties and catalytic activities, *J. Chem. Soc., Perkin Trans. 2* (1998), 2395-2404.
59. Sasaki, T. and Kaiser, E. T.: Helichrome: Synthesis and enzymatic activity of a designed hemeprotein, *J. Am. Chem. Soc.* **111** (1989), 380-381.
60. Mihara, H., Haruta, Y., Sakamoto, S., Nishino, N., and Aoyagi, H.: Chiral assembly of porphyrins regulated by amphiphilic-helix peptides, *Chem. Lett.* (1996), 1-2.
61. Mihara, H., Tomizaki, K., Fujimoto, T., Sakamoto, S., Aoyagi, H., and Nishino, N.: Artificial membrane protein functionalized with electron transfer system, *Chem. Lett.* (1996), 187-188.
62. Nastri, F., Lombardi, A., Morelli, G., Maglio, O., D'Auria, G., Pedone, C., and Pavone, V.: Hemoprotein models based on a convalent helix-heme-helix sandwich: 1. Design, synthesis, and characterization, *Chem. Eur. J.* **3** (1997), 340-349.
63. Zhou, N. E., Zhu, B.-Y., Kay C. M., and Hodges, R. S.: The two-stranded α-helical coiled coil is an ideal model for studying protein stability and subunit interactions, *Biopolymers* **32** (1992), 419-426.
64. Scholtz, J. M., Qian, H., York, E. J., Stewart, J. M., and Baldwin, R. L.: Parameters of helix-coil transition theory for alanine-based peptides of varying chain lengths in water, *Biopolymers* **31** (1991), 1463-1470.
65. Tanford, C.: The interpretation of hydrogen ion titration curves of proteins, *Adv. Protein Chem.* **17** (1961), 69-165.
66. Engelman, D. M., Steitz, T. A., and Goldman, A.: Identifying nonpolar transbilayer helices in amino acid sequences of membrane proteins, *Ann. Rev. Biophys. Biomol. Struct.* **15** (1986), 321-353.

MODEL SIGNAL PEPTIDES: PROBES OF MOLECULAR INTERACTIONS DURING PROTEIN SECRETION

A. MILLER, L. WANG, AND D. A. KENDALL
Department of Molecular and Cell Biology
University of Connecticut, Storrs Connecticut 06269 USA

Abstract

A signal peptide is required for the entry of a protein into the secretory pathway but how it functions in concert with the other transport components to achieve protein localization is not known. In *Escherichia coli*, SecA is a component of the transport machinery which may play a role in targeting the preprotein to membrane-bound translocation sites and then utilize the energy of ATP hydrolysis to initiate membrane insertion of the preprotein. This model requires that the signal peptide interact specifically with SecA and that features of the signal peptide promote binding. These issues were examined using a wild type synthetic signal sequence derived from *E. coli* alkaline phosphatase and several model signal peptides that differ in amino-terminal charge, core region hydrophobicity, and the ability to form an α-helical structure. Using a SecA/lipid ATPase assay as an indicator of binding, we observe maximum activity with the functional wild type peptide, 3K7L and 1K7L; these have very hydrophobic core regions and a high propensity for α-helix formation, while no significant reactions were noted for the non-functional peptides, 3K2L and 1K2L. Although peptides of intermediate hydrophobicity, 3K4L and 1K4L, both stimulated the SecA ATPase activity to an intermediate extent, the level of stimulation was more marked with the 3K4L peptide. This is consistent with *in vivo* analyses which indicate that for signal peptides of intermediate hydrophobicity, the amino terminal basic residues also play a key role in enhancing transport activity. The data suggest that signal peptide core region hydrophobicity, amino-terminal charge, and α-helicity contribute to the modulation of SecA/lipid ATPase activity by altering the binding affinity of the peptide for SecA. Separately, a competition binding assay was employed to establish that signal peptides also interact with SecA in aqueous solution. Furthermore, the interaction of functional signal peptides with SecA alters the SecA conformation sufficiently, for both soluble and membrane-associated forms, to cause a marked change in its V8 protease sensitivity.

1. Introduction

The structural characteristics of signal peptides and their interactions with the various components of the translocation pathway have been extensively studied and reviewed. Briefly, in *E. coli*, a typical signal sequence is comprised of a positively charged amino-

A. Aggeli et al. (eds.), Self-assembling Peptide Systems in Biology, Medicine and Engineering, 207–220.

terminus, followed in succession by an uncharged central hydrophobic core region, and a more polar carboxyl terminus (von Heijne, 1985). It is generally acknowledged that protein transport across the cytoplasmic membrane requires an initial interaction between the preprotein and SecA, an ATPase found in the cytoplasm and inner membranes of prokaryotes. We have established that the interaction involves specific binding of the signal peptide region of the preprotein and that signal peptide binding stimulates SecA/lipid ATPase activity (Miller et al., 1998). The energy derived from the hydrolysis of ATP by SecA drives the translocation process (Schiebel et al., 1991; Wickner and Leonard, 1996). Other protein components involved in transport are SecB, a cytosolic species which functions like a molecular chaperone, maintaining a transport-competent (unfolded) state for some exported proteins (Collier et al., 1988; Hardy and Randall, 1991; Randall et al., 1997) and SecY, E, G, D, F, and yajC (Hartl et al., 1990; Ito, 1992; Matsuyama et al., 1993; Taura et al., 1993; Wickner and Leonard, 1996; Duong and Wickner, 1997). Transport also has an absolute requirement for anionic phospholipids (Lill et al., 1990; Hendrick and Wickner, 1991). The SecYEG heterotrimer forms a translocation channel which may be activated by signal sequence recognition (Kaufmann et al., 1999). SecD and SecF may stabilize membrane-associated SecA in an active conformation (Duong and Wickner, 1997) or they may be involved in either the release of the transported protein into the periplasmic space (Matsuyama et al., 1993) or the maintenance of the protonmotive force across the cytoplasmic membrane (Arkowitz and Wickner, 1994). In addition, SecDFyajC interacts with the SecYEG complex (Duong and Wickner, 1997). A model of some of the components of the transport machinery in *E. coli* is shown in Figure 1.

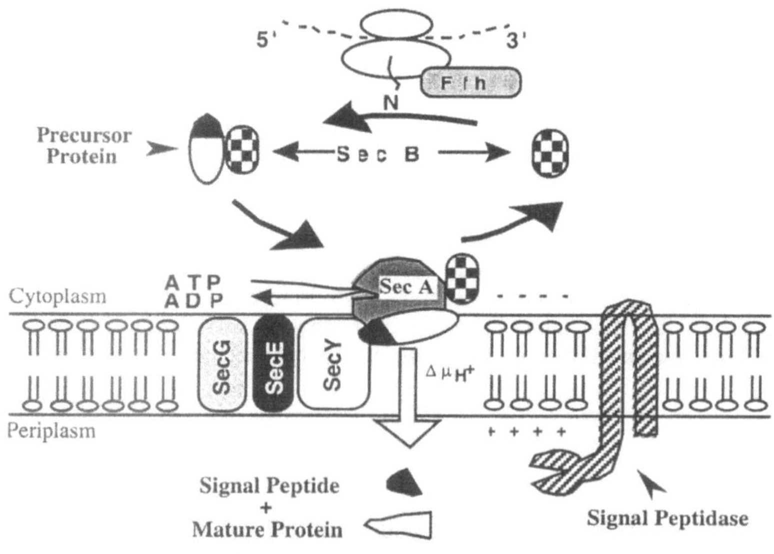

Figure 1. Model of *E. coli* protein transport machinery.

Since there is a lack of sequence homology among signal peptides, other factors such as amino-terminal charge, core region hydrophobicity, and secondary structure have been implicated in SecA binding and protein transport. In this study we have determined the relative level of synthetic signal peptide hydrophobicity required for binding to SecA and for maximal peptide-stimulated SecA/lipid ATPase activity, and have demonstrated that ionic interactions and α-helicity can also play significant roles in peptide binding to SecA.

2. Results and Discussion

In order to evaluate the relative roles of signal peptide amino-terminal charge, core region hydrophobicity and α-helicity on interactions with SecA, the series of synthetic signal peptides given in Table 1 were generated.

TABLE 1. Sequences and properties of model signal peptides

Peptide	Sequence [a]	Hydrophobicity[b]	Retention Time (min)	Net Charge[c]	*in vivo* Functionality [d]
wild type[e]	MKQST **IALALLPLLF** TPVTKAC-NH_2	0.77 (1.34)	22.1	+2	+
1K7L	MKQQQ **LLALLALALL** SSSASAC-NH_2	0.67 (1.28)	24.1	+1	+
3K7L	MKQKK **LLALLALALL** SSSASAC-NH_2	0.59 (1.28)	22.3	+3	+
1K4L	MKQQQ **LALAAAALAL** SSSASAC-NH_2	0.47 (0.87)	18.7	+1	+/-
3K4L	MKQKK **LALAAAALAL** SSSASAC-NH_2	0.40 (0.87)	16.6	+3	+/-
1K2L	MKQQQ **AALAAAALAA** SSSASAC-NH_2	0.35 (0.54)	14.7	+1	-
3K2L	MKQKK **AALAAAALAA** SSSASAC-NH_2	0.28 (0.54)	12.9	+3	-

[a] The hydrophobic core regions of the peptides are represented in bold face and the amidated carboxyl-termini are designated by the NH_2 groups linked to the cysteine residues.
[b] The hydrophobicity was calculated as described in Fauchère J.L. et al. (1988). Numbers in parentheses represent the hydrophobicity of the core region.
[c] Number of basic amino acid residues.
[d] (+) indicates the ablility to support efficient preprotein translocation.
[e] *E. coli* wild type alkaline phosphatase signal sequence.

These peptides were designed so that the hydrophobic core regions of 3K7L and 1K7L, containing seven leucines and three alanines, are above the threshold hydrophobicity required for efficient processing of an otherwise wild type signal peptide, while 3K2L and 1K2L, with two leucines and eight alanines, represent non-functional signal peptides due to their low hydrophobicity (Doud et al., 1993; Izard et al., 1995). An intermediate hydrophobicity is represented by 3K4L and 1K4L.

2.1. EFFECT OF SIGNAL PEPTIDE CORE REGION HYDROPHOBICITY ON SecA/LIPID ATPASE ACTIVITY

SecA was isolated and purified from *E. coli* BL21.14/pCS1 S300 extracts by affinity chromatography, using reactive blue 4 agarose essentially as described by Mitchell and Oliver (1993) with minor revisions. All ATPase reactions were performed in 50 mM HEPES-KOH (pH 7.0), containing 30 mM KCl, 30 mM NH_4Cl, 0.5 mM $Mg(OAc)_2$, 1 mM DTT, 4 mM ATP, BSA (0.5 mg/ml), SecA (40 μg/ml), and either *E. coli* phospholipids or the artificial lipids dioleoylphosphatidylethanolamine (DOPE), dioleoylphosphatidylcholine (DOPC), and dioleoylphosphatidylglycerol (DOPG). P_i released was determined as described by Lanzetta et al. (1979). The effects of the above synthetic peptides (Table 1) on SecA/lipid ATPase activity in the presence of *E. coli* phospholipids are shown in Figure 2.

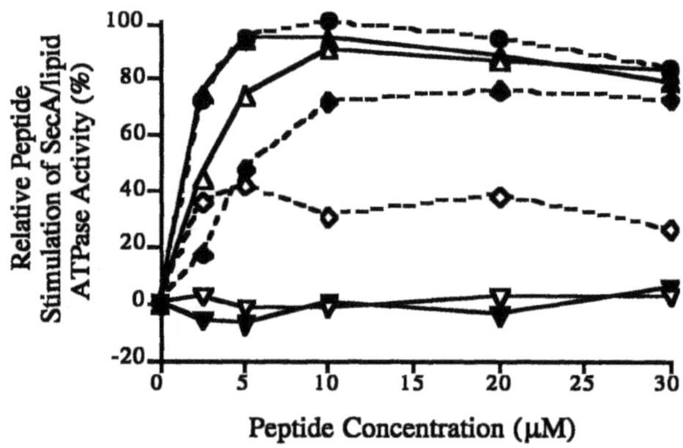

Figure 2. Differential effects of model signal peptides on SecA/lipid ATPase activity. Maximum peptide stimulated P_i release, 175 pmol/min/μg SecA, by wild type peptide at 10 μM was equated to 100% relative peptide stimulation. ●, wild type alkaline phosphatase signal peptide; ▲, 3K7L; △, 1K7L; ◆, 3K4L; ◇, 1K4L; ▼, 3K2L; ▽, 1K2L.

The peptide-stimulated SecA/lipid ATPase activity is reported as the difference in the rate of ATP hydrolysis in the presence and absence of peptide (Miller et al., 1998) and each data point represents an average of triplicate assays. Optimal conditions for the enzyme assay were determined and are described later in the text. Maximum stimulation of SecA/lipid ATPase by the wild type peptide, 3K7L and 1K7L was approximately 175, 165 and 160 pmol of P_i released/min/μg of SecA, expressed as relative % activity of 100%, 94% and 91%, respectively, and occurred at a peptide concentration of 10 μM; half-maximal stimulation was in the 1.5 to 2.0 μM range. Since no significant increases in ATPase activity were observed for non-functional 1K2L and 3K2L, and intermediate values were evident for 1K4L and 3K4L, the data suggest that peptide-stimulated SecA/lipid ATPase activity increases with higher hydrophobicity (Table 1). The retention times of the charge- and hydrophobicity- modified signal peptides determined by reverse-phase HPLC agree well with the calculated hydrophobicities (Table 1). Therefore, based on both the hydrophobicity of the signal peptide core region and the extent of stimulation of ATPase activity, the peptides may be ranked as follows: wild type peptide > 3K7L and 1K7L > 3K4L and 1K4L > 3K2L and 1K2L.

The results correlate with our *in vivo* analyses in which we demonstrated an interrelationship between signal peptide hydrophobicity and the length of the core region required for signal peptide function (Chou and Kendall, 1990). A high net hydrophobicity contributed by increased length partly restored function that was diminished by a lower mean hydrophobicity per residue. However, a high mean hydrophobicity per residue must be attained within the absolute bounds of length limitations. In other words, an ideal signal sequence requires a threshold 'hydrophobic density' to ensure rapid and efficient protein translocation (Chou and Kendall, 1990; Doud et al., 1993)

2.2. EFFECT OF SIGNAL PEPTIDE AMINO-TERMINAL CHARGE ON SecA/LIPID ATPASE ACTIVITY

In addition to the effect of signal peptide hydrophobicity on SecA/lipid ATPase activity noted above, it was also possible to demonstrate an influence of signal peptide charge. An elevated ATPase activity of 3K4L relative to 1K4L (Figure 2) was observed which suggests the participation of electrostatic, in addition to hydrophobic, interactions. At a peptide concentration of 10 μM, the ATPase activity induced by 3K4L and 1K4L was about 71% and 31%, respectively, relative to the wild type. A similar effect was noted for 3K7L and 1K7L but only at concentrations lower than 10 μM. The values calculated for 1K7L relative to 3K7L (100%) at 2.5 μM and 5 μM were 60% and 74%, respectively (Figure 2). As the hydrophobicity of the signal peptide increases (Table 1), the influence of electrostatic interactions appears to diminish.

The importance of the positive charge at the amino-terminus of the signal peptide in protein translocation has been demonstrated *in vivo* (Inouye et al., 1982; Vlasuk et al., 1983; Iino et al., 1987) and *in vitro* (Yamane and Mizushima, 1988; Akita et al., 1990; Sasaki et al., 1990; Mori et al., 1997). However, it should be emphasized that a positively charged amino-terminal signal is not an absolute requirement for protein transport (Phoenix et al., 1993) and signal peptides with highly hydrophobic core regions can support rapid rates of transport even in the presence of a negatively charged amino-terminus (Izard et al., 1996).

2.3. EFFECT OF SIGNAL PEPTIDE α-HELICITY ON SecA/LIPID ATPASE

The potential of the synthetic signal peptides to form α-helical structures was determined by CD spectroscopy (Izard et al., 1995) and the results are summarized in Table 2.

TABLE 2. α-Helical content of model signal peptides

Peptide	Helical Content (%)	
	water	40% TFE
wild type	15	31
1K7L	ND	93
3K7L	19	100
1K4L	ND	ND
3K4L	ND	ND
1K2L	0	13
3K2L	8	53

In the structure promoting solvent, trifluoroethanol (TFE), both 3K7L and 1K7L exhibit high α-helical contents (Izard et al., 1995) which correlate well with their stimulatory affect on SecA/lipid ATPase activity *in vitro* (Figure 2) and may explain the unusually high transport efficiency of preproteins with these signal sequences observed *in vivo* (Doud et al., 1993). The α-helical potential for the wild type peptide, though significant, was considerably lower which may be explained by the presence of proline residues in the wild type sequence (Table 1) that could limit the helix-forming potential of the peptide in solution (Izard et al., 1995). Although the formation of an α-helical structure may not, in itself, be sufficient for protein secretion (McKnight et al., 1989), the overall hydrophobicity of the signal peptide core region is enhanced by shielding the polar amide groups through intrastrand hydrogen bonding, and such a hydrophobic α-helix may provide an important recognition element (Kendall et al., 1986).

The data presented thus far agree well with results previously reported for *in vivo* transport (Izard et al., 1995) and suggest that the core region of the signal peptide could be an important recognition element for interaction with the transport machinery. Specificity may be contributed by the extent of hydrophobicity, the length of the core segment (Chou and Kendall, 1990), the amino-terminal charge, and the propensity for α-helix formation (Jones et al., 1990; Izard et al., 1995) which may be attenuated by the properties of the flanking signal peptide segments and mature protein.

In an attempt to optimize the peptide-stimulated SecA/lipid ATPase assay, the following parameters were further studied: SecA:lipid molar ratio, lipid concentration and type, ATP and Mg^{2+} concentrations, and ionic strength. The results are summarized below.

2.4. EFFECT OF SecA:LIPID MOLAR RATIO, LIPID CONCENTRATION AND COMPOSITION ON PEPTIDE-STIMULATED SecA/LIPID ATPASE ACTIVITY

Near optimal peptide-stimulated ATPase activity occurred at a SecA to lipid molar ratio of about 1:900, calculated using a molecular weight of approximately 100-kDa for the SecA monomer and the following concentrations of SecA and phospholipid: 40 μg/ml and 300 μg/ml, respectively. If one assumes that each small unilamellar vesicle (SUV) is comprised of 3.6 x 10^3 phospholipid molecules, then at the above molar ratio, 1:900, approximately four SecA molecules can associate with each SUV, as illustrated in Figure 3A.

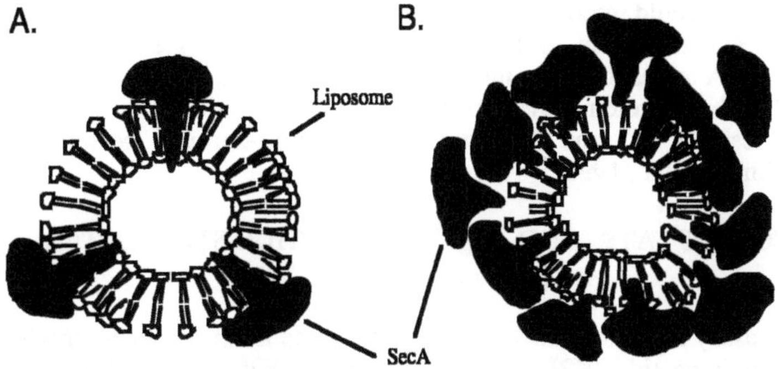

Figure 3. Importance of SecA:lipid molar ratio on SecA/SUV association. (A) SecA:lipid molar ratio 1:900. (B) SecA:lipid molar ratio 1:100.

As the SecA:lipid molar ratio decreases, the number of SecA molecules associated with each vesicle increases accordingly, resulting in a non-uniform incorporation of SecA (Figure 3B) and decreased peptide-stimulated ATPase activity. For example, at a SecA:lipid molar ratio of 1:100, there are approximately 32 SecA molecules associated with each SUV.

In addition to the SecA:lipid molar ratio, liposomes enriched in anionic phospholipids, such as DOPG, as well as the lipid concentration (120 μg/ml to 320 μg/ml) have a decided affect on the 3K7L-stimulated SecA/lipid ATPase activity (Table 3; Miller et al., 1998).

TABLE 3. Effects of lipid type and concentration on 3K7L-stimulated SecA/lipid ATPase activity

Lipid Concentration (μg/ml)	3K7L-stimulated SecA/lipid ATPase activity (pmol/min/μg·SecA)		
	E. coli lipid	DOPE/DOPG (3:1)	DOPC/DOPG (3:1)
120	6.5	14.3	3.3
160	32.2	58.6	41.1
200	36.9	51.8	60.5
240	85.1	45	57
280	117.8	83.4	49.4
320	181.6	118.7	91.9

At peptide and lipid concentrations of 40 μg/ml and 320 μg/ml, respectively (corresponding to a SecA:lipid molar ratio of 1:900), SecA/lipid ATPase activity was greater in the presence of *E. coli* phospholipids followed by DOPE/DOPG and

DOPC/DOPG (3:1). Enzyme activity was generally inhibited at lipid concentrations below 120 μg/ml. It is suggested that a lipid induced conformational change in SecA (Ulbrandt et al., 1992) may expose a previously buried hydrophobic region or groove in the protein which could interact with the projecting hydrophobic side-chains in the core region of the signal peptide. A high concentration of lipid relative to SecA may be required for SecA to become fully integrated in the bilayer, assume its active topology, and interact with preprotein. The proposed preprotein binding site is located between amino acid residues 267 and 340, within the amino-terminal portion of the SecA sequence (Kimura et al., 1991)

2.5. EFFECT OF ATP AND Mg^{2+} ON PEPTIDE-STIMULATED SecA/LIPID ATPASE

The ATPase assay conditions are essentially the same as those stated previously except that the synthetic peptide 3K7L (20 μM) and DOPC/DOPG 1:1 (150 μg/ml) were used. As shown in Figure 4, high concentrations of both ATP (8 mM) and Mg^{2+} (10 mM) markedly diminished the 3K7L-stimulated SecA/lipid ATPase activity.

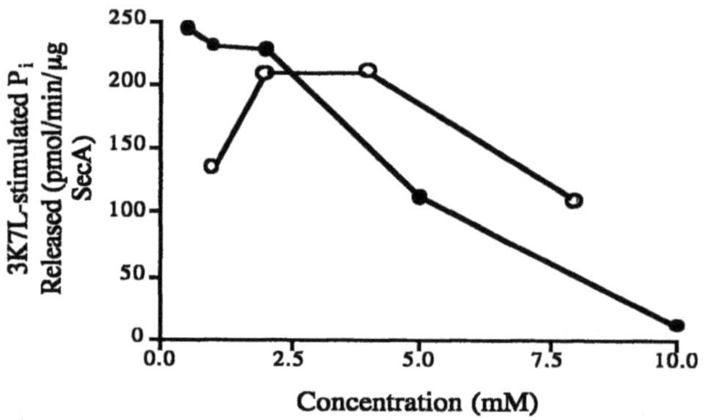

Figure 4. Effect of ATP and Mg^{2+} on 3K7L-stimulated SecA/lipid ATPase activity. ○, ATP; and ●, Mg^{2+}.

Optimal stimulation occurred at low Mg^{2+} concentrations (0.5 mM) and ATP concentrations ranging from 2 mM to 4 mM. This is consistent with studies by Lill et al. (1990) which indicate that preprotein-induced SecA ATPase activity can be stimulated greatly in the presence of anionic phospholipids and low concentrations of Mg^{2+}. Low concentrations of ATP and Mg^{2+} are also essential for specific SecA-RNA binding (Dolan and Oliver, 1991). ATP binding to SecA and subsequent hydrolysis may be associated with cycles of precursor protein binding and release (Schiebel et al., 1991).

2.6. EFFECT OF HIGH SALT CONCENTRATIONS ON SecA/LIPID ATPASE

At high ionic strengths, e.g. 200 mM to 500 mM, both KOAc and KCl significantly reduced 3K7L-stimulated SecA/lipid ATPase activity. High salt concentrations may decrease liposome stability as well as promote and strengthen hydrophobic-hydrophobic interactions among peptides, resulting in extensive peptide aggregation. The possible affect of high salt concentrations on the secondary structure of SecA was not evaluated.

Interactions between SecA and various cellular components such as nucleotides, lipids, and signal peptides were studied and evaluated using V8 protease sensitivity, and chemical cross-linking reactions; the results obtained are discussed below.

2.7. APDP CROSS-LINKING OF RADIOLABELLED WILD TYPE SIGNAL PEPTIDE TO SecA

A schematic representation of the signal peptide-SecA APDP cross-linking reaction is shown in Figure 5.

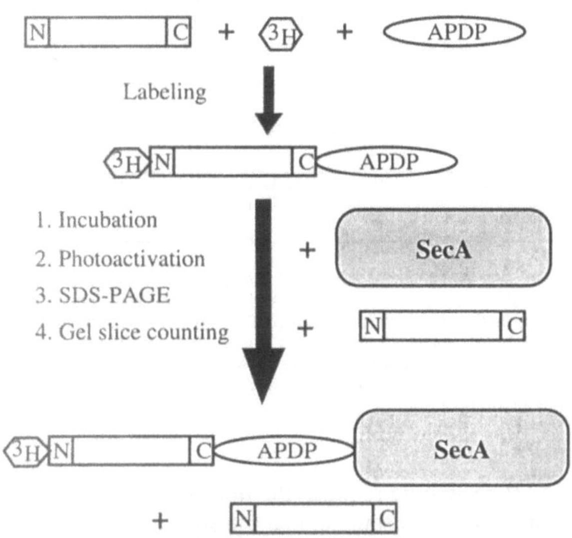

Figure 5. Signal peptide-SecA APDP cross-linking reaction scheme.

The amino-terminal group of the wild type peptide, is radiolabelled with ^{3}H-acetic anhydride in N-methylpyrrolidine as described by Seligman (1993). The cross-linking of APDP to the tritiated peptide was carried out as described in Figure 5 and as suggested by the manufacturer (Pierce Chemical Co., Rockford, IL). A Kwik Sep polyacrylamide 1800 desalting column, also from Pierce, was used for final purification.

The cross-linking reactions and competition experiments were conducted as described previously (Miller et al., 1998) but the liposomes were omitted except when specifically stated otherwise. The radiolabelled, cross-linked complexes were detected by SDS-PAGE and subsequent liquid scintillation counting of the solubilized gel slices. As shown in Figure 6A, radioactivity was primarily associated with the monomeric 102-kDa subunit of SecA; a small amount was also detected with the SecA dimer (204-kDa) which was not dissociated by SDS-PAGE.

A.

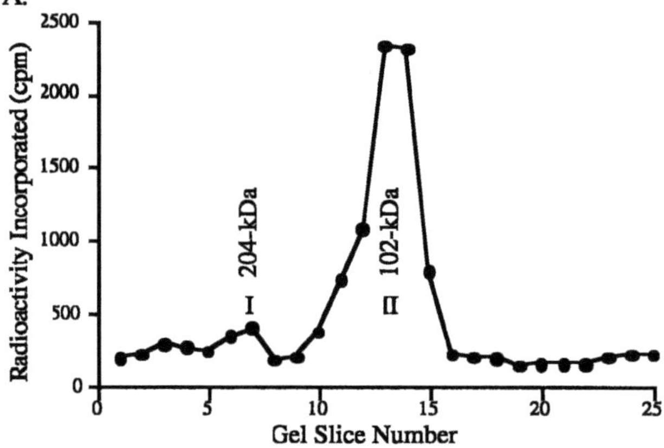

B.

Peptide	WT	3K7L	3K2L	lipid	lipid+WT
% Displacement	56	84	18	5	69

Figure 6. (A) APDP cross-linking of radiolabelled wild type peptide to SecA. Peak I corresponds to the SecA dimer (MW 204-kDa) and peak II represents the monomeric protein (MW 102-kDa). Radioactivity from solubilized gel slices (1 mm) was quantified by liquid scintillation counting. (B) Relative radioactivity associated with SecA monomer displaced by ten-fold unlabelled peptide, lipid alone, or lipid plus ten-fold unlabelled peptide.

Since the ATPase activity requires the liposome bound form of SecA, we used the competition binding assay to establish that the signal peptide binds SecA in aqueous solution. Competition assays, in the presence of a 10-fold excess of either unlabelled wild type peptide or the functional 3K7L revealed displacement values of 56% and 84%, respectively. In contrast, only 18% displacement was obtained with the nonfunctional 3K2L. Separately, liposomes were included in the reaction to test the relative affinity of peptide for SecA and lipid. In either the absence or presence of artificial lipids (DOPC/DOPG 1:1, 150 μg/ml) or *E. coli* phospholipids (150 μg/ml), the APDP-mediated cross-linking of the labelled wild type peptide to SecA was similar, indicating that lipid did not significantly influence the peptide-protein interaction. For example, when liposomes composed of DOPC/DOPG were added with no unlabelled peptide, only 5% displacement of the radioligand occurred (Figure 6B). The data indicate a direct interaction between functional signal peptide and SecA (Miller et al., 1998).

2.8. SecA SENSITIVITY TO PROTEOLYSIS BY *Staphylococcus aureus* V8 PROTEASE

Digestion of SecA by V8 protease, which cleaves at the carboxylic side of glutamate and aspartate residues, confirmed our earlier observation (Miller et al., 1998), reported herein, that functional signal peptides in the absence of mature protein, can interact directly and specifically with SecA. Controlled proteolytic digestions of SecA were conducted at 37°C essentially as described by Shinkai et al. (1991) and Blanco et al. (1998). All reaction mixtures (total volume 25 μl) contained SecA (10 μg), 50 mM Tris-HCl (pH 8.0), 50 mM KCl, 5 mM $MgCl_2$, and 1 mM DTT. After the addition of ATP, peptides and/or lipids, all reactions were preincubated at 37°C for 20 min after which V8 protease (125 ng to 500 ng/25 μl) was added. Incubation was continued at 37°C and aliquots (5 μl) were collected at 0, 5, 15, 30 and 60 min followed by adding an equal volume of 2X gel loading buffer to terminate the reaction. These were analyzed by 10% SDS-PAGE as described by Laemmli (1970).

The 95-kDa amino-terminal fragment of SecA, which contains the nucleotide binding sites (Mitchell and Oliver, 1993) and the proposed peptide binding site (Kimura et al., 1991), was very resistant to V8 hydrolysis in the presence of either ATP (2 mM), ADP (2 mM), or the non-hydrolyzable analog of ATP, AMP-PNP (2 mM), in agreement with Shinkai et al. (1991). The effects of ATP, wild type peptide and 1K2L on the V8 proteolysis of SecA in aqueous solution are shown in Figure 7.

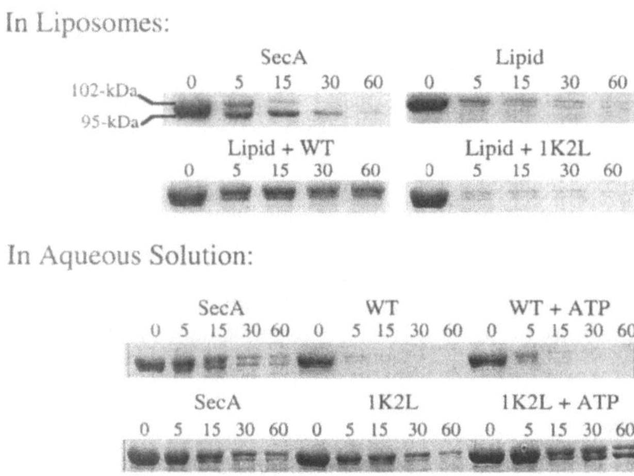

Figure 7. **Effects of ATP, lipid, wild type peptide (WT) and 1K2L on the sensitivity of SecA to V8 protease digestion.**

The addition of wild type peptide (SecA:peptide molar ratio 1:50) markedly increased the V8 sensitivity of SecA, in the absence or presence of nucleotide, suggesting a partial unfolding of the SecA molecule which would expose previously masked protease-sensitive sites. As expected, V8 protease had little or no affect on SecA digestion in the presence of the non-functional peptide 1K2L. The accumulation of the 95-kDa fragment in the gel profile shown for 1K2L plus ATP was due primarily to the affect of the nucleotide and not to any conformational change induced by the peptide. The V8 resistance observed in the presence of ATP and the increased sensitivity noted upon addition of either the wild type peptide or wild type peptide plus ATP indicate the induction of two different SecA conformations. As suggested by den Blaauwen et al. (1996), the ATP-bound conformation may correspond to membrane-inserted SecA while the ADP-bound conformation may represent the deinserted state of the protein.
Also shown in Figure 7, an increased sensitivity of SecA (4 μM) to V8 digestion was readily apparent in the presence of lipid vesicles (DOPC/DOPG 2:1; 3 mg/ml). The addition of wild type peptide to SecA in a lipid environment resulted in a SecA molecule which was decidedly more resistant to V8 digestion while 1K2L had no effect. However, at the high peptide concentration (240 μM) used in this particular experiment, the effects of either peptide-peptide aggregation or peptide-induced liposome aggregation have to be carefully considered. To address this potential problem, V8 proteolysis of SecA was also carried out under more dilute conditions; for example, SecA (0.8 μM instead of 4 μM), wild type peptide (60 μM instead of 240 μM), and lipid (3-7.5 mg/ml), keeping the SecA:peptide molar ratio at 1:75 and maintaining total reaction volume 125 μl instead of 25 μl. Since the same proteolysis pattern, although less intense, was readily apparent using the latter conditions, the observed effect cannot be attributed solely to aggregation of the peptide in either the aqueous phase or lipid bilayer, or to a peptide-induced aggregation of liposomes. In any case, the data seem reasonable, since SecA (MW 102-kDa) is believed to insert into the membrane at SecYEG and the 30-kDa and 65-kDa fragments, which are derived from the C- and N-domains, respectively, and represent over 90% of the molecule, become protease-inaccessible (Economou and Wickner, 1994; Eichler and Wickner, 1997). Synders et al. (1997) have demonstrated a direct interaction between SecA and SecY and, more recently, Kaufmann et al. (1999) have suggested that the conformational change induced in SecA by a functional signal peptide or precursor protein may be required to open the SecYEG channel to allow subsequent protein translocation.

The V8 proteolysis data strongly support the cross-linking results (Miller et al., 1998) and demonstrate that a functional synthetic signal peptide can interact directly and specifically with SecA, in the absence of mature protein or phospholipids, and illustrate the adaptability of SecA to switch between aqueous and lipid-associated forms.

3. Conclusions

In these studies we have used an assay based on the SecA ATPase activity, a competition binding assay, and a V8 protease sensitivity assay to establish that signal peptides, in the absence of the mature protein, specifically bind SecA. In each case, the results indicate that only signal peptides which are functional *in vivo* interact with SecA; nonfunctional ones do not. A clear correlation is observed between the degree of hydrophobicity of the signal peptide and the extent of interaction. For signal peptides of intermediate hydrophobicity, the amino terminal charge also influences the SecA/lipid ATPase activity. This finding is in good agreement with *in vivo* studies in which an interplay between the properties of the signal peptide amino-terminus and core region was found to impact the extent of preprotein transport (Izard et al., 1996). Indeed, the strong parallel observed between the hierarchy of signal peptide characteristics which promote transport activity *in vivo* and those which enhance interactions with SecA *in vitro*, suggests that the signal peptide-SecA interaction is fundamental to the kinetics of transport overall. Furthermore, the data indicate that signal peptides interact with SecA in aqueous and in lipid environments. This finding is consistent with the possibility that, *in vivo*, SecA can interact with preproteins in the cytoplasm and facilitate the targeting of these to translocation sites in the membrane.

4. References

Akita, M., Sasaki, S., Matsuyama, S., and Mizushima, S. (1990) SecA interacts with secretory proteins by recognizing the positive charge at the amino terminus of the signal peptide in *Escherichia coli*, *J. Biol. Chem.* **265**, 8164-8169

Arkowitz, R. A., and Wickner, W. (1994) SecD and SecF are required for the proton electrochemical gradient stimulation of preprotein translocation, *EMBO J.* **13**, 954-963

Blanco, J., Driessen, J. M., Coque, J. J. R., and Martin, J. F. (1998) Biochemical characterization of the SecA protein of *Streptomyces lividans* interaction with nucleotides, binding to membrane vesicles and *in vitro* translocation of proAmy protein, *Eur. J. Biochem.* **257**, 472-478

Chou, M. M., and Kendall, D. A. (1990) Polymeric sequences reveal a functional interrelationship between hydrophobicity and length of signal peptides, *J. Biol. Chem.* **265**, 2873-2880

Collier, D. N., Bankaitis, V. A., Weiss, J. B., and Bassford, P. J. Jr. (1988) The antifolding activity of SecB promotes the export of the *E. coli* maltose-binding protein, *Cell* **53**, 273-283

den Blaauwen, T, Fekkes, P., de Wit, J. G., Kuiper, W., and Driessen, A. J. M. (1996) Domain interactions of the peripheral preprotein translocase subunit, *Biochemistry* **35**, 11994-12004

Dolan, K. M., and Oliver, D. B. (1991) Characterization of *Escherichia coli* SecA protein binding to a site on its mRNA involved in autoregulation, *J. Biol. Chem.* **266**, 23329-23333

Doud, S. K., Chou, M. M., and Kendall, D. A. (1993) Titration of protein transport activity by incremental changes in signal peptide hydrophobicity, *Biochemistry* **32**, 1251-1256

Duong, F., and Wickner, W. (1997) Distinct catalytic roles of the SecYE, SecG, and SecDFyajC subunits of preprotein translocase holoenzyme, *EMBO J.* **16**, 2756-2768

Economou, A., and Wickner, W. (1994) SecA promotes preprotein translocation by undergoing ATP-driven cycles of membrane insertion and deinsertion, *Cell* **78**, 835-843

Eichler, J., and Wickner, W. (1997) Both an N-terminal 65-kDa domain and a C-terminal 30-kDa domain of SecA cycle into the membrane at SecYEG during translocation, *Proc. Natl. Acad. Sci. U.S.A.* **94**, 5574-5581

Fauchère, J. L., Charton, M., Kier, L. B., Verloop, A., and Pliska, V. (1988) Amino acid side chain parameters for correlation studies in biology and pharmacology, *Int. J. Pept. Prot. Res.* **32**, 269-278

Hardy, S. J. S., and Randall, L. L. (1991) A kinetic partitioning model of selective binding of nonnative proteins by the bacterial chaperone SecB, *Science* **251**, 439-443

Hartl, F.-U., Lecker, S., Schiebel, E., Hendrick, J., and Wickner, W. (1990) The binding cascade of SecB to SecA to SecY/E mediates preprotein targeting to the *E. coli* plasma membrane, *Cell* **63**, 269-279

Hendrick, J. P., and Wickner, W. (1991) SecA protein needs both acidic phospholipids and SecY/E protein for functional high-affinity binding to the *Escherichia coli* plasma membrane, *J. Biol. Chem.* **266**, 24596-24600

Iino, T., Takahashi, M., and Sako, T. (1987) Role of amino-terminal positive charge on signal peptide in staphylokinase export across the cytoplasmic membrane of *Escherichia coli*, *J. Biol. Chem.* **262**, 7412-7417

Inouye, S., Soberon, X., Franceschini, T., Nakamura, K., Itakura, K., and Inouye, M. (1982) Role of positive charge on the amino-terminal region of the signal peptide in protein secretion across the membrane, *Proc. Natl. Acad. Sci. U.S.A.* **79**, 3438-3441

Ito, K. (1992) SecY and integral membrane components of the *Escherichia coli* protein translocation system, *Mol. Microbiol.* **6**, 2423-2428

Izard, J. W., Doughty, M. B., and Kendall, D. A. (1995) Physical and conformational properties of synthetic idealized signal sequences parallel their biological function, *Biochemistry* **34**, 9904-9912

Izard, J. W., Rusch, S. L., and Kendall, D. A. (1996) The amino-terminal charge and core region hydrophobicity interdependently contribute to the function of signal sequences, *J. Biol. Chem.* **271**, 21579-21582

Jones, J. D., McKnight, C. J., and Gierasch, L. M. (1990) Biophysical studies of signal peptides: Implications for signal peptide functions and the involvement of lipid in protein transport, *J. Bioenerg. Biomembr.* **22**, 213-222

Kaufmann, A., Manting, E. H., Veenendaal, A. K. J., Driessen, A. J. M., and van der Does, C. (1999) Cysteine-directed cross-linking demonstrates that helix 3 of SecE is close to helix 2 of SecY and helix 3 of a neighboring SecE, *Biochemistry* **38**, 9115-9125

Kendall, D. A., Bock, S. C., and Kaiser, E. T. (1986) Idealization of the hydrophobic segment of the alkaline phosphatase signal peptide, *Nature* **321**, 706-708

Kimura, E., Akita, M., Matsuyama, S.-I., and Mizushima, S. (1991) Determination of a region in SecA that interacts with presecretory proteins in *Escherichia coli*, *J. Biol. Chem.* **266**, 6600-6606

Laemmli, U. K. (1970) Cleavage of structural proteins during the assembly of the head of bacteriophage T4, *Nature* **227**, 680-685

Lanzetta, P. A., Alvarez, L. J., Reinach, P. S., and Candia, O. A. (1979) An improved assay for nanomole amounts of inorganic phosphate, *Anal. Biochem.* **100**, 95-97

Randall, L. L., Topping, T. B., Hardy, S. J., Pavlov, M. Y., Freistroffer, D. V., and Ehrenberg, M. (1997) Binding of SecB to ribosome-bound polypeptides has the same characteristics as binding to full-length, denatured proteins, *Proc. Natl. Acad. Sci. U.S.A.* **94**, 802-807

Lill, R., Dowhan, W., and Wickner, W. (1990) The ATPase activity of SecA is regulated by acidic phospholipids, SecY, and the leader and mature domain of precursor proteins, *Cell* **60**, 271-280

Matsuyama, S., Fujita, Y., and Mizushima, S. (1993) SecD is involved in the release of translocated secretory proteins from the cytoplasmic membrane of *Escherichia coli*, *EMBO J.* **12**, 265-270

McKnight, C. J., Briggs, M. S., and Gierasch, L. M. (1989) Functional and nonfunctional LamB signal sequences can be distinguished by their biophysical properties, *J. Biol. Chem.* **264**, 17293-17297

Miller, A., Wang, L., and Kendall, D. A. (1998) Synthetic signal peptides specifically recognize SecA and stimulate ATPase activity in the absence of preprotein, *J. Biol. Chem.* **273**, 11409-11412

Mitchell, C., and Oliver, D. (1993) Two distinct ATP-binding domains are needed to promote protein export by *Escherichia coli* SecA ATPase, *Mol. Microbiol.* **10**, 483-497

Mori, H., Araki, M., Hikita, C., Tagaya, M., and Mizushima, S. (1997) The hydrophobic region of signal peptides is involved in the interaction with membrane-bound SecA, *Biochim. Biophys. Acta* **1326**, 23-36

Phoenix, D. A., Kusters, R., Hikita, C., Mizushima, S., and de Kruijff, B. (1993) OmpF-Lpp signal sequence mutants with varying charge hydrophobicity ratios provide evidence for a phosphatidylglycerol-signal sequence interaction during protein translocation across the *Escherichia coli* inner membrane, *J. Biol. Chem.* **268**, 17069-17073

Sasaki, S., Matsuyama, S., and Mizushima, S. (1990) *In vitro* kinetic analysis of the role of the positive charge at the amino-terminal region of signal peptides in translocation of secretory protein across the cytoplasmic membrane in *Escherichia coli*, *J. Biol. Chem.* **265**, 4358-4363

Schiebel, E., Driessen, A. J. M., Hartl, F.-U., and Wickner, W. (1991) $\Delta\mu H^+$ and ATP function at different steps of the catalytic cycle of preprotein translocase, *Cell* 64, 927-939

Shinkai, A., Mei, L. H., Tokuda, H., and Mizushima, S. (1991) The conformation of SecA, as revealed by its protease sensitivity, is altered upon interaction with ATP, presecretory proteins, everted membrane vesicles, and phospholipids, *J. Biol. Chem.* **266**, 5827-5833

Seligman, S. J. (1993) Radiolabelling of synthetic peptides by acetylation of the N-terminal amino group, *Anal. Biochem.* **211**, 324-325

Synders, S., Ramamurthy, V., and Oliver, D. (1997) Identification of a region of interaction between *Escherichia coli* SecA and SecY proteins, *J. Biol. Chem.* **272**, 11302-11306

Taura, T., Baba, T., Akiyama, Y., and Ito, K. (1993) Determinants of the quantity of the stable SecY complex in the *Escherichia coli* cell, *J. Bacteriol.* **175**, 7771-7775

Ulbrandt, N. D., London, E., and Oliver, D. B. (1992) Deep penetration of a portion of *Escherichia coli* SecA protein into model membranes is promoted by anionic phospholipids and by partial unfolding, *J. Biol. Chem.* **267**, 15184-15192

Vlasuk, G. P., Inouye, S., Ito, H., Itakura, K., and Inouye, M. (1983) Effects of the complete removal of basic amino acid residues from the signal peptide on secretion of lipoprotein in *Escherichia coli*, *J. Biol. Chem.* **258**, 7141-7148

von Heijne, G. (1985) Signal sequences. The limits of variation, *J. Mol. Biol.* **184**, 99-105

Wickner, W., and Leonard, M. R. (1996) *Escherichia coli* preprotein translocase, *J. Biol. Chem.* **271**, 29514-29516

Yamane, K., and Mizushima, S. (1988) Introduction of basic amino acid residues after the signal peptide inhibits protein translocation across the cytoplasmic membrane of *Escherichia coli*, *J. Biol. Chem.* **263**, 19690-19696

STRUCTURE, FOLDING AND ASSEMBLY OF ADENOVIRUS FIBERS

ANNA MITRAKI[1]*, MARK VAN RAAIJ[2], ROB RUIGROK[2], STEPHEN CUSACK[2], JEAN-FRANÇOIS HERNANDEZ[1] AND MARY LUCKEY[3]
[1]Institut de Biologie Structurale, 41 rue Jules Horowitz, 38027 Grenoble Cedex 1, France
[2]European Molecular Biology Laboratory, Grenoble Outstation, c/o Institut Laue Langevin, BP 156, 38042 Grenoble Cedex 9, France
[3]Department of Chemistry and Biochemistry, San Francisco State University, 1600 Holloway avenue, 94132 San Francisco, CA, USA
*correspondence to this author: mitraki@ibs.fr

Abstract
Adenovirus fibers are trimeric proteins that protrude from the 12 five-fold vertices of the virion. They consist of three segments: a N-terminal tail, a thin shaft carrying 15-amino acid pseudo-repeats, and a C-terminal globular head (or knob) which recognizes the primary cell receptor. Our recent folding studies have demonstrated that the fiber unfolds in SDS through a stable intermediate in which the C-terminal head and five repeats of the shaft remain folded and trimeric. This stable domain has been cloned and expressed in *Escherichia coli* and its structure has been solved at 2.4 Å resolution. The structure reveals a novel triple β-spiral fibrous fold for the shaft. In order to assemble into the correct triple β–spiral conformation, the shaft needs to be brought into the correct registration. The globular head appears to act as the registration signal, and in its absence synthetic peptides corresponding to shaft sequences fail to assemble correctly. Instead, they aggregate and form amyloid-like fibrils. In this chapter we discuss the fiber as a model system to address the interplay between folding, assembly and misassembly of β-sheet proteins. We also discuss potential implications for materials science since this protein has been used as a model for synthetic fiber design.

Morphology of the fiber

Adenoviruses have a wide cell tropism, infecting a variety of human and animal cell types. The fiber proteins are located at the fivefold vertices of the icosahedral capsid, and are responsible for the interaction with the primary cell receptor (fig. 1a) [1-3]. The human serotype 2 fiber is a trimer of 582 residues assembled with an asymmetrical topology. It is composed of three distinct segments: the N-terminal tail of approximately 45 residues

A. Aggeli et al. (eds.), Self-assembling Peptide Systems in Biology, Medicine and Engineering, 221–234.

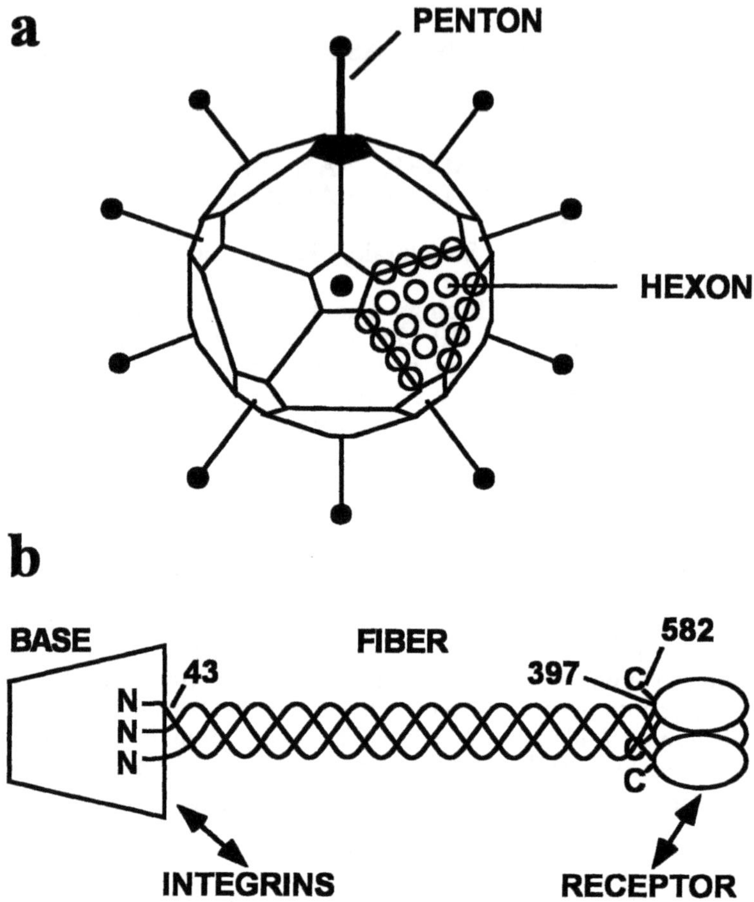

Figure 1. **a.** The adenovirus virion. Hexon is the major capsid protein, and the penton complex consists of the penton base plus the fiber. The fibers emanate from the twelve five-fold vertices of the icosahedral capsid. **b.** Morphology and domain organization of the fiber protein. The N-terminal tail of 43 residues is noncovalently embedded in the penton base. The fibrillar shaft (residues 43-397) consists of 22 pseudo-repeats of about 15 amino acids each. The C-terminal globular head (residues 397-582) interacts with the primary receptor. Virus internalization is subsequently promoted by interaction of the penton base with integrins.

interacts with the penton base protein of the capsid, a long thin shaft comprising pseudo-repeats of approximately 15 amino acids, and finally the C-terminal globular head (or knob) which binds to the cell receptor (fig. 1b) [4,5]. This cell receptor has been identified to be the Coxsackievirus and Adenovirus Receptor (CAR), a human protein of unknown function [6,7].

The sequence of the shaft reveals pseudo-repeats of 15 amino acids containing conserved hydrophobic residues and a conserved proline or glycine (fig. 2) [8]. The fiber of human adenovirus type 2 has 22 such repeats making up a total length of the shaft of 29 nm [9].

The fiber as a model system for folding and assembly

Viral fibers have been proven excellent model systems for understanding protein folding and assembly both in the test tube and inside prokaryotic cells [10-14]. Quite extensive studies have also been carried out on the folding and quality control of membrane glycoproteins in the endoplasmic reticulum of eukaryotic cells using viral proteins as models [15-16]. However, not many studies exist on the folding of proteins assembled in the eukaryotic cytoplasm, with the exception of reovirus sigma protein [17].

Adenovirus capsid proteins are synthesized in the cytoplasm and subsequently transported to the nucleus where assembly of the viral particles takes place. Fiber trimerization occurs in the cytoplasm of the infected cells, as deletion of the nuclear targeting signal (amino acids 2 to 5, K-R-A-R) which is sufficient to abolish translocation of the fiber to the nucleus, results in accumulation of adenovirus type 2 fiber trimers in the cytoplasm of vaccinia-infected cells [18]. *In vitro* translation experiments also suggest that fiber assembly might occur in the cytoplasm of the infected cell [19].

The native fiber is resistant to SDS at low temperatures (4° C). In the presence of 2% SDS without boiling, native trimers do not dissociate into monomers, and migrate at approximately 180 kDa when run in SDS-PAGE gels with refrigeration [20]. Dissociation into monomers and complete denaturation can be achieved after boiling at 100° C for 3 minutes, after which the fiber polypeptide chains migrate in the monomer position (62 KDa; fig. 3).

Unfolding intermediates can also be visualized in gels as well as with electron microscopy (fig. 3, 4) [20]. Furthermore, monoclonal antibodies that specifically recognize the trimeric form of the head exist [21]. The existence of the above tools allows easy distinction of native trimers from partially folded and unfolded states. Thus, adenovirus fibers might serve as a model system for studying protein folding, misfolding or misassembly both *in vitro* and inside eukaryotic cells.

		a	b	c	d	e	f	g	h					i	j	k	l	m	n	o	
			β	β	β	β	β	β	β							β	β	β	β		
				tail			*P*	*P*	*T*	*V*	*P*	*F*	*L*	*T*	*P*	*P*	*F*	*V*	*S*	*P*	*N G F Q E S P P*
1	45	G	V	L	S	L	R	V	S	.	.	.	.	E	P	L	D	T	S	H	. . . .
2	60	G	M	L	A	L	K	M	G	.	.	.	.	S	G	L	T	L	D	K	A . . . .
3	76	G	N	L	*T*	*S*	*Q*	*N*	*V*	*T*	*T*	*V*	*T*	*Q*	*P*	*L*	*K*	*K*	*T*	*K*	. . . .
4	95	S	N	I	S	L	D	T	S	.	.	.	.	A	P	L	T	I	T	S	. . . .
5	110	G	A	L	T	V	A	T	T	.	.	.	.	A	P	L	I	V	T	S	. . . .
6	125	G	A	L	S	V	Q	S	Q	.	.	.	.	A	P	L	T	V	Q	D	. . . .
7	140	S	K	L	S	I	A	T	K	.	.	.	.	G	P	I	T	V	S	D	. . . .
8	155	G	K	L	A	L	Q	T	S	.	.	.	.	A	P	L	S	G	S	D	S . . . .
9	171	D	T	L	T	V	T	A	S	.	.	.	.	P	P	L	T	T	A	T	. . . .
10	186	G	S	L	G	I	N	M	E	.	.	.	.	D	P	I	Y	V	N	N	. . . .
11	201	G	K	I	G	I	K	I	S	.	.	.	.	G	P	L	Q	V	A	Q	N S . . . . . .
12	218	D	T	L	T	V	V	T	G	.	.	.	.	P	G	V	T	V	E	Q	. . . .
13	233	N	S	L	R	T	K	V	A	.	.	.	.	G	A	I	G	Y	D	S	S . . . .
14	249	N	N	M	E	I	K	T	G	.	.	.	.	G	G	M	R	I	N	N	. . . .
15	264	N	L	L	I	L	D	V	D	.	.	.	.	Y	P	F	D	A	Q	.	. . . .
16	278	T	K	L	R	L	K	L	G	Q	.	.	.	G	P	L	Y	I	N	A	S . . . .
17	295	H	N	L	D	I	N	Y	N	.	.	.	.	R	G	L	Y	L	F	N	A S N N T . . .
18	315	K	K	L	E	V	S	I	K	K	S	.	.	S	G	L	N	F	D	N	. . . .
19	332	T	A	I	A	I	N	A	G	.	.	.	.	K	G	L	E	F	D	T	N T S E S P D I
20	355	N	P	I	K	T	K	I	G	.	.	.	.	S	G	I	D	Y	N	E	N . . . .
21	371	G	A	M	I	T	K	L	G	.	.	.	.	S	G	L	S	F	D	N	S . . . .
22	387	G	A	I	T	I	G	*N*	*K*	*N*	*D*	*D*	*K*	*L*	*T*	*L*	*W*	*T*	*T*	*P*	*D P head*
P		X	X	φ	X	φ	X	φ	X	.	.	.	.	X	P	φ	X	φ	X	X	. . . .
G		X	X	φ	X	φ	X	φ	G	.	.	.	.	X	G	φ	X	φ	X	X	. . . .
		a	b	c	d	e	f	g	h					i	j	k	l	m	n	o	

Figure 2. Alignment of the 22 repeats in the adenovirus serotype 2 fiber shaft. The repeating residues are labelled a to o. Residues belonging to β–strands are indicated with β. Residues involved in the main hydrophobic core (columns c, e, k) and in the peripheral hydrophobic patches (columns g and m) are shown. The conserved glycines or prolines in the tight turn are boxed in bold (column j). Residues 23-44 of the tail domain are shown in italics at the top. Repeat 3 (also shown in italics) cannot be aligned reliably and might adopt a different structure. The flexible linker (residues 393-397) connecting the shaft to the head as well as the first nine residues of the head domain are also shown in italics. The consensus sequences of the proline and glycine repeats (P and G respectively) are shown at the bottom (φ is hydrophobic). Adapted from ref. 22 with permission.

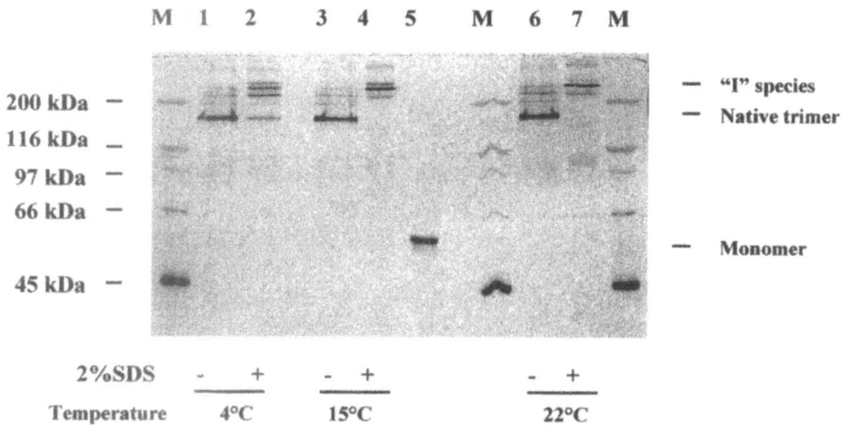

Figure 3. Conformational states of the fiber as a function of incubation time at 2% SDS in various temperatures. Purified fibers were incubated for 100 min at 4, 5 or 22° C either in buffer (lanes 1, 3 and 6) or in 2% SDS (lanes 2, 4 and 7). At the end of icubation, samples were quenched at 4° C with ice-cold sample buffer and immediately separated by SDS gel electrophoresis at 4° C. The denatured control (lane 5) was boiled for 3 min at 100° C. Lane M, molecular mass markers. Reproduced from [20] with permission.

The fiber unfolds via a stable intermediate comprising the head and part of the shaft

In an attempt to understand stability of the trimeric fibers, the unfolding pathway of Ad2 fibers induced by SDS and temperature has been investigated [20]. Presence of SDS at moderate temperatures (4°-22° C) induces partial unfolding starting from the N-terminus and a stable intermediate accumulates that has the C-terminal head and part of the shaft structured as seen by electron microscopy (fig. 4). This unfolding intermediate displays a characteristic mobility in SDS-PAGE gels, running slower than the native trimer. It is stable and long-lived at 22° C (fig. 3).

Upon further rising of the temperature, it dissociates to the monomeric, SDS-sensitive state. The unfolded parts of the stable intermediate have been digested using limited proteolysis and the stable domain has been identified to span residues 319-582 [20]. It still remains correctly folded and trimeric, as probed by reactivity to antibodies specific for the trimeric form of the head, and change of mobility on SDS-PAGE upon boiling. The observations on the unfolding pathway of the fiber can be summarized in the cartoon depicted in figure 5 [20].

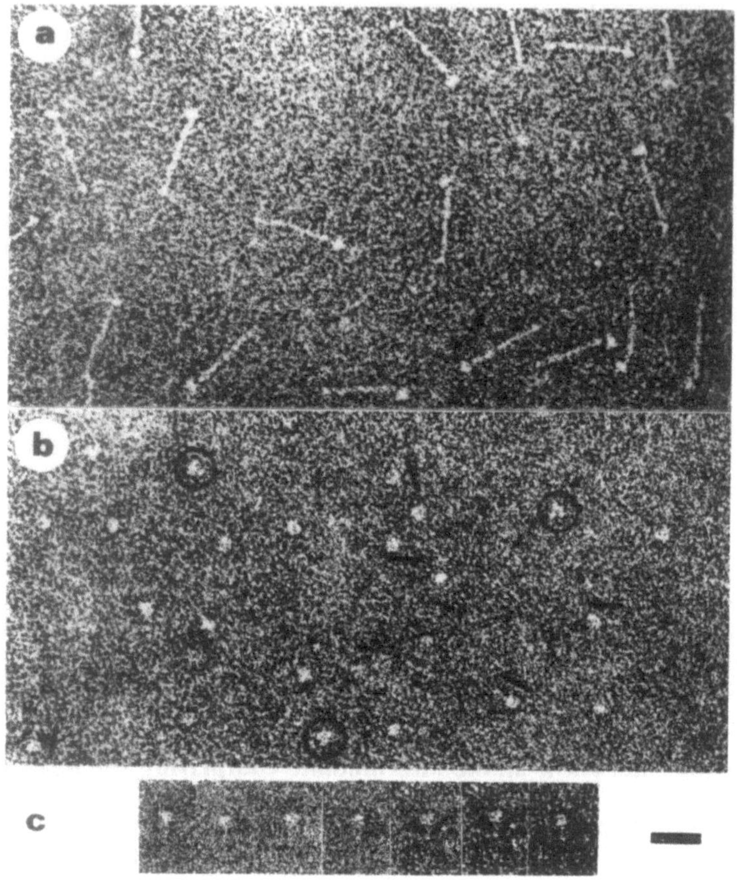

Figure 4. Electron micrographs of negatively stained native fibers (**a**) and partially unfolded (SDS-treated) fibers (**b**). Fibers in side-views are indicated by a small bar and trimeric fiber heads in end-on view are encircled. **c.** Gallery of side-views of SDS-treated fibers. The bar represents 20 nm. Reproduced from [20] with permission.

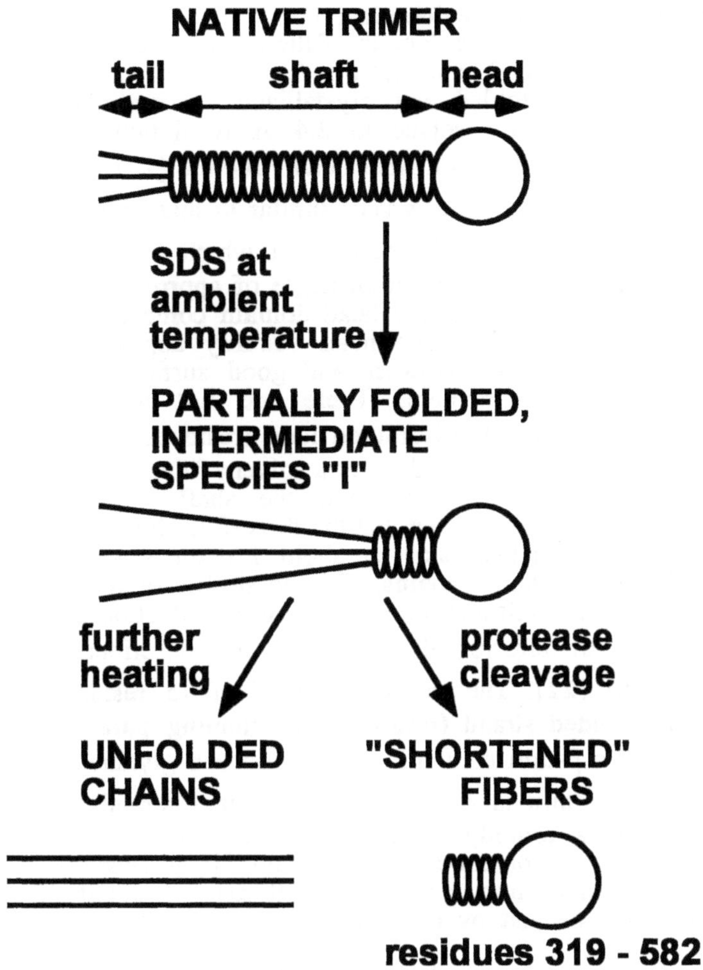

Figure 5. Unfolding pathway of the fibers. In the presence of SDS, in the range of temperatures from 4 to 22°C, unfolding of the native trimer starts from the N-terminus and goes through an unfolding intermediate that has the head plus five shaft repeats structured. This partially folded intermediate is stable at ambient temperatures; further rising of the temperature causes the complete unfolding of the fiber. The unfolded parts of the stable unfolding intermediate « I » can be digested, giving rise to a stable, « shortened » fragment, comprising residues 319-582.

Crystal structure of the stable domain

The identification of the stable domain of the adenovirus fiber prompted us to undertake its cloning and expression in *E. coli*. The recombinant domain was trimeric, properly folded and crystallized successfully. The crystal structure was subsequently solved at 2.4 Å resolution [22] using the recently obtained high resolution fiber head structure [23].

The fiber head structure is very similar to that of serotype 5 [24]. Each monomer consists of a β-sandwich in which two four-stranded β-sheets are packed so that they are at an angle of approximately 30 degrees to each other. The monomers of the head domain come together to form a stable three-bladed propeller. The head stability appears to be mainly caused by a hydrogen-bond network and good surface-complementarity, holding together the monomers. There are not many hydrophobic interactions [23]. A relatively disordered loop (residues 393-397) connects the head domain to the shaft and there is a solvent-filled volume around the three-fold axis between the head and the shaft (Figure 6a). These observations, together with the fact that the three-fold shaft axis is tilted with respect to the three-fold head axis by about 2 degrees, suggest flexibility between the head and shaft domains. This possibility of relative movement between the head and shaft may be required in the infection process.

The shaft structure reveals a new fold, best described as a triple β-spiral (Figure 6b) [22]. The repeating motif of 15 residues (Figure 2) comprises an extended strand (residues b-h) running parallel to the fiber axis, followed by a β-turn containing the conserved glycine (or proline in other repeats). The turn is followed by another β-strand (residues k-n), running backwards at roughly 45 degrees to the shaft axis. The repeats are joined together by a solvent-exposed variable loop (residues o-a). One repeat leads to a translation along the shaft axis of about 13 Å and a clockwise rotation of just over 50 degrees, as seen from the N- to C-terminus.

In contrast to the head domain, the shaft domain contains an extensive hydrophobic core (Figure 7a). Three conserved hydrophobic residues in each repeat (c, e, and k) point towards the central three-fold axis forming a longitudinal hydrophobic core with the equivalent residues from the other chains. Two further residues (g and m) form stabilizing hydrophobic patches at greater radius. The framework of the structure is formed by conserved intra- and inter-chain hydrogen bonds (Figure 7b). The resulting structure (Figure 6b) is a complex, yet regular, highly cross-linked structure which, together with a high proportion of buried surface, accounts for the high rigidity and stability of the shaft. The fold of the adenovirus shaft is unique in protein structures solved until now and different from both the cross-β model [8] and the triple-helical model [25]. Therefore, with the triple β-spiral adenovirus shaft fold, we have discovered a new fold for a fibrous protein; it remains to be seen if this structure will be found in other fibrous proteins.

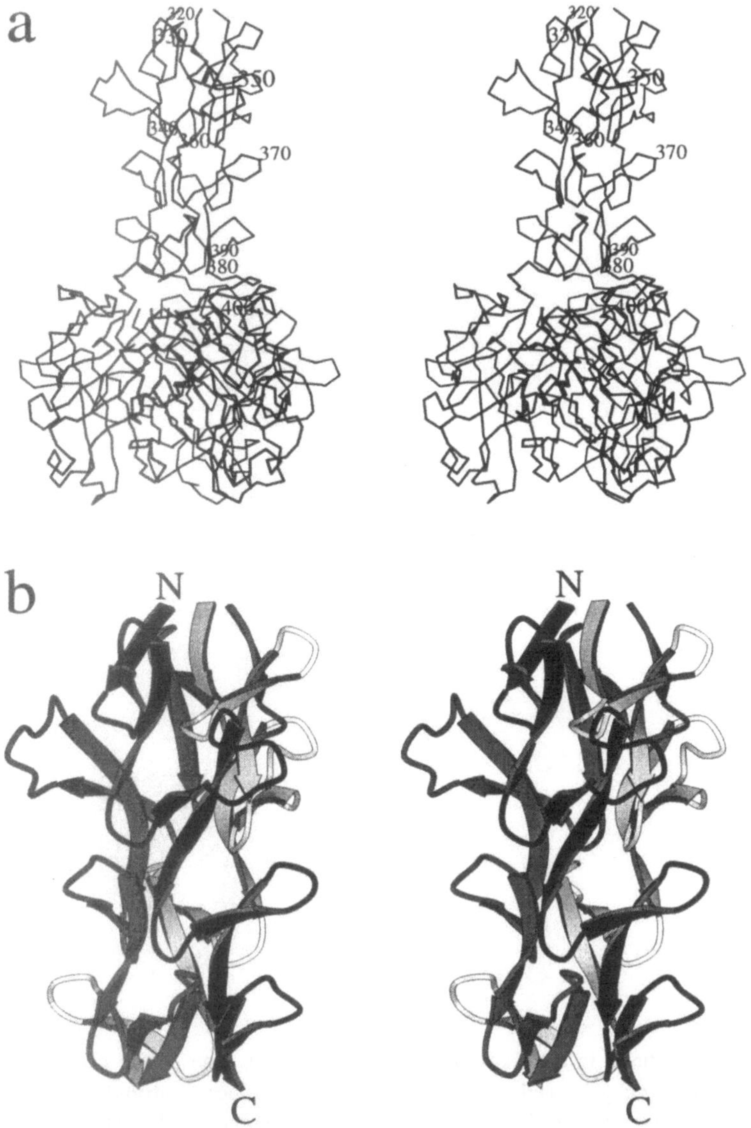

Figure 6. The adenovirus fiber fold. The two parts of the figure are shown in stereo. **a.** Chain trace of the Cα atoms of one of the trimers present in the asymmetric unit. Every tenth residue of the shaft domain is numbered. **b.** The shaft domain as present in the crystal. One of the chains is shown in black and the other two in gray. The N- and C-termini of one chain are labelled. Adapted from ref. 22 with permission.

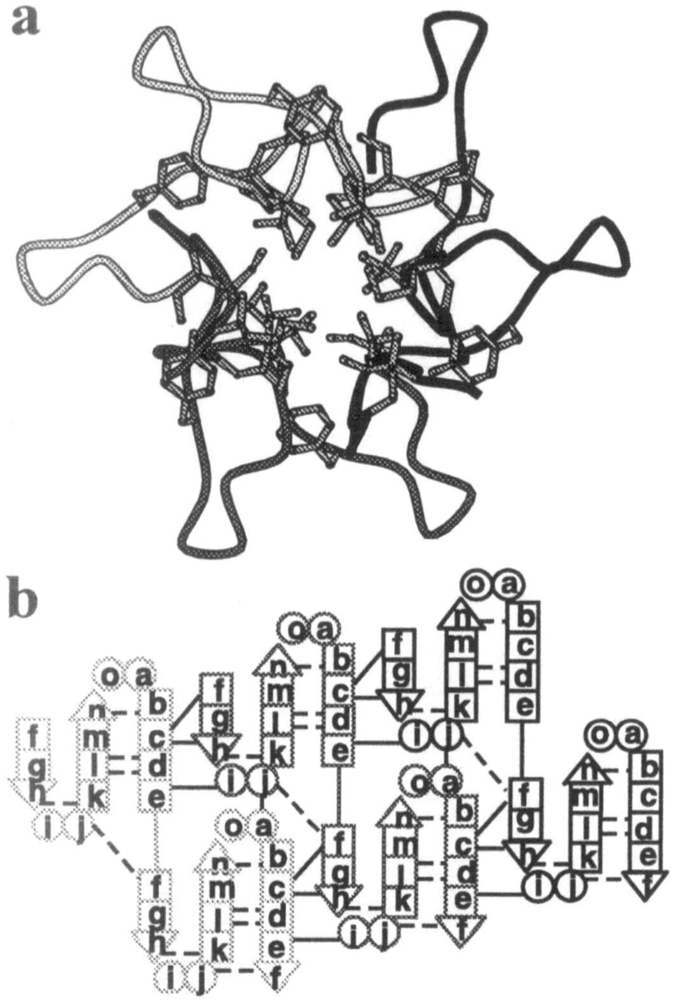

Figure 7. Interactions in the shaft. **a.** hydrophobic core. One of the chains is shown in black, one in dark grey and one in light grey. Residues contributing to the central hydrophobic core (c, e, k) and to the peripheral hydrophobic patches (g and m) are shown. **b.** diagram of the conserved hydrogen bonding network within one monomer and between adjacent monomers. Residues are labelled according to figure 2. Intrachain (dotted lines) and interchain (solid black lines) hydrogen bonds are shown. Adapted from ref. 22 with permission.

Assembly *versus* misassembly of the fibers

Our unfolding studies provide evidence that the head domain is very stable and is the last part of the fiber to unfold. Stable intermediates populated in an unfolding pathway are not necessarily populated in the refolding pathway, and therefore the fiber head would not necessarily be the first part of the fiber to refold [11,26]. However, the head is also known to be essential for trimerization as deletions or mutations in this part of the protein hinder the trimerization process *in vivo* [27,28]. This domain might act as a registration signal necessary for the three chains to align and fold together, thus playing a role analogous to that of the procollagen peptides in collagen folding [29,30]. Thus, it is reasonable to speculate that the head could initiate the folding and/or assembly process during refolding of the fiber *in vitro* or folding *in vivo*.

Synthetic peptides corresponding to the part of the shaft which is immediately adjacent to the head fail to assemble correctly. Instead, they form amyloid-like fibrils (fig. 8) that are birefringent when stained with the diagnostic dye Congo red (unpublished results). Thus, in the absence of the head to provide registration, the peptides aggregate, probably through out-of register interactions to form the amyloid-like fibrils. This observation provides further support for an essential role of the head domain in fiber assembly. It also raises the question of possible involvement of chaperone proteins during the *in vivo* folding and trimerization of the fiber, in the sense that the N-terminal part will certainly have to remain protected from aggregation during the time required for synthesis and folding of the head. Fiber folding and trimerization do not seem to involve other viral proteins, since fiber or fiber portions can be expressed and correctly trimerize in various prokaryotic and eukaryotic expression systems in the absence of any other viral components [27,28,31,32]. In extracts of cells infected with the human serotype 5 the fiber can be co-immunoprecipitated with the heat-shock protein hsp70 [33]. However, only 5% of the newly synthesized protein was found complexed with hsp70. Thus, elucidation of folding and assembly mechanisms of the fiber will certainly require further detailed studies.

The adenovirus fiber as a model for synthetic fiber design

Natural polymers increasingly serve as models for the design of new materials. The adenovirus fiber with its repetitive structural motifs has also served as model for synthetic fiber design. O'Brien et al. [34] have bacterially expressed recombinant polymers with fiber repeats as building blocks which were produced as inclusion bodies. These were solubilized in denaturants, refolded, purified and spun into fibers. The results have been encouraging, demonstrating the feasibility of obtaining highly aligned and ordered fibers with mechanical properties comparable to those of commercial textile fibers. Our structural information on the fiber shaft will hopefully stimulate further efforts on rational design of synthetic fibers with desired mechanical and functional properties.

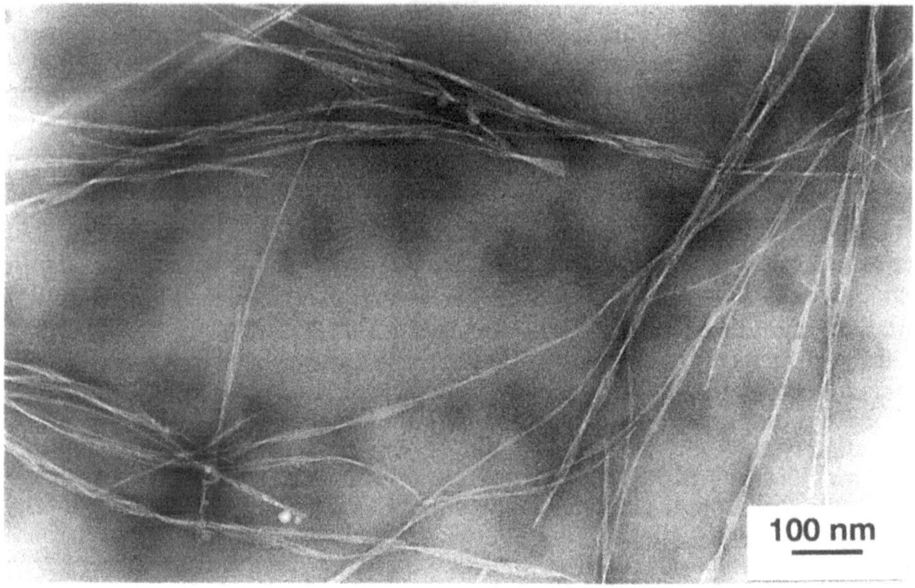

Figure 8. Electron micrograph of amyloid-like fibrils formed by synthetic peptides corresponding to fiber shaft sequences. Note the twisted ribbon morphology of the fibrils.

Perspectives

Due to its importance in cell attachment the adenovirus fiber has been the subject of intense investigation during the last 20 years. The discovery of the primary receptor for most adenovirus subgroups and the emergence of structural information on the fiber are certainly important steps towards the understanding of its function. However, many important aspects remain unresolved, such as the nature of the interaction between fiber and penton base, possible conformational changes and flexibility associated with the infection process and structures of fibers that use a different primary receptor.

A deeper understanding of fiber structure, folding and assembly is not only needed on fundamental grounds, but also for practical reasons. Adenoviruses infect a variety of human tissues and are currently used as gene transfer vectors. Future developments in new generation adenovirus gene therapy vectors might require engineering of fiber variants to specifically target certain tissues. The presence of the variable loop in the shaft structure opens up the possibility for insertion of sequences potentially useful for gene therapy. However, we still do not know if these loops play an important role during folding and/or assembly, and if any inserted sequences will perturb the overall folding of the fiber.

Future folding studies combined with structural information will bring further insight to the assembly and stabilization mechanisms of the fiber.

Acknowledgements

We would like to thank Gilles Lavigne, Pierrette Goeltz, Nancy Cohet, Annie Barge, Jean-Pierre Andrieu and Gerard Arlaud for their contributions to this research. We are grateful to Jean Gagnon for support and critical reading of the manuscript.

References

1. Valentine, R.C. and Pereira, H.G. (1965) *J. Mol. Biol.* **13**, 13-20.
2. Ginsberg, H.S., Pereira, H.G., Valentine, R.C. and Wilcox, W.C. (1966) *Virology* **28**, 782-3.
3. Philipson, L., Lonberg-Holm, K. and Pettersson, U. (1968) *J. Virol.* **2**, 1064-1075.
4. Devaux, C., Caillet-Boudin, M.L., Jacrot, B. and Boulanger, P. (1987) *Virology* **161**, 121-128.
5. Devaux, C., Adrian, M., Berthet-Colominas, C., Cusack, S. and Jacrot, B. (1990) *J. Mol. Biol.* **215**, 567-588.
6. Bergelson, J.M., Cunningham, J.A., Droguett, G., Kurt-Jones, E.A., Krithivas, A., Hong, J.S., Horwitz, M.S., Crowell, R.L. and Finberg, R.W. (1997) *Science* **275**, 1320-1323.
7. Roelvink, P.W., Lizonova, A., Lee, J.G., Li, Y., Bergelson, J.M., Finberg, R.W., Brough, D.E., Kovesdi, I. and Wickham, T.J. (1998) *J. Virol.* **72**, 7909-7915.
8. Green, N.M., Wrigley, N.G., Russell, W.C., Martin, S.R. and McLachlan, A.D. (1983) *EMBO J.* **2**, 1357-65.
9. Ruigrok, R.W., Barge, A., Albiges-Rizo, C. and Dayan, S. (1990) *J. Mol. Biol.* **215**, 589-596.
10. Goldenberg, D.P., Smith, D.H. and King, J. (1983) *Biopolymers* **22**, 125-129.
11. Fuchs, A., Seiderer, C. and Seckler, R. (1991) *Biochemistry* **30**, 6598-604.
12. Mitraki, A., Fane, B., Haase-Pettingell, C., Sturtevant, J. and King, J. (1991) *Science* **253**, 54-58.
13. Mitraki, A., Danner, M., King, J. and Seckler, R. (1993) *J. Biol. Chem.* **268**, 20071-20075.
14. Betts, S. and King, J. (1999) *Structure Fold. Des.* **7**, R131-R139.
15. Doms, R.W., Lamb, R.A., Rose, J.K. and Helenius, A. (1993) *Virology* **193**, 545-62.
16. Tatu, U. and Helenius, A. (1997) *J. Cell. Biol.* **136**, 555-565.
17. Lee, P.W. and Leone, G. (1994) *Bioessays* **16**, 199-206.
18. Hong, J.S. and Engler, J.A. (1991) *Virology* **185**, 758-767.
19. Novelli, A. and Boulanger, P.A. (1991) *J. Biol. Chem.* **266**, 9299-9303.
20. Mitraki, A., Barge, A., Chroboczek, J., Andrieu, J.P., Gagnon, J. and Ruigrok, R.W. (1999) *Eur. J. Biochem.* **264**, 599-606.
21. Fender, P., Kidd, A.H., Brebant, R., Oberg, M., Drouet, E. and Chroboczek, J. (1995) *Virology* **214**, 110-117.

22. van Raaij M.J., Mitraki, A., Lavigne, G. and Cusack, S. (1999) *Nature* **401**, 935-938.
23. van Raaij M.J., Louis, N., Chroboczek, J. and Cusack, S. (1999) *Virology* **262**, 333-343.
24. Xia, D., Henry, L.J., Gerard, R.D. and Deisenhofer, J. (1994) *Structure* **2**, 1259-1270.
25. Stouten, P.F., Sander, C., Ruigrok, R.W. and Cusack, S. (1992) *J. Mol. Biol.* **226**, 1073-1084.
26. Chen, B. and King, J. (1991) *Biochemistry* **30**, 6260-6269.
27. Novelli, A. and Boulanger, P.A. (1991) *Virology* **185**, 365-376.
28. Hong, J.S. and Engler, J.A. (1996) *J. Virol.* **70**, 7071-7078.
29. Engel, J. and Prockop, D.J. (1991) *Annu. Rev. Biophys. Biophys. Chem.* **20**, 137-152.
30. McLaughlin, S.H. and Bulleid, N.J. (1998) *Matrix Biol.* **16**, 369-377.
31. Louis, N., Fender, P., Barge, A., Kitts, P. and Chroboczek, J. (1994) *J. Virol.* **68**, 4104-4106.
32. Henry, L.J., Xia, D., Wilke, M.E., Deisenhofer, J. and Gerard, R.D. (1994) *J. Virol.* **68**, 5239-5246.
33. Macejak, D.G. and Luftig, R.B. (1991) *Virology* **180**, 120-125.
34. O'Brien, J.P., Hoess, R.H., Gardner, K.H., Lock, R.L., Wasserman, Z.R., Weber, P.C., and Salemme, F.R. (1993) Design, synthesis, and fabrication of a novel self-assembling fibrillar protein, in D. Kaplan, W.W. Adams, B. Farmer and C. Viney (eds.), *Silk Polymers, Materials Science and Biotechnology*, American Chemical Society, Washington D.C., pp. 104-117.

SOLVING THE STRUCTURE OF COLLAGEN

A. RICH
Department of Biology
Massachusetts Institute of Technology
Cambridge, MA, 02139-4307, USA

This article was presented at the Crete workshop. The story was first told as a telephone-interview by the Editor of *Nature Structural Biology* and was written in three parts as the history by the Editor for August, September and October issues in 1998 [1-3].

Follow that fiber

Nature magazine came on Saturday morning. Francis Crick normally read it in bed and we would discuss it over breakfast. On the morning of August 29, 1955, I asked Francis "Anything in the Nature?" "No", he said, nothing very interesting...Oh, but there was this paper by the Courtaulds people." [4].

This was the group that included C.H. Bamford and A. Elliot from the Courtaulds Research Laboratory in Maidenhead, UK. They have made an insoluble form of polyglycine (II) that gave an X-ray power pattern which was different from the usual form, polyglycine I which has a fully extended beta-polypeptide chain. Although they had a lot of infrared observations on polyglycine II [5], they did not know what its structure was, except that it was different from polyglycine I.

So eating our way through breakfast we said, you know, it should not be too difficult to solve the structure because, after all, there are not too many degrees of freedom; let's try and work on it when we go to the [Cavendish] Laboratory. To work on it meant to make molecular models from these brass skeletal rods, 5 cm to the Angstrom, and see if we could build something that has the properties described for polyglycine II in the Courtaulds paper.

By the time we arrived at the Cavendish we had forgotten about the problem, but shortly thereafter remembered about it again. This was around 10 a.m. In solving the structure our thinking went like this. In polyglycine I the glycyl residues are fully extended, having a 2_1 screw axis. Perhaps the chain is not fully extended. So we decided to try model a 3_1 screw axis. As soon as we did this, we found that the chains formed a large three-dimensional lattice, in which all the hydrogen bonds were satisfied. The planes of the peptide units were stacked in such a way that we could

A. Aggeli et al. (eds.), Self-assembling Peptide Systems in Biology, Medicine and Engineering, 235–241.

predict, immediately and just by looking at the structure, that they would give very intense reflections with the spacings 4.15 Angstroms and 3.1 Angstroms, as seen in the X-ray pictures. We could see that the 3.1 Angstrom reflections derived from the extension of the glycyl residues along the fiber axis, one every 3.1 Angstrom, and the 4.15 Angstrom reflection was the spacing perpendicular to the neighboring planar peptide units. We measured the coordinates from the model, and calculated the intensity of the reflections and the model was clearly right. By noon, we had solved the structure.

We then had the devilish idea, why don't we write it up, have [Lawrence] Bragg submit it to *Nature* and have it come out next week. Then, we thought about this a little more and Francis said, "Well, you know, maybe we should wait because I think the Courtaulds people, if they knew it was this simple, might unhappy, and the situation could be complicated. Let's invite them to come and look at the structure."

So we invited them over and Bamford and Elliot came by. It turned out to be a very good thing to do. They gave us more information about their infrared studies and showed us the original X-ray photos. It was clear that the data was consistent with the structure. We wrote the manuscript up and sent it into Nature on September 20, thanking Bamford and Elliot for useful discussion---a much more diplomatic way of proceeding [6].

Chasing Collagen

What passed for frontier structural biology in those days was to deduce the structure you were interested in, using clues, intuition and whatever wit you had, and then see whether it agreed with the X-ray diffraction data, or whatever other data were available. It was a different animal entirely from the kind of work done today. For one thing, in today's structure determinations you have redundancy – you have more data points then you have unknowns – and thus when you get a solution you know that it is almost certainly correct. In those days you had more unknowns, that is, the coordinates, than data points.

For fibrous molecules random rotations around the fiber axis yielded a structure that produced the continuous transform in its diffraction pattern, this was produced generally in biological molecules, including the alpha helix and collagen. If the proposed structure predicted the observed layer line spacings and the intensity maximum then you could feel confident that the structure was consistent with the data and therefore might be correct.

In September 1953, Linus Pauling organized a conference on protein structure, which was held at the California Institute of Technology. This is where I first met Francis Crick. I learned from Francis that they had an X-ray machine with a high intensity beam at Cambridge that Toni Broad, who was there technical assistant, had developed. Francis said, "Look, if you get something that looks interesting why don't you come over, spend some time and collect some data on our rotating anode."

In 1954 I went to the National Institute of Health, in Maryland, and set up a section on physical chemistry, dealing with the molecular structures of biological materials. We started working on RNA fibers. Severa Ochoa and Marianne Grunberg-Manago had recently discovered an enzyme – polynucleotide phosphorylase--which would make RNA polymers without a template. Ochoa sent me some of the polymer material (polyadenylic acid, and so on) and, using this material, we pulled a polyA fiber that gave an oriented and very interesting diffraction pattern. But, as usual, this and the other RNA fibers we were looking at were small (0.1-0.4 mm in diameter), thus the X-ray beam at the Cavendish lab was potentially a useful tool for us. So I wrote to Francis and shortly thereafter he invited me to Cambridge.

I arrived in England with my fibers in July 1955. Francis said, "Why don't you come and stay with us, we have a spare room." I accepted, as it looked as though I would only be staying for a two – or – three week visit. In fact, it was more like the movie of The Man Who Came to Dinner – I was still there in 1956, slightly over a half year later. Upon my arrival I shared a desk with Francis and had roughly 28 inches of bench space, Hugh Huxley had the next 28 inches. Nevertheless, or perhaps because of this it was a very productive lab. Then, shortly after my arrival, perhaps a week or two later, Jim Watson came over and joined us in the polyA business; and, in fact, we wrote a paper on the subject [7].

One of the more civilized things about research in Britain is that you can have tea in the afternoon and sort of reflect. We were working in the Cavendish Laboratory (this was before Max Perutz's hut had been built) and there was a tea-room, but on this particular afternoon we decided not to have tea there. Rather, we found a tea-room in town and, while sitting on the sidewalk, I said, "You know, Francis, I have been looking at that model of polyglycine and C-H bond of glycine is virtually coplanar with N-H bond. That means you could make a pyrrolidine ring there without any distortion." I told him I had been wondering whether, if you look three cyclically hydrogen bond chains out of the lattice and simply put pyrrolidine rings in positions two and three, you would have a model for collagen, because at the time we knew that collagen had sequences Gly-Pro and Gly-Pro-Hypro and Hypro-Gly. It was also known that collagen gave an ordered diffraction pattern.. Furthermore, I though, if we just twisted three molecules together into the form of a right handed helix (polyglycine itself is a left handed helix) to make, literally, a coiled coil, then you could bring the repeat length down to 2.86 angstroms, which is the main collagen meridional reflection, instead of 3.1 angstroms for the polyglycine, and that did not distort the polypeptide chain very much. So then we said, let's work on it.

I should say that there had been many proposals for the structure of collagen. Linus Pauling and Robert Cory had proposed a three-chained model [8], which then they subsequently withdrew because it did not fit the data. (This was part of the great flood of papers Pauling published in the *Proceedings of National Academy of Sciences of the USA* while he was the editor--the model of the alpha-helix, the model of the beta-sheets, along with the structure of collagen; these papers help establish the *Proceedings* as an important journal – until then it had been unimportant). Lots of people were wrong, though: Francis himself had proposed a two-chained model for collagen and then withdrew it [9], before we revisited the problem. The idea was at that time, that each person was entitled to one published guess per molecule, otherwise you were spinning your wheels, which was considered inappropriate. Nevertheless, Linus said that because he and Cory had been co-authors on the collagen model paper he was entitled to one more guess, so Francis felt that since he was going to be a co-author on our modeling paper he was also entitled to one more guess.

Structure of collagen

Molecular models of biological polymers—reliable, hand-built model—were not widely spread at the time. Linus Pauling built models, John Kendrew at the Cavendish in Cambridge built models, most people did not and therefore, were essentially guessing at their structural proposals. But with models, we could get coordinates and make Fourier transforms, and calculations. This provided a technological edge. It sounds primitive nowadays, but the difference was that we had these soldered, tetrahedral-shaped wires for the carbon atom, and we had a planar peptide unit based on the most recent work that Linus and his colleagues had done on the peptide bond.

Thus armed, we wen to work on the collagen model in late September, 1955. We took three of the cyclically hydrogen bonded polypeptide chains from the polyglycine II structure, built them and after rotating the chains to make a coiled coil, we found that the model came together easily—there was no conflict, in terms of bad coordinates. In essence, what we were using was the organization of polyglycine and its hydrogen bonding systems as the core of the collagen molecule. This turned out to be quite reasonable because it was known polyproline formed a helix [10], of three residues per turn, which had the same backbone twist but, of course, did not have the potential for hydrogen bonding. The problem with most other models at the time was that they did not have good van der Waals contacts, bond angles and distances. Since we stared with polyglycine model, which had an excellent stereochemistry, we could get to collagen without any problems in terms of atomic crowding.

The model we built of poly (GlyProPro) was about six feet high, at a scale of 5 cm to the Angstrom. There was a very long corridor in the basement of the Cavendish, it was about 60-70 feet long. We placed an extremely strong light at the one end of the corridor (we wanted roughly the parallel light, of course), put the model at the other end and placed a large sheet of a white paper on the wall behind the model and simply drew the projection (circles and labels) on the paper. Since the model had a ten fold symmetry (there are ten GlyProPro triples per turn of the helix), a single projection consisted ten views of the structure. We then rotated the model 18o and repeated the process, superimposing the projections to give a view of the structure, randomly rotated about the fiber axis. We were, of course, making an optical transform, which is an optical analog of the X-ray diffraction fiber pattern.

Once we had this projection, which was covered with dots representing the positions of the atoms from the corridor experiments, we used a reducing pantograph to transfer the information onto a much smaller aluminum plate, and drilled a hole everywhere there was an atom, such that as you held this 6 by 3 inch piece of metal up to the light, you saw a reduced image of a whole repeat of the six-foot collagen model. We took the template to the laboratories at Courtaulds, in Maidenhead, UK, because they had a very good optical diffractometer. There we could look at the optical diffraction pattern—there was no phase problem with light—and what we got, using the optical diffractometer, was the continuous Fourier transform. In this way we could compare the experimental data—a fiber X-ray diffraction pattern of a collagen fiber—with the data obtained from the model and could see that there were no steric problems with the model.

We wrote up the manuscript. However, we then discovered, much to our embarrassment, that although the polyglycine chains in the polyglycine II structure pack in what looks like a hexagonal array, the actual unit cell is trigonal, not hexagonal. What that meant was there were two ways we could make a cluster of three chains out of the polyglycine II lattice. Thus we could built two different collagen conformers, collagen I and collagen II. This had escaped us initially so we had then to rewrite the paper before submitting it as a brief note to *Nature* [6]. Around about this time, (24 September, to be exact) a paper came out by Ramachandran and Kartha [11] which proposed a coiled coil structure for collagen. But it was not possible to build it satisfactorily because of bad van der Waals contacts—the structure was wrong.

After the second World War, a number of biophysics units had been set up around the UK by the Medical Research Council and many of these decided to go into biology, including those at King's College, London, under John Randall (who had a very distinguished career and who discovered the cavity magnetron during the War), and at Cambridge. It had been decided early on that Cambridge MRC biophysics unit would concentrate on crystalline proteins and King's College MRC unit would concentrate on DNA and collagen. Well, you know the story of the determination the structure of DNA. Sensitivities were already heightened and now we had gone ahead and solved the structure of collagen—it was a case of déjà vu over again.

Of course I did not know the background to all this but Francis said, "Well, we have to let the people at King's know." I said, "let' em read in *Nature*." Francis was, once again, a little sober about the matter; "No, no, we better let them know." I said "OK". John Kendrew happened to be going down to London to visit King's MRC unit so he said he would casually mention we had a structure. On hearing this from Kendrew they became very upset, having built a large number of models, all of which are wrong. They just hadn't tumbled to the connection between polyglycine (which in fairness, had just appeared in the literature) and collagen. They subsequently published a structure of collagen, which is very similar to ours [12].

A year later, in 1956, the Gelatin and Glue Research Association of Great Britain held a conference on recent advances in gelatin and glue research, for which we wrote a paper, outlining our structures [13]. (It was possible, from the collagen model, to understand how gelatin and glue form). By this time, we understood the collagen II was the correct model, not collagen I, both on experimental and stereochemical grounds. By measuring the intensity of the ten experimental layer lines, and where the peaks occurred, it was clear that the collagen II fit the data better. Furthermore, collagen I made a bad van der Waals contact if the chain was deformed ever so slightly, suggesting the structure would be less energetically favorable.

Around this time, we used to talk a lot about whether we should go on and do more work on collagen. After some discussion we decided that, since collagen is found in all animals but there is no collagen in plants, it was not really fundamental to life. It would make more sense to stay with the nucleic acids, which are.

This was a very exciting period and for that reason I stayed on with Francis until January 1956, and even brought my wife, Jane. Things evened up in 1959, though. Francis was coming over as a visiting professor at Harvard and he and his family were planning to rent the house of Bill Moffatt, who was going on sabbatical. Tragically, a week before they were scheduled to arrive, Bill had a coronary infarct and died. We had a large house in Cambridge, Massachusetts, and we simply took the Cricks, their two daughters and their French maid and had a very pleasant time together.

There is a very gratifying postscript to the story. Helen Berman, at Rutgers University, New Jersey, and colleagues determined the single crystal X-ray structure of a collagen-like peptide [14]. Her comparison with a half-a-dozen or so models for collagen revealed that our original coordinates were, after all, the closest to the real one.

References

1. Riddinhough, G. (1998) *Nature Structural Biology* **5**, 675.

2. Riddinhough, G. (1998) *Nature Structural Biology* **5**, 760.

3. Riddinhough, G. (1998) *Nature Structural Biology* **5**, 858-859.

4. Bamford, C.H. et. al. (1955) *Nature* **176**, 396-397.

5. Elliot A. & Malcolm, B. R. (1954) *Farad. Soc. Discuss*. (Maidenhead; 1954).

6. Crick, F.H. C. & Rich, A. (1955) *Nature* **176**, 915-916.

7. Rich, A. Davies, D. R., Crick, F.H. C. & Watson, J. D. (1961) *J. Mol. Biol.* **3**, 71-86.

8. Pauling, L. & Cory, R. B. (1951) *Proc. Natl. Acad. Sci. USA* **37**, 272-281.

9. Crick, F. H. C. (1954) *J. Chem. Phys*. **22**, 347.

10. Cowan, P. M., & McGavin, S. (1955) *Naturè* **176**, 501-503.

11. Ramachandran, G. N. & Kartha, G. (1955) *Nature* **176**, 593-595.

12. Cowan, P. M., McGavin, S. & North, A.C. T. (1955) *Nature* **176**, 1602-1604.

13. Rich, A. & Crick, F. H.C. (1957) *Recent advances in gelatin and glue research.* (Pergamon Press, London) pp.20-24.

14. Bella, J. Eaton, M. Brodsky, B. Berman, H.M. (1994) *Science* **266,** 75-81.

DISULFIDE BOND BASED SELF-ASSEMBLY OF PEPTIDES LEADING TO SPHEROIDAL CYCLIC TRIMERS

MIRIAM ROYO, MIQUEL ÀNGEL CONTRERAS, JUAN CEBRIÁN, ERNEST GIRALT, FERNANDO ALBERICIO, MIQUEL PONS
Departament de Química Orgànica
Universitat de Barcelona
Martí i Franquès, 1. E-08028-Barcelona, Spain

Abstract

Oxidation in the presence of TFE of several [19]- or [20]-peptides with two cysteine residues in positions i, i+17 and with two serine residues at positions i+8 and i+9 leads to the spontaneous formation, in high yields, of cyclic disulfide bonded trimers in which two of the chains are parallel and the third one is antiparallel. The serine residues in the center of the sequence are a necessary requirement for trimer formation which results from destabilization of the helical parallel cyclic dimer.

The cyclic structure is incompatible with a cylindrical coiled-coil helix bundle but is consistent with a quasi-spherical topology in which individual peptides, in an α-helical conformation, occupy non-intersecting edges of an octahedron and the disulfide bonds complete the cyclic structure along three of the remaining edges.

The easy synthetic access to this peptide scaffold opens the way to a number of possible applications, including aster molecules, dendrimer cores or immunological carriers. By using peptide mixtures, trimer libraries can be prepared. Disulfide bond scrambling under thermodynamic control also opens the way to the preparation of virtual trimer libraries.

1. Covalent peptide self-assembly

The self assembly concept is changing the way chemists face the quest for molecules of increasing complexity [1]. The synthetic effort associated with the preparation of non-trivial large molecules raises steeply with the size of the target, but the information content that can be included in the molecular structure also increases with the molecular complexity. Thus, structure-directed spontaneous association ("self-assembly") of properly designed intermediates can be used to provide molecules of larger size and complexity at a fraction of the cost. This is illustrated schematically in Figure 1.

Classically, self assembly is associated with non-covalent bonds. Weak, reversible bonding ensures thermodynamic control of product formation . Additionally, the

A. Aggeli et al. (eds.), Self-assembling Peptide Systems in Biology, Medicine and Engineering, 243–256.

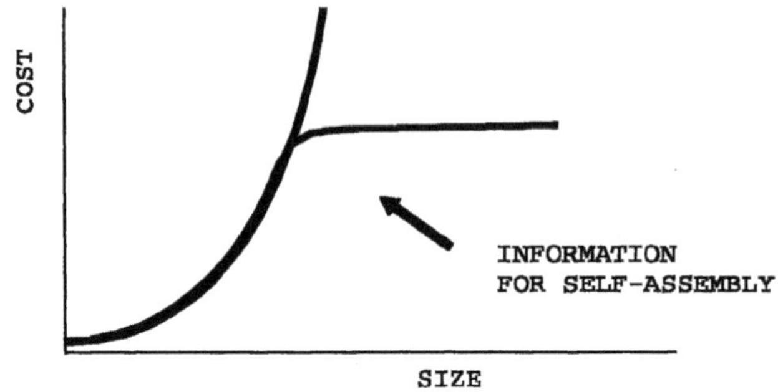

Figure 1. The self assembly concept. The information content of synthetic intermediates is used to direct the spontaneous assembly of the complete molecule at no extra cost.

requirement that several weak interactions are simultaneously present to stabilize the final product, introduces strict structural requirements that direct the formation of a unique product under a given set of experimental conditions (solvent, temperature, pH). However, non-covalent bonding may hinder the use of self-assembled products in real situations where the environmental conditions may change. Also, the structural requirements for non-covalent self-assembly complicate the search for new compounds with improved properties, as small variations in the structure of the final product may require the complete redesign of the synthetic strategy for its preparation.

Covalent stabilization of the assembly may lead to more robust systems with a larger potential for application. Of the possible alternatives, disulfide bonds have a number of desirable properties as connecting units between individual constituents in supramolecular assemblies: i) Disulfide bonds are easily formed under oxidizing conditions from the free thiols. ii) Disulfide bonds are stable at low pH. iii) The sulfur-sulfur bond has a rather long bond-distance and the C-S-S-C dihedral angle is restricted to values close to ±90° with an interconversion barrier of ca. 40 KJ mol^{-1} , intermediate between that of a single bond and a partial double bond, such as the peptide bond. iv) The entropic cost of immobilization of a S-S bond has been estimated to be around 3.5 J mol^{-1} K^{-1} in CH_3-S-S-CH_3 and is the lowest in the series of molecules with the formula CH_3-X-Y-CH_3 with X,Y = C, Si, S, O, N, P [2].

Peptides are starting to play a key role in different fields outside the strict study of natural biological systems. The knowledge accumulated on the structural requirements for secondary structure formation and packing allow for the de novo design of peptide scaffolds that may be conveniently decorated to provide new functionality [3]. In addition, peptide materials may be conveniently prepared either biotechnologically or by standarized chemical methods.

Cysteine containing peptides can be oxidized to form inter- or intramolecular disulfide bonds [4]. Disulfide bond scrambling can take place at high pH in the presence of excess free thiols. Thus, when alternative bonding patterns are possible, the thermodynamically controlled product is formed. On the other hand, disulfide bonds are rather stable at low pH in the absence of strong nucleophiles. Thus, intermolecular disulfide bond formation may lead to a particular class of self-assembly that leads to products that are covalently linked and thus stable under environmental conditions (e.g. solvents, pH...) completely different from those that favored their spontaneous formation.

Disulfide bonds are flexible and compatible with a range of local conformations. Thus, disulfide bonded peptide assemblies are best characterized by their topological properties, rather than by a unique conformation that, if exists, is determined by additional non-covalent interactions. The conceptual similarities between disulfide bonded peptide oligomers and non-covalently assembled systems are summarized in the following chart.

NON COVALENT SELF-ASSEMBLY	DISULFIDE BONDED PEPTIDE OLIGOMERS
Reversible Cooperative effects Strict structural requirements	Reversibly formed at high pH Multiple disulfide bonds introduce selectivity Maintain the topological integrity High flexibility
Preparing similar molecules may require complete redesign	Allow for sequence variability
Sensitive to environment	Stable at low pH

2. Spontaneous cyclic trimer formation by bis-cysteine peptides

In previous work from our group [5], we had studied the spontaneous oxidation of peptide **1**.

1: Ac-CXKLHAELSSLEAHLKXCG-NH_2 (X= Aib)

Peptide **1** had been designed to have a high propensity to adopt an amphipathic α-helical conformation by placing hydrophobic leucine residues separated by three or four residues along the sequence. Oppositely charged residues in i, i+4 relative positions [6], and α-aminoisobutiric acid (Aib) residues near both ends of the sequence were also used to increase the overall helix propensity [7]. The original sequence was chosen to be a palindrome as it was part of experimental efforts to differentiate electrostatic effects arising from the peptide backbone and the side chains. Cysteine residues in positions 1 and 18, that correspond to hydrophobic sites as defined by the leucine pattern, were allowed to form disulfide bonds at pH 8 at different concentrations and in the presence of increasing amounts of the α-helix inducing solvent, trifluoroethanol (TFE).

Only three products were formed in any of the conditions studied: a cyclic monomer, with an intramolecular disulfide bond, an antiparallel cyclic dimer, and a cyclic trimer. The products were characterized by their mass spectra and, in the case of the dimer, by comparison with compounds prepared univocally using regioselective disulfide bond forming methods [8]. The topology of the trimer was determined from the analysis of the products of tryptic hydrolysis. Tryptic fragments containing Cys1-Cys1' bonds as well as Cys1-Cys18' disulfides were detected proving that the trimer has two peptide chains parallel to each other and antiparallel to the third one (Figure 2).

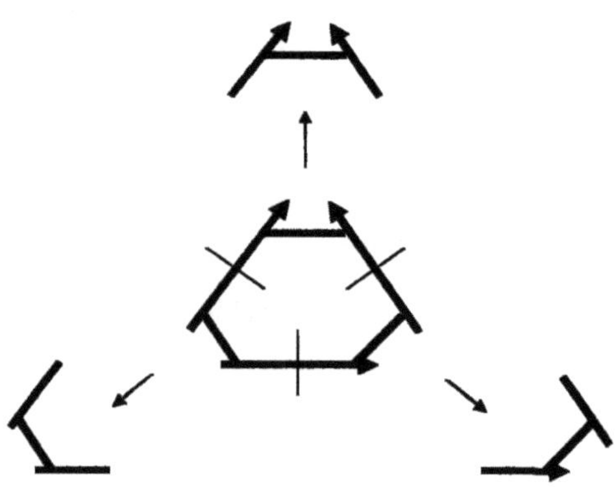

Figure 2. Enzymatic hydrolysis of a parallel-antiparallel trimer provides three different disulfide bonded fragments that were characterized by mass spectrometry.

Circular dichroism and NMR studies showed that the trimer is nearly 100% helical in TFE but has no tertiary structure.

The yields of monomer, antiparallel dimer and trimer obtained under different oxidation conditions are plotted in Figure 3.

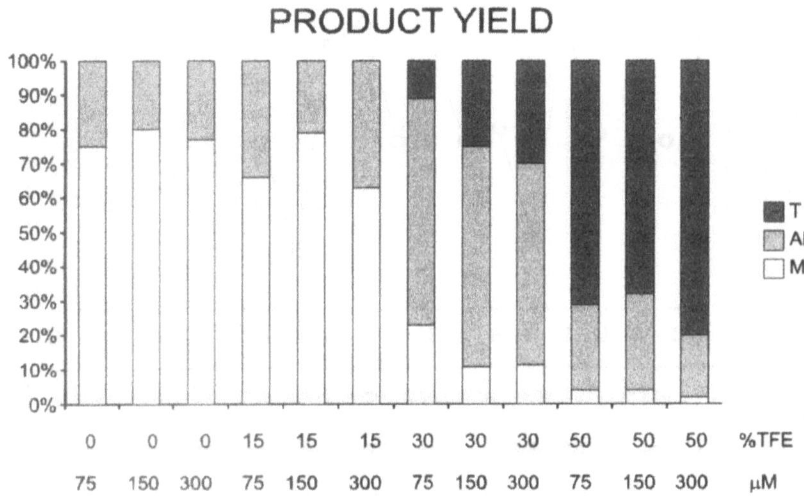

Figure 3. Yield of cyclic monomer (M), antiparallel dimer (AD) and trimer (T) obtained by spontaneous oxidation of **1** at different peptide concentrations and in the presence of different percentages of TFE.

Trimer formation becomes the dominant process in the presence of 50% TFE and, at lower percentages of TFE, when peptide concentration is high. The monomer is the main product in the absence of TFE independently of the concentration of peptide used.

The conformational preferences of peptide **1** were studied by NMR using a model peptide with the same sequence but with the two cysteine residues replaced by leucine. Representative results are shown in Figure 3.

A helical conformation can be detected by a series of strong sequential NN(i, i+1) and medium intensity αN(i, i+3) and αN(i, i+4) NOEs. CHα conformational shifts, i.e. the difference between the observed chemical shift and the one expected from the same residue in a disordered peptide provide an additional indication of the presence of secondary structures. For α-helices conformational shifts are negative. According to these criteria, the model peptide dissolved in a helix inducing solvent, displays two helical regions separated by a central region that contains the two serine residues and has a low helicity. According to the conformational shifts, the *N*-terminal half of the molecule appears to be less helical than the *C*-terminal moiety.

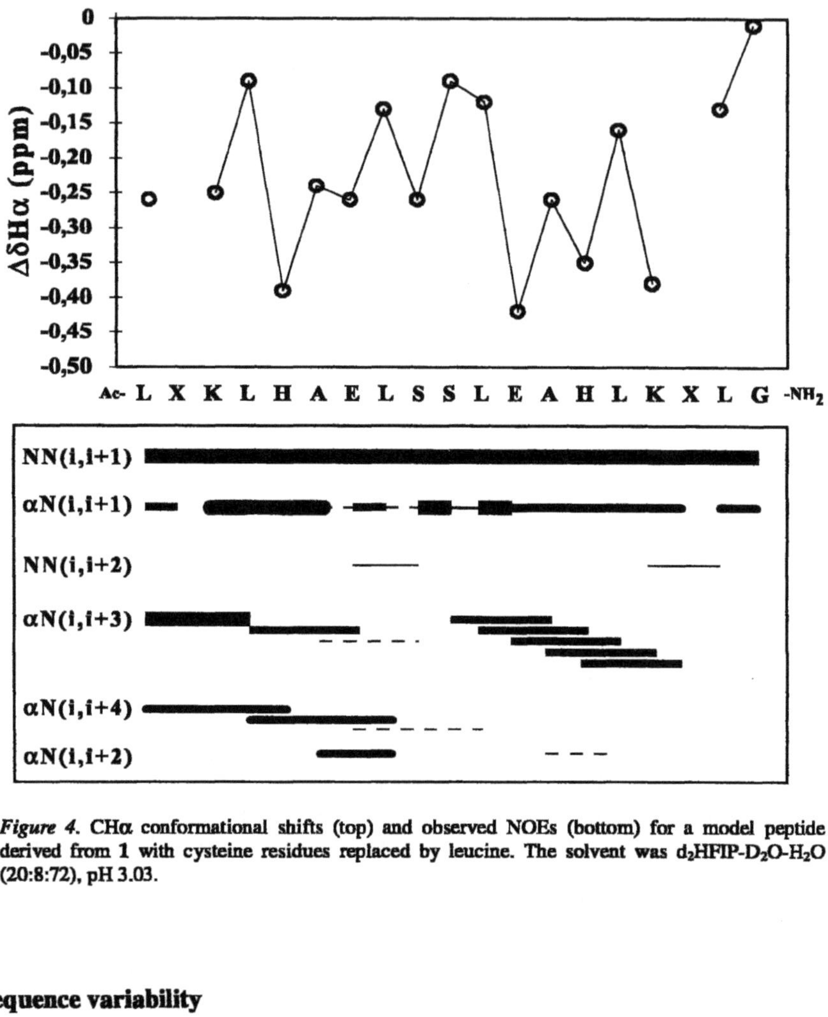

Figure 4. CHα conformational shifts (top) and observed NOEs (bottom) for a model peptide derived from **1** with cysteine residues replaced by leucine. The solvent was d_2HFIP-D_2O-H_2O (20:8:72), pH 3.03.

3. Sequence variability

In order to investigate the robustness of the spontaneous trimerization process to changes in peptide sequence we have prepared a number of analogs of **1** in which a) the propensity of the *N*-terminal half of the molecule was modified, c) hydrophobic residues were replaced by phenylalanine, isoleucine or alanine, and c) the central serine residues were replaced by alanine. Additionally, in some of the peptides Aib residues were replaced by alanine, in order to produce peptides containing only natural aminoacids. The sequences of the peptides studied are presented in Table 1.

TABLE 1. Sequence variants of peptide **1**. Modified residues are shown in boldface

Peptide	Sequence	Fraction of trimer[a]
1	Ac-CXKLHAELSSLEAHLKXCG-NH_2	80%
AG	Ac-**G**C**A**KLHAELSSLEAHLK**A**CG-NH_2	75%
XN	Ac-**N**CXKLHAELSSLEAHLKXCG-NH_2	77%
XEK	Ac-CX**E**LHA**K**LSSLEAHLKXCG-NH_2	61%
XR	Ac-CX**R**LHAELSSLEAHL**R**XCG-NH_2	87%
XQ	Ac-CXKLHA**Q**LSSL**Q**AHLKXCG-NH_2	54%
XF4	Ac-CXK**F**HAELSSLEAHLKXCG-NH_2	73%
XF15	Ac-CXKLHAELSSLEAH**F**KXCG-NH_2	64%
AF15	Ac-C**A**KLHAELSSLEAH**F**K**A**CG-NH_2	69%
AF11	Ac-C**A**KLHAELSS**F**EAHLK**A**CG-NH_2	77%
AF4F15	Ac-C**A**K**F**HAELSSLEAH**F**K**A**CG-NH_2	85%
AF8	Ac-C**A**KLHAE**F**SSLEAHLK**A**CG-NH_2	45%
AI8	Ac-C**A**KLHAE**I** SSLEAHLK**A**CG-NH_2	78%
AA8	Ac-C**A**KLHAE**A**SSLEAHLK**A**CG-NH_2	29%
AGA10A11	Ac-**G**C**A**KLHAEL**AA**LEAHLK**A**CG-NH_2	0%

a) Oxidation was carried out at 300 μM peptide concentration in 50% TFE pH 8 until complete conversion of the starting product. The fraction of trimer was determined by integration of the HPLC peaks of trimer, dimer and cyclic monomer. Oxidation of some of the peptides yielded also polymeric material. Total yield of trimer ranged from 87% (**XR**) to 19% (**AA8**).

The different helical content of the two halves of the molecules can be rationalized by the different orientation of the salt bridge between lysine and glutamic acid side chains. Additionally, one of the serine residues could act as an *N*-cap [9] for the second helix, but the *N*-terminal helix does not have this possibility. In order to increase the helix propensity, an asparagine residue that could act as an *N*-cap for the first helix was introduced in peptide **XN** and the first i, i+4 lysine-glutamic acid pair was inverted in peptide **XEK**[6]. In this arrangement the negatively charged side chain is expected to interact with the positive part of the helix macrodipole. In both cases, spontaneous oxidation in the presence of 50% TFE afforded the expected trimer in yields comparable to those obtained using peptide **1**.

Interestingly, oxidation of **XEK** produced 38% of antiparallel dimer in the presence of 50% TFE and 64% in aqueous buffer. Yields of antiparallel dimer from peptide **1** are 18% and 23% under the same conditions. This suggests that electrostatic interactions between the charged side chains stabilize the antiparallel dimer and cause a reduction in the fraction of trimer [10].

The possible role of charged residues was further explored by replacing the glutamic acid residues by glutamine in peptide **XQ** or by substituting arginine residues for lysine in peptide **XR**. The substitution by glutamine eliminates the possible contribution of a salt bridge, although glutamine can still participate in hydrogen bond formation with the charged lysine side chain. In 50% TFE **XR** gives the corresponding cyclic trimer in very good yield but **XQ** has a reduced tendency to form trimer. The removal of the two salt bridges present in **1** may explain the lower trimer yield. However, we have found that changes in position 8, adjacent to one of the sites modified in peptide **XQ**, had a strong effect in the oxidation products (see below). Therefore, the reasons for the lower trimerization propensity of **XQ** remain ambiguous.

Finally, Aib residues, that stabilize the helix ends in water, were replaced by alanine in several peptides without any drastic effect in the formation of the corresponding cyclic trimers.

We can conclude that sequence variations that modify the helix propensity of **1** have only minor effects in the formation of cyclic trimer in the strong helix forming conditions provided by 50% TFE. In aqueous buffer, a cyclic monomer is the main product in all cases, except for peptide **XEK** that gives preferentially antiparallel dimer.

We next explored the hydrophobic positions. Flexible leucine side chains are known to give unstructured hydrophobic cores in coiled coil peptides [11], and were replaced by phenylalanine in different positions. Peptides modified at positions 4, 11 and/or 15 give a product distribution similar to the one observed for **1**. However, replacing leucine in position 8 by other hydrophobic residues has a strong effect in the relative yields of oxidation products. Peptides **AF8** and **AA8**, with phenylalanine or alanine respectively produce substantially less trimer than **1**. Peptide **AI8**, with isoleucine, and peptide **1** form the corresponding trimers in comparable amounts. However, no antiparallel dimer was observed when **AI8** was oxidized. Also, dimer formation was completely suppressed in **AF8**. The results are summarized in Figure 5.

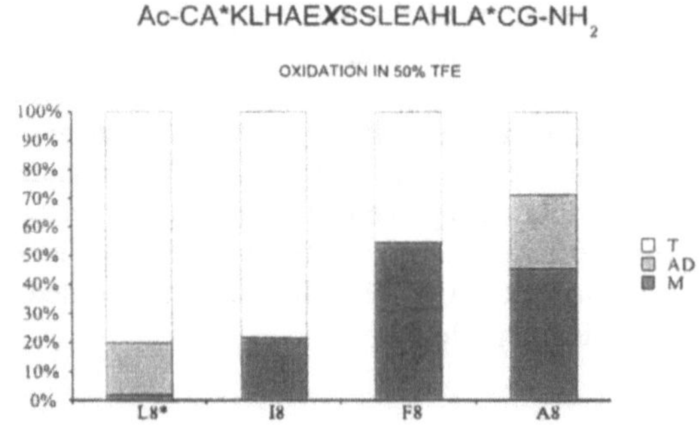

Figure 5. Oxidation products in 50% TFE of peptides containing different hydrophobic residues in position 8. All peptides have alanine residues in positions 2 and 16 except for **1** that has Aib in these positions. Peptide concentration was 300 μM.

The different trimers were examined by NMR. All of them gave a single set of signals. Also, no aromatic induced shifts could be observed suggesting that they do not have an structured hydrophobic core.

4. Serine residues in the central positions are essential for trimer formation

Qualitatively, the same oxidation pattern was observed for all peptides included in table 1 with the exception of **AGA10A11**: a cyclic monomer is formed in the absence of TFE, a trimer is the main product in 50% TFE and no parallel dimer is obtained.

Peptide **AGA10A11** with alanine in the two central positions shows a completely different behavior. Yields of cyclic monomer are very low, even in the absence of TFE and no trimer is formed in any of the conditions studied. On the other hand, both parallel and antiparallel dimers are formed, and their relative proportions depend on peptide concentration and in the percentage of TFE added. The results are shown in Figure 6.

In 50% TFE the antiparallel dimer is formed in 70-80% yield, independently of the peptide concentration used. In the absence of TFE, the antiparallel dimer is the main product (around 60%) when the oxidation is carried out at low concentration (75 μM) but the parallel dimer dominates, with also ca. 60% yield, in the oxidation products when the reaction is carried out at a peptide concentration of 300 μM.

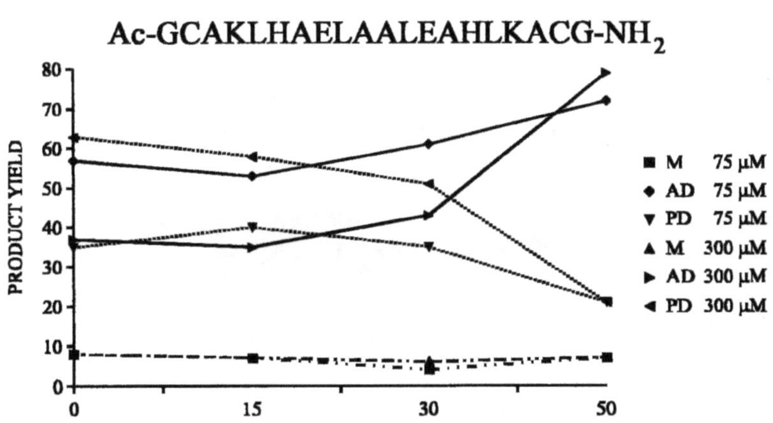

Figure 6. Oxidation products of **AGA10A11** under different experimental conditions. The topologies of the two dimers, parallel (PD), and antiparallel (AD), were determined by enzymatic digestion. No trimer was formed.

5. Trimer formation arises from frustrated parallel dimers.

These results indicate that serine residues in the central position of the sequence are required for trimer formation and are also responsible for the absence of parallel dimer. We believe that the two observations are related.

NMR results using a model of peptide **AGA10A11** with cysteine residues replaced by leucine (Figure 7) show a number of medium range NOEs that expand the complete peptide sequence, including the central part. This is an evidence indicating that **AGA10A11** forms a continuous helix in which the two cysteine residues are aligned along one side of the helix. Thus, both parallel and antiparallel dimers can be formed. In peptide **1** a discontinuity of the helix in the central region prevents a correct alignment of the two cysteines. In the antiparallel topology, formation of the cyclic dimer is still possible as the distortions in each peptide chain compensate each other. On the other hand, in a parallel arrangement, distortions in both chains add together preventing the formation of the second disulfide bond

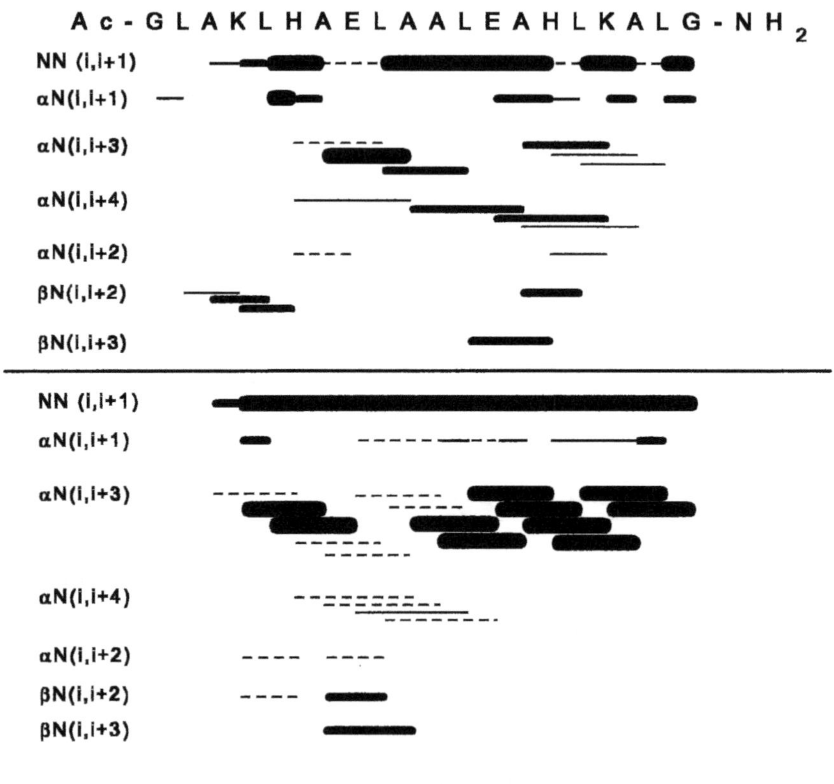

Figure 7. NOE results for **AGA10A11** in water (top) and in 20% HFIP (bottom).

This is supported by independent evidence showing that parallel and antiparallel "open" dimers of **1**, i.e. with a single disulfide bond, are formed in statistical yields. Also in agreement with this hypothesis, we found that in the parallel dimer, synthesized by regioselective disulfide bond formation, the helix conformation is destabilized as compared to the open dimers or the antiparallel cyclic dimer [5]. Thus, after formation of the first disulfide bond in a parallel arrangement, formation of the second disulfide bridge, leading to a cyclic parallel dimer is disfavored and the open intermediate reacts with a third peptide molecule giving the cyclic trimer (Figure 8) . This explanation also accounts for the observed parallel-antiparallel topology

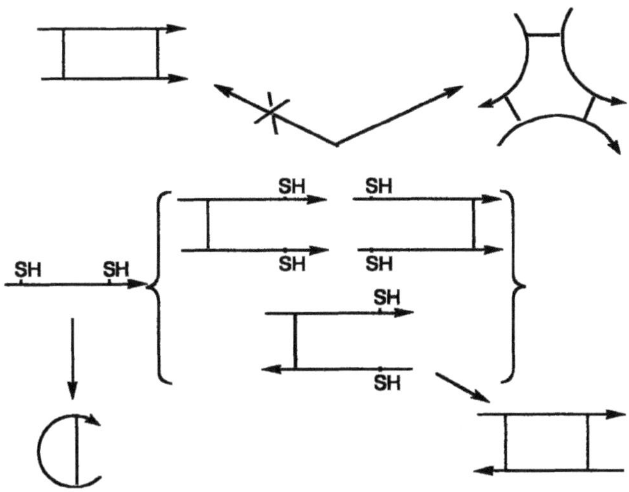

Figure 8. Schematic representation of the processes leading to the formation of parallel-antiparallel cyclic trimers.

The topological constraints imposed by the three disulfide bonds in the cyclic trimer give a geometry very different than the one found in classical coiled-coil three helix bundles [12]. Coiled-coils have an overall cylindrical geometry and the axes of the helices typically form a small angle. This cylindrical arrangement is not compatible with the formation of three disulfide bonds between the ends of helical peptides. In contrast, disulfide bonded cyclic trimers probably have an overall quasi-spherical arrangement with helix axes crossing at a wider angle, as suggested by the polyhedron model of Murzin and Finkelstein [13], where three cylindrical objects (the three individual peptide helices) pack together with their long axes along the non-intersecting vertices of an octahedron. (Figure 9).

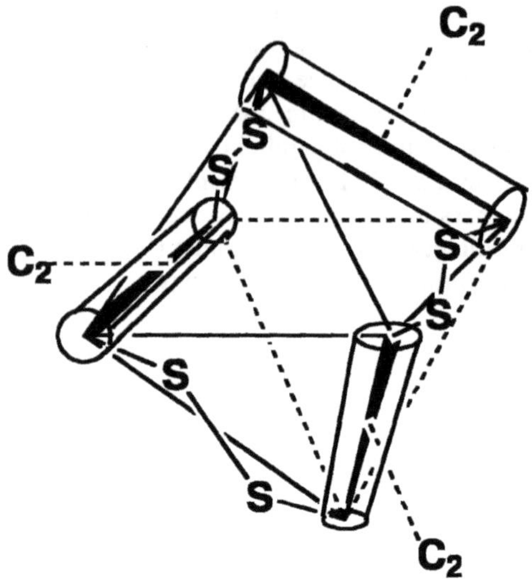

Figure 9. Octahedral model for the quasi-spherical packing of three helices. An enantiomorphic arrangement is also possible.

6. Applications prospects

The two distinctive features of the octahedron model for the structure of the trimer are the hydrophobic cavity defined by the three amphipathic helices and the radial disposition of the hydrophilic side chains.

The hydrophobic interior of the trimer readily accommodates the substitution of at least six phenylalanine residues for leucines but, even in this case, the resulting trimer is highly flexible. On the other hand, the six lysine or the six glutamic acid side chains can be made to react with other molecules, including other peptide chains, leading to aster molecules, sixth order dendrimer cores, or carriers for immunization.

An specially appealing aspect of the self assembly process leading to peptide trimers is the possibility of generating virtual combinatorial libraries [14]. (Figure 10).

We have shown that peptides containing different polar residues or even hydrophobic ones except in position 8, form trimers in similar yields. When a mixture of two bis-cysteine peptides is allowed to react under conditions favoring the formation of trimers, up to eight (2^3) different molecules can be formed because the parallel-antiparallel topology makes the three chains in a trimer, non-equivalent. In a mixture of n-peptides, n^3 molecules can be formed, although not all of them are necessarily observed. At high pH, disulfide bond scrambling allows the interconversion of the different molecules potentially present in the library. Therefore, if one particular species is favored, e.g. because it binds to a particular target, this species will predominate. However, in the presence of a different target, the same starting mixture is likely to produce a different predominating species. We are presently exploring the attachment of additional peptide chains to the surface of trimers to be used as immunogens. This could open the way to a combinatorial approach to the development of synthetic vaccines based on discontinuous epitopes.

Figure 10. Virtual combinatorial libraries using self assembling trimers.

7. Acknowledgements

This project was supported by the Dirección General de Enseñanza Superior (DGES, PB97-0933) and by the Generalitat de Catalunya (Centre de Referència de Biotecnologia). M.R. holds a RED postdoctoral fellowship. We acknowledge the use of the facilities of Serveis Cientific-Tècnics de la Universitat de Barcelona.

8. References

1. a) Ungaro, R. and Dalcanale, E. (eds.) (1999) *Supramolecular Science: Where It Is and Where It Is Going*. Kluwer, Dordrecht. b) Lehn, J.-M. (1995) *Supramolecular Chemistry – Concepts and Perspectives*, VCH, Weinheim. c) Rebek, J, Jr. (1996) *Chem. Soc. Rev.* 255. (d) Rebek, J. Jr. (1996) *Acta Chim. Scand.* **50**, 707.
2. Mammen, M., Shakhnovich, E.I. and Whitesides, G.M. (1998) *J. Org. Chem.* **63**, 3168.
3. a) Handel, T. and DeGrado, W.F. (1990) *J. Am. Chem. Soc.* **112**, 6710. b) Choma, C.T., Lear, J.D., Nelson, M.J., Dutton, P.L., Robertson, D.E. and DeGrado, W.F. (1994) *J. Am. Chem. Soc.* **116**, 856. c) Gibney, B.R., Rabanal, F., Skalicky, J., Wand, J.A. and Dutton, P.L. (1997) *J.Am. Chem. Soc.*, **119** 2323.
4. For a review on disulfide bond formation see: Andreu, D., Albericio, F., Solé, N., Munson, M., Ferrer, M. and Barany, G. (1994) In *Methods in Molecular Biology, Vol 35: Peptide Synthesis Protocols*; Pennington, M.W., Dunn, B.M., Eds.; Humana Press: Totowa, NJ,; pp 91-169.
5. Royo, M., Contreras, M.A., Giralt, E., Albericio, F. and Pons, M. (1998) *J. Am. Chem. Soc.* **120**, 6639.
6. Marqusee, S. and Baldwin, R.L. (1987) *Proc. Natl. Acad. Sci. USA* **84**, 8898.
7. Karle, I.L. and Balaram, P. (1990) *Biochemistry* **29**, 6748.
8. a) Ruiz-Gayo, M., Albericio, F., Pons, M.,Royo, M., Pedroso, E. and Giralt, E. (1988) *Tetrahedron Lett.* **29**, 3845. b) Ruiz-Gayo, M., Royo, M., Fernandez, I., Albericio, F., Giralt, E. and Pons, M., (1993) *J. Org. Chem.* **58**, 6319.
9. a) Harper, E.T. and Rose, G.D. (1993) *Biochemistry*, **32**, 7605. b) Chakrabartty, A., Kortemme, T. and Baldwin, R.L. (1994) *Protein Sci.* **3**, 843.
10. Monera, O.D., Zhou, N.F., Kay, C.M. and Hodges, R.S. (1993) *J. Biol. Chem.* **268**, 19218.
11. Betz, S.F., Raleigh, D.P. and DeGrado, W.F., (1993) *Curr. Opin. Struct. Biol.* **3**, 601.
12. Lovejoy, B., Choe, S., Cascio, D., McRorie, D.K., DeGrado, W. F. and Eisenberg, D. , (1993) *Science* **259**, 1288.
13. Murzin, A.G. and Finkelstein, A.V. (1988) *J. Mol. Biol.* **204**, 749.
14. (a) Sanders, J.K.M (**1998**) *Chem. Eur. J.* *4*, 1378. (b) Crego-Calama, M., Hulst, R., Fokkens, R., Nibbering, N.M. M., Timmerman, P. and Reinhoudt, D.N. (1988) *Chem. Commun.* **1998**, 1021.

A NEW CIRCULAR HELICOID-TYPE SEQUENTIAL OLIGOPEPTIDE CARRIER FOR ASSEMBLING MULTIPLE ANTIGENIC PEPTIDES

MARIA SAKARELLOS-DAITSIOTIS, VASSILIOS TSIKARIS AND CONSTANTINOS SAKARELLOS
Department of Chemistry, University of Ioannina, P.O.BOX 1186, 54110 Ioannina, Greece

Abstract

A novel class of oligopeptide carriers termed Sequential Oligopeptide Carriers (SOCs) is presented. SOCs are formed either by the Lys-Aib-Gly or the Aib-Lys-Aib-Gly sequential motif. By varying the number of the repeating units, carriers of the general formula either SOC_n-I, $(Lys\text{-}Aib\text{-}Gly)_n$, or SOC_n-II, $(Aib\text{-}Lys\text{-}Aib\text{-}Gly)_n$, where n=2-7, are prepared. The carriers adopt a predetermined secondary structure of 3_{10}-helix, which is given by the presence of Aib, an unnatural amino acid with a known propensity to induce ordered helicoid backbone. This helicoidal structure contributes to the reduction of steric hindrance and conformational restrictions of the carrier, and thus allows antigenic peptides to retain their original structure as confirmed by 1HNMR and molecular modeling studies. Four antigenic sequences were selected to assess the utility of SOCs-conjugates, either as antigens in solid-phase immunoassays, or as immunogens in eliciting specific anti-peptide antibodies and producing an immune spreading:
(i) the [Ala^{76}] decapeptide derivative of the main immunogenic region (MIR, Trp^{67}-Asn-Pro-Ala-Asp-Tyr-Gly-Gly-Lys^{76}) of the Torpedo nicotinic acetylcholine receptor (AChR),
(ii) the gp63-SRYD octapeptide fragment (Ile^{250}-Ala-Ser-Arg-Tyr-Asp-Gln-Leu^{257}) of the gp63, the major surface glycoprotein of Leishmania,
(iii) the B-cell epitopes of the La/SSB antigen, TLHKAFKGSIFVVFDSIESA (145-164), ANNGNLQLRNKEVTWEVLEG (289-308), VTWEVLEGEVEKEALKKI (301-318) and GSGKGKVQFQGKKTKF (349-364), major targets of anti-La/SSB antibodies in primary Sjögren's syndrome (pSS) and systemic lupus erythematosus (SLE), and
(iv) the PPGMRPP heptapeptide, found in several copies in the Sm and U1RNP antigens, main targets of anti-Sm/U1RNP antibodies in SLE. The spatial arrangement and the regular secondary structure of the multiepitopic SOCs-conjugates result in such an orientation and accessibility of the antigens that favour their recognition by antibodies in immunoassays and by the cells of the immune process.

A. Aggeli et al. (eds.), Self-assembling Peptide Systems in Biology, Medicine and Engineering, 257–271.

1. Introduction

Short synthetic peptides can mimic linear, and, to a certain extent, conformational epitopes and they can be used in solid-phase immuno assays or vaccination instead of complex antigens that are often difficult to purify and prepare in large quantities. It is also possible that peptides, rather than whole isolated proteins, resemble more closely short regions accessible at the surface of macromolecular complexes, which may be involved in the immune response. Following the conventional approaches the peptide fragment (epitope or antigenic determinant) must be conjugated to a protein carrier, as for example bovine serum albumin or tetanus toxin. However, several disadvantages, such as immune response to the carrier and ambiguous composition and structure, frequently result from this conjugation [1].

Construction of an artificial carrier for inducing an immune response must fulfill a variety of requirements, as for example, the simultaneous assembling of B and T-cell epitopes, as well as a linkage between class II MHC and T-cell receptors. A central issue is the size and topographical nature of the antigenic parts anchored to the carrier, since B-cells recognize antigens in a highly conformation-dependent fashion [2]. The lysine core matrix (multiple antigenic peptide MAP), the Pam_3-Cys-Ser (a palmitoyloxy lipopeptide) and the template assembled synthetic protein (TASP) concept are some artificial carriers appeared the last decade [3-5]. For review see references [6,7], where the physico-chemical and biological properties of peptide-protein conjugates, MAPs, TASPs and MAP-SOCs are given, for comparison.

2. Concept and design of the Sequential Oligopeptide Carriers (SOCs)

A new class of carriers named Sequential Oligopeptide Carriers (SOCs) has been recently modeled from our group with the aim to optimize epitope presentation and help in the reconstruction and/or mimicking of a native epitope [8-10]. One of our major guidelines was to obtain constructions with predetermined three-dimensional structure, so that the attached antigenic peptides would obtain a defined spatial orientation. Our support, formed by the repetitive Lys-Aib-Gly moiety, incorporates lysine for anchoring of the antigen, the α-aminoisobutyric residue for inducing a helicoid structure of the peptide backbone and glycine for its small stereochemical volume. The Sequential Oligopeptide Carrier Ac-(Lys-Aib-Gly)$_n$ (n=2-7), named SOC$_n$-I, analyzed in the following, adopts a regular secondary structure allowing thus, the antigenic peptides to retain their natural «active» conformation without interacting with each other or with the carrier (Fig. 1a). The structure regularity of the carrier (distorted 3_{10} helix), and the absence of conformational restrictions and steric hindrances for the constructed conjugates favour the antibody recognition and the generation of potent immunogens [8-11].

A second carrier of antigenic peptides, belonging to the same class with the precedent one, is also presented (Fig 1b). It incorporates two Aib residues in each repetitive moiety: Ac-(Aib-Lys-Aib-Gly)$_n$, (SOC$_n$-II) (n=2-4) [12,13]. Previous studies have correlated the high content (≥50%) of Aib in a peptide sequence, with pronounced 3_{10}-helical structures [14]. The Sequential Oligopeptide Carriers (SOC$_n$-II) assume a

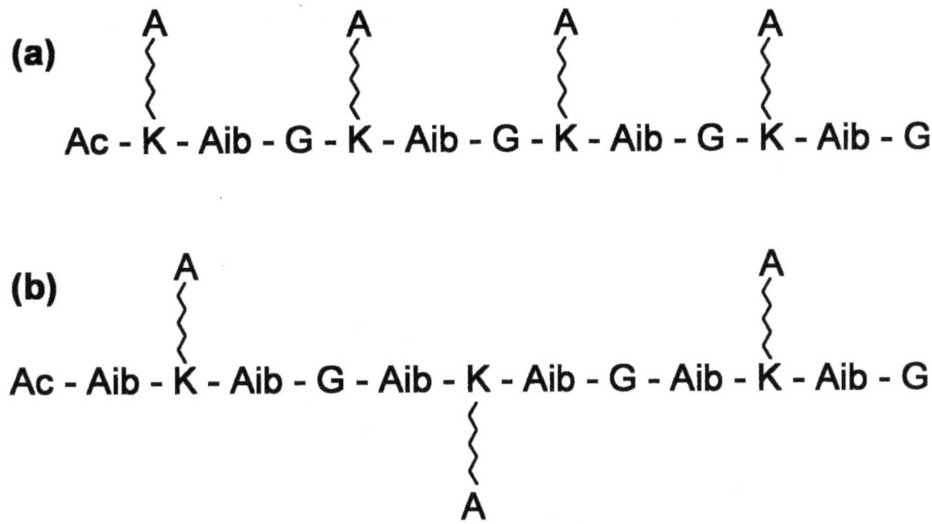

Figure 1. Schematic representation of the sequential oligopeptide carriers SOC_n-I (**a**), SOC_n-II (**b**) with the anchored antigenic/immunogenic peptides A: [A^{76}]-MIR, gp63 - SRYD, PPGMRPP and La/$SSB_{349-364}$

more rigid, compared to the previous one, and regular 3_{10}-helical secondary structure without steric hindrances between the attached antigens. The consecutive lysines bearing the antigens are located in every fourth position (i, i+4, i+8, ..), whereas in SOC_n-I were positioned in every third one (i, i+3, i+6, ...). This construction compared to SOC_n-I provides also information about the significance of the lysine positions along the oligopeptide backbone for an optimal antibody recognition of the anchored antigens.

3. Selected applications of SOC_n-I and SOC_n-II.

Four peptide antigens have been selected for applying the SOCs:

(i) The [Ala^{76}] analogues of the Main Immunogenic Region (MIR, Trp^{67}-Asn-Pro- Ala-Asp-Tyr-Gly-Gly-Ile-Lys^{76}) of the α-subunit of the Torpedo nicotinic acetylcholine receptor (AChR), against which is directed the majority of the anti-AChR autoantibodies from myasthenic patients [11].

(ii) The Ile^{250}-Ala-Ser-Arg-Tyr-Asp-Gln-Leu^{257} fragment (gp63-SRYD) of the major surface glycoprotein of Leishmania, gp63, which efficiently inhibits parasite attachment to the macrophage receptors [15].

(iii) The B-cell epitopes of the La/SSB antigen, TLHKAFKGSIFVVFDSIESA (145-164), NNGNLQLRNKEVTWEVLEG (289308), VTWEVLEGEVEKEALKKI (301-318) and GSGKGKVQFQGKKTKF (349-364), which are a major target of anti-La/SSB antibodies in primary Sjögren's syndrome (pSS) and SLE [16].

(iv) The PPGMRPP sequence, found in several copies in the Sm and U1RNP autoantigens, is the main target of anti-Sm autoantibodies, which are marker antibodies for systemic lupus erythematosus (SLE) [17]. This epitope was coupled to both carriers (SOC_n-I and SOC_n-II) separately and used as antigenic substrate.

Scheme 1. Step-by-step protocol for the synthesis of the sequential oligopeptide Carrier (SOC).

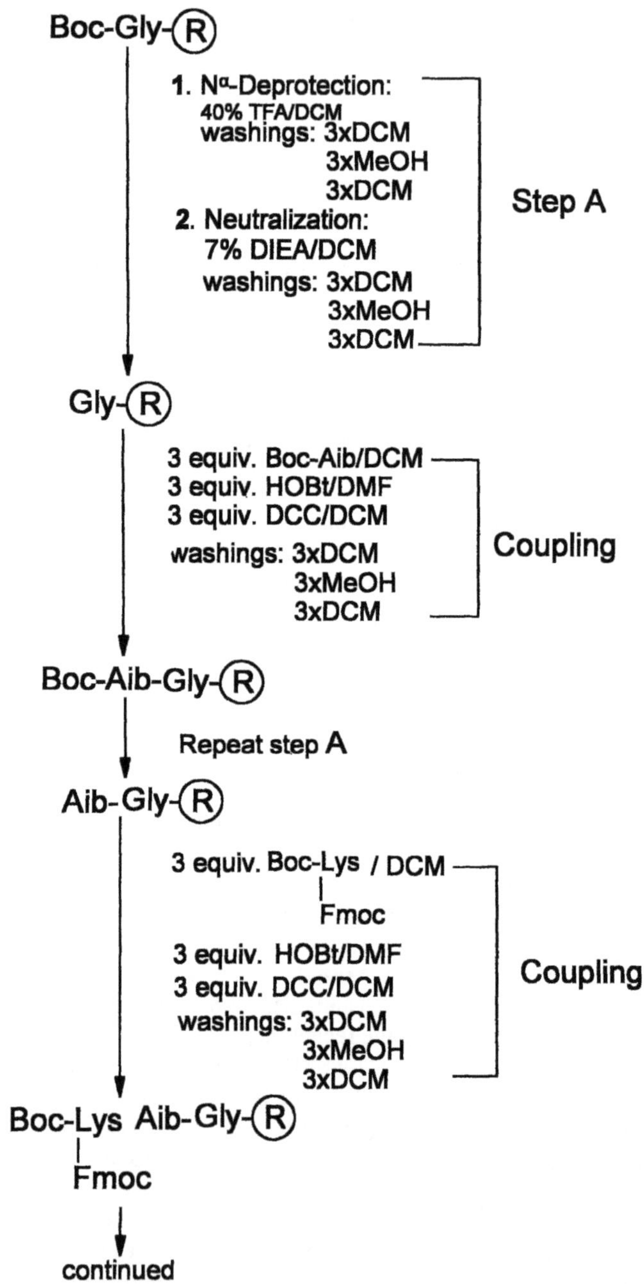

continued

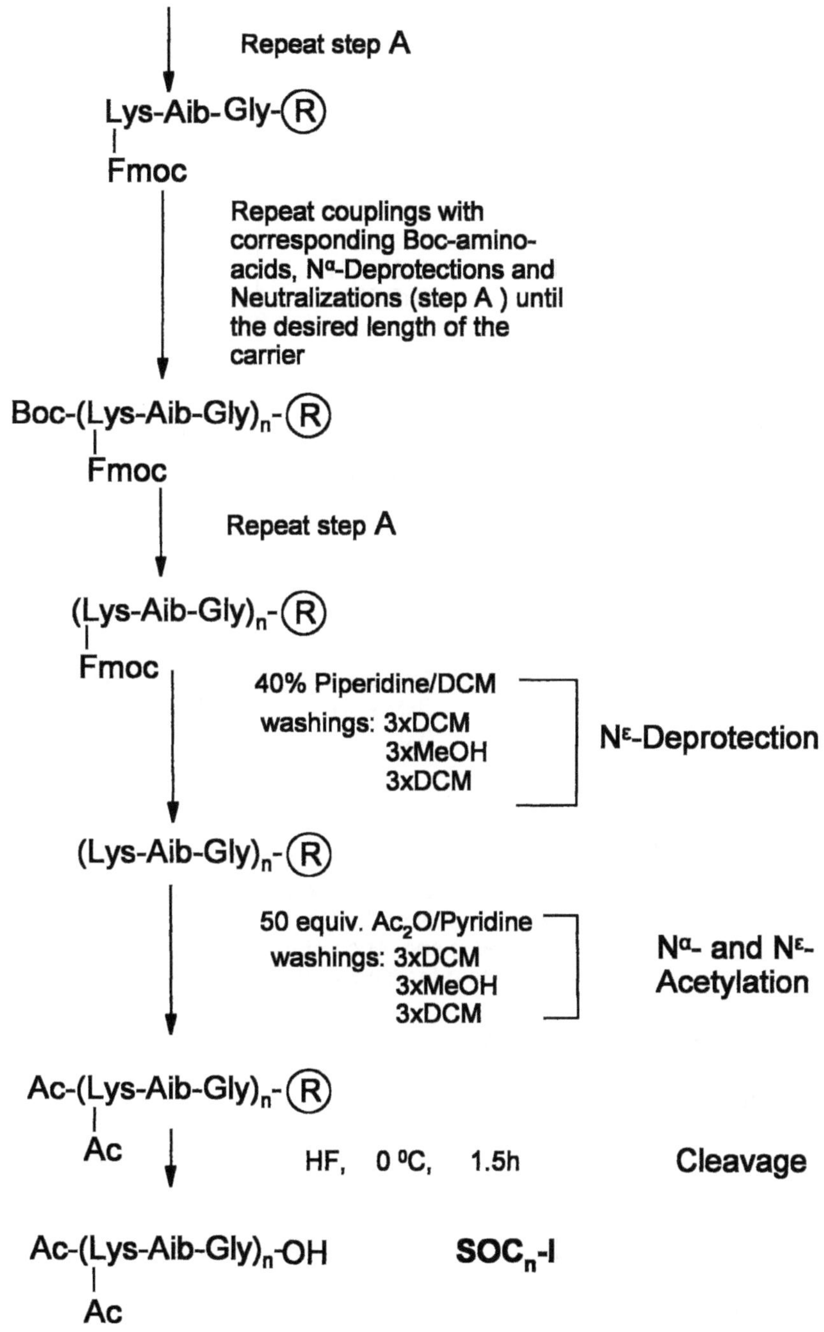
Repeat step A
Lys-Aib-Gly-(R)
Fmoc
Repeat couplings with corresponding Boc-amino-acids, Nα-Deprotections and Neutralizations (step A) until the desired length of the carrier
Boc-(Lys-Aib-Gly)n-(R)
Fmoc
Repeat step A
(Lys-Aib-Gly)n-(R)
Fmoc
40% Piperidine/DCM
washings: 3xDCM
3xMeOH
3xDCM
Nε-Deprotection
(Lys-Aib-Gly)n-(R)
50 equiv. Ac2O/Pyridine
washings: 3xDCM
3xMeOH
3xDCM
Nα- and Nε-Acetylation
Ac-(Lys-Aib-Gly)n-(R)
Ac
HF, 0 °C, 1.5h
Cleavage
Ac-(Lys-Aib-Gly)n-OH
Ac
SOCn-I

Scheme 2. **Step-by-step protocol for the synthesis of the Sm heptapeptide conjugate (PPGMRPP)$_n$-SOC$_n$-I**

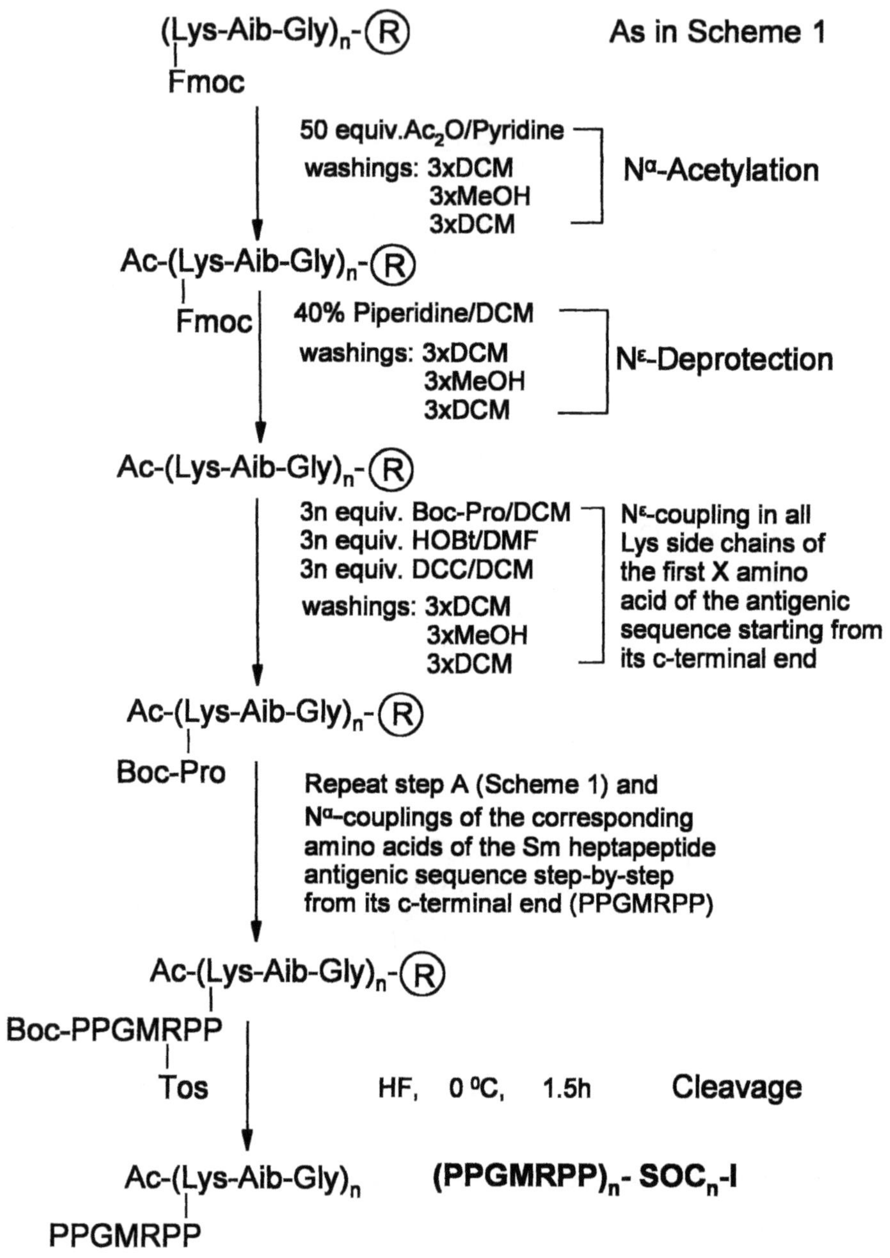

4. Synthetic aspects of SOCs and conjugates

The SOCs (either SOC_n-I or SOC_n-II) were synthesized by a stepwise solid-phase procedure, using the Boc-Gly-OCH_2-Pam resin (Boc: tert-butyloxycarbonyl) and the protected Boc-Aib, Boc-Gly and Boc-L-Lys-(N^{ε}-Fmoc)-OH (Fmoc: 9-fluorenylmethyloxycarbonyl) for the couplings. After elongating the carrier to the desired length (repeating units from 2 to 7), using the Boc-chemistry, the N^{ε}-Fmoc protecting groups were cleaved and the antigenic peptides were built simultaneously, according to the Boc-chemistry, step by step, with attachment to the SOCs by the Lys-N^{ε}-H_2 groups. The final construct, SOCs-conjugate, was cleaved from the resin using HF and subjected to suitable purification.

4.1. SYNTHETIC PROTOCOL OF SOCs

The synthesis of the sequential oligopeptide carriers SOC_n-I (SOC_5) was carried out manually (Scheme 1) by a stepwise solid phase procedure using the Boc-Gly-OCH_2-Pam resin [8,11]. After Boc removal by 40% trifluoroacetic acid (TFA) in dichloromethane (DCM) and neutralization of the resulting salt by diisopropylethylamine (DIEA), the Boc-Aib was coupled using a ratio in mmol of amino acid/HOBT/DCC/Resin of 3/3/3/1 (HOBT: 1-hydroxybenzotriazole, DCC: dicyclohexylcarbodiimide). Cleavage of the Boc group was followed by coupling of the Boc-Lys(Fmoc)-OH. The same protocol was used for the sequential propagation of the Boc-Lys(Fmoc)-Aib-Gly moiety, which resulted in the formation of the desired oligopeptide. Deprotection of the N^{α}-terminal lysine (cleavage of Boc group) and the Lys-N^{ε}-side-chains groups (40% solution of piperidine in dimethylformamide, DMF) was followed by acetylation using $(CH_3CO)_2O$ in DCM (1:10 v/v) and a ratio in mmol of $(CH_3CO)_2O$/resin (30/1). The crude material obtained from the resin by hydrogen fluoride (HF) (yield ranged from 80 to 85%) was subjected to preparative high pressure liquid chromatography, HPLC (A, H_2O/0.1% TFA; and B, CH_3CN/0.1% TFA) and the yield of the purified product ranged from 15 to 20% of the crude material. The composition of the purified product was confirmed by amino acid analysis and 1HNMR spectroscopy.

In a similar way SOC_n-II was also synthesized on a Boc-Gly-OCH_2-Pam resin. After cleavage of Boc, three coupling cycles were carried out for the attachment, in the following order, of Boc-Aib, Boc-Lys (Fmoc)-OH and Boc-Aib. The composition of the final purified product Ac-(Aib-Lys-Aib-Gly)$_n$ was confirmed by amino acid analysis and 1HNMR spectroscopy.

4.2. SYNTHETIC PROTOCOL OF SOCs-CONJUGATES

As an example for the synthesis of the SOCs-conjugates the (PPGMRPP)$_5$-SOC_5-I [18] derivative is described in Scheme 2. On a Boc-Gly-OCH_2-Pam resin was synthesized, stepwise, the Ac-[Lys(N^{ε}-Fmoc)-Aib-Gly]$_5$-Pam compound (see 4.1). The Lys- N^{ε}-protecting groups of the pentameric carrier (SOC_5) were removed using 40% piperidine in DMF and then Boc-Pro was coupled simultaneously to each lysine using a ratio in mmol of Boc-Pro/HOBt/DCC/Resin 15/15/15/5. Cleavage of the Boc group was followed by a second synthetic cycle for the coupling of the following Boc-Pro residue. Similarly, in the following order, Boc-Arg (Tos)-OH, Boc-Met, Boc-Gly, Boc-Pro and Boc-Pro were coupled.

After completion of the synthesis, the SOC_5-conjugate was cleaved from the resin by hydrogen fluoride (HF) in the presence of anisole and phenol as scavengers. HF was evaporated under vacuum, the resin was washed with diethyl ether and the peptide was extracted with 2M aqueous acetic acid. The yield of the crude $(PPGMRPP)_5$-SOC_5 was 70%. The obtained conjugate was dialyzed against water using dialysis tubes with molecular mass cut-off of ca. 1500. The lyophilized material was subjected to preparative HPLC on C-18 column and a programmed gradient elution (3ml/min) was applied (A, H_2O/0.1% TFA; and B, CH_3CN/0.1 TFA). The yield of the purified product was 15%. The purified material gave satisfactory amino acid analysis and its composition was confirmed by amino acid analysis and 1HNMR spectroscopy.

5. Conformational study of SOCs and conjugates.

The conformational study of SOCs was performed by proton magnetic resonance (1HNMR), circular dichroism (CD) and infrared (IR) spectroscopy. Molecular dynamics (MD) calculations were applied in the case of SOC_n-I. The NMR samples were prepared by lyophilization of aqueous solutions adjusted at pH 5, and weighed amounts were dissolved in DMSO-d_6 at concentrations 5-7 mM. Dilution experiments in the range of 3-10 mM proved that molecular associations were excluded. The NMR spectra were run on a Bruker ARX-400 spectrometer, using the standard COSY, HOHAHA and NOESY microprograms.

CD spectra were recorded on a Jobin Yvon CD6 spectrophotometer at sodium dodecylsulfate (SDS) and phosphate buffer ranged from 10^{-3} - 10^{-5} M. All spectra were reported in terms of ellipticity units per mole of peptide residue ($[\Theta]_R$). FT-IR spectra were recorded on a Bruker IFS-85 spectrometer at concentrations 0.7 - 0.3 mM in DMSO and CH_3CN. Pellets of KBr were used for solid state IR.

5.1. CONFORMATIONAL CHARACTERIZATION OF SOC_n-I AND SOC_n-II.

The NH region of the 2D spectrum of SOC_n-I, Ac-[Lys(Ac)-Aib-Gly]$_4$-OH, gave intense NOE connectivities, between successive amide protons, compatible either with a helical conformation or with a random structure. This latter was excluded due to the following: (i) the low absolute temperature coefficient values of all the LysNHs indicate that they are involved in intramolecular interactions, (ii) the variation of the $J_{N\alpha}$ and $J_{N\alpha'}$ coupling constants for the glycines, except the C-terminal one, versus the torsional angle Gly-Φ denotes that the Gly-Φ assumes a defined value about ±70° [19] and (iii) a remarkable similarity between the amide proton chemical shifts of the same repeating residue, when increasing the carrier length, indicates that the Lys-Aib-Gly segments share a common repetitive conformation initiated from the carboxy end of SOC_n-I. It has been concluded that the carrier adopts a rigid conformation with some regularity [8,11].

The NMR data were introduced in Molecular Dynamics (MD) calculations in order to refine the SOC_n structure. Considering the main conformational angles of the time-averaged structure and the average interproton distances, estimated from NOE measurements, a time-averaged structure, stabilized by a network of hydrogen bonds was obtained. The most frequent ones are of the i+3 → i type, and effectively involve the NHs having the lower absolute temperature coefficients. The time-averaged structure of SOC_n is a distorted 3_{10}-helix with somewhat curved axis and rms deviation between every atom pair of a canonical 3_{10}-helix and the SOC_n of less than 1.4 Å [11].

The NH and $C^{\alpha}H$ resonances, the amide proton temperature coefficient values and the $^3J_{N\alpha}$ coupling constants of SOC_n-II, Ac-(Aib-Lys-Aib-Gly)$_n$ (n=3,4) were estimated. Except of the intense NOE connectivities between successive amide protons, some medium range d_{NN} (i, i+2), $d_{\alpha N}$ (i+2) and $d_{\alpha N}$ (i, i+3) were also detected for SOC_n-II (n=3,4). These later peaks argue in favour of a helical structure in both compounds, whereas the appearance of $d_{\alpha N}$ (i, i+2) are diagnostic for a 3_{10}-helix. One may note that $d_{\alpha\beta}$ (i, i+3) cross peaks can not exist in SOC_n-II due to the presence of Aib. Further contribution to the helical structure constitutes the low absolute temperature coefficient values of almost all of the AibNH, suggesting that these NHs are not entirely exposed to the solvent and they are possibly involved in intramolecular interactions [12,13].

The helical structure of SOC_n-II was also confirmed by the CD spectra. Upon addition of SDS, even at concentrations below the critical micelle concentration (8 mM), a positive band at 192 nm and two negative bands at 208 and 222 nm appeared, typical of helical structures. Their intensities increased progressively with SDS concentration and reached a plateau around 30 mM of SDS. A similar CD profile was observed for SOC_3-II and SOC_2-II. The helical content of SOC_n-II (n=2-4), in various concentrations of SDS, was calculated on the basis of the $[\theta]_{222nm}$ values. It is concluded that the molar fraction of the SOCn-II (n=2-4), found in helicoid structure, increases as a function of the carrier's length. Further confirmation of the 3_{10} helix comes from the presence of the amide I band at ~1660 cm^{-1} in the solid state and organic solvents, since it has been demonstrated, using model compounds incorporating Aib, that the main amide I band of fully stable 3_{10}-helices occurs at this frequency [12,13].

5.2 CONFORMATIONAL CHARACTERIZATION OF SOCs-CONJUGATES

Detailed conformational study of the SOCs conjugates was performed, by 1HNMR spectroscopy, with the combined use of COSY, HOHAHA and NOESY experiments [11,12,15]. Taking in consideration the chemical shift values, the temperature coefficients and the NOE effects, it was concluded that the antigenic peptides ([Ala^{76}]MIR and gp63-SRYD) covalently linked to SOCs do not interact with each other or with the carrier. Thus, the antigenic peptides are available either for an optimal recognition with antibodies, or for offering the correct «processed» fragment.

The 1HNMR data indicated also that the peptide epitopes after coupling to SOCs ratain their initial «active» conformation of their free form, preserving thus their topological characteristics. More precisely, the very small perturbations between the chemical shifts of the [Ala^{76}]MIR free and bound to SOC_4-I indicate a conservative structure of the antigenic peptides in both states [11]. Also, the low absolute temperature coefficient for the Asp^{71}-NH proton resonance argues in favour of a β-folded Asn-Pro-Ala-Asp sequence, which was shown to be a prerequisite for optimal antibody recognition [19,20]. Further confirmation of the preserved «active» conformation, after coupling to SOCs, came also from the 1HNMR study of gp63-SRYD-SOC_5-I and gp63-SRYD-SOC_6-I. In particular, an ionic interaction between the guanidinium of arginine and the β-carboxylate group of aspartic acid, as well as the formation of a type I β-turn involving GlnNH → ArgCO of the Ile^{250}-Ala-Ser-Arg-Tyr-Asp-Gln-Leu^{257} of gp63 were identified in both free and bound states [15,21].

The example of $(PPGMRPP)_5$-SOC_5-I and $(PPGMRPP)_4$-SOC_4-II deserves to be noted. The 1HNMR analysis of PPGMRPP indicated the presence of at least three conformers due to the cis-trans isomerism of the X-Pro bond [17]. Grafting of PPGMRPP to SOCs resulted in the prevalence of one peptide's conformer (extended form) which, in turn, enhanced the anti-Sm/U1RNP specific recognition, as it is shown in the following section 6 for biological studies as well as in references [17,22].

6. Biological studies

The SOCs-conjugates were used as antigens and/or immunogens. In brief, in the case of antigenic substates, an appropriate quantity of the SOCs-conjugate, dissolved in buffer, was added on ELISA plates and incubated overnight at 4°C. After washing with phosphate buffer, the plates were incubated with bovine serum albumin, washed with buffer and incubated either with mAbs solution or diluted sera, following the experiment under investigation. In the case of a radioimmunoassay (RIA test), rabbit anti-rat γ-globulin in buffer was added followed by ^{125}I-labelled protein-A. The bound radioactivity was removed by 1% sodium dodecyl sulfate, transferred to test tubes and measured in α γ-counter [23]. When an ELISA immunoassay was performed, the sera bound to the antigenic substrate were detected with alkaline phoshatase conjugated anti-human IgG. After washings a solution of p-nitrophenyl phosphate disodium was added and the absorbance was read at 405 nm [22].

Immunization experiments were performed using New Zealand white rabbits, which were immunized subcutaneously with the SOCs-conjugate, complete and incomplete Freund's adjuvant [10,15]. Experiments were also realized with SOCs alone and without conjugate. Production of anti-SOCs-conjugate antibodies was tested by ELISA assays.

6.1. THE SOCs-CONJUGATES AS ANTIGENS

The use of SOCs-conjugates for the development of specific diagnostic immunoassays was proved very successful. Assuming that recognition of an antigen by B-cells is higly dependent on the three-dimensional conformation of the antigen, it is very probable that the antigenic peptides presented on SOCs adopt a topological pattern that mimics the native epitopes. Obviously the optimal presentation of the epitopes is directly related to the helicoid structure of SOCs.

Anchorage of the $[Ala^{76}]$MIR analogue to the SOC_n-I carrier resulted to a construct that efficiently inhibited after pre-incubation, anti-MIR mAbs from binding to the immobilized MIR peptide [11]. The SOC_n-I carrier bearing four or five copies of the $[Ala^{76}]$MIR analogue, showed up to a tenfold increase of the MIR binding capacity, suggesting a clear advantage in using MIR-SOCs rather than MIR peptides, as potent antigens in diagnostic assays or in elaborating specific resins for antibody depletion in myasthenic sera.

ELISA and dot blot tests based on recombinant La/SSB and $(La/SSB_{349-364})_4$-SOC_4-I exhibited similar reactivities (88%), whereas the specificity of the later SOC_4-I conjugate was 90% compared to recombinant La/SSB (rLa/SSB), which gave 63% only (sera tested: anti-Ro/SSA positive but anti-La/SSB negative). These results suggest that $(La/SSB_{349-364})_4$-SOC_4-I displays high sensitivity and specificity and can be used in convenient and reproducible immunoassays for detecting anti-La/SSB antibodies [24].

A sensitive, highly reproducible and rapid ELISA was developed as a valuable screening method to test anti-Sm/U1RNP reactivities using $(PPGMRPP)_5$-SOC_5 as antigenic substrate [22]. The sensitivity and specificity of the method were found 98% and 68% for the detection of anti-Sm antibodies and 82% and 86% respectively, for the detection of antibodies to small nuclear ribonucleoproteins (snRNPs), either of anti-Sm or anti-U1RNP specificity. Taking in consideration that in ELISA experiments, using the Sm/U1RNP purified complex (snRNPS), the sensitivity in detecting anti-snRNP was 74%, we conclude that $(PPGMRPP)_5$-SOC_5 is a better alternative as an antigenic substrate in ELISA tests.

The PPGMRPP epitope coupled to the SOC_n-II carrier was also employed for the detection of anti-Sm/U1RNP antibodies [12,13]. Surprisingly the binding capacity of the tested sera was found rather limited compared to $(PPGMRPP)_n$-SOC_n-I. One may note that the functional groups of all lysines, which occupy the i, i+4 positions of SOC_n-II, are located all around the axis in every 3-residues turn of the 3_{10}-helix, while in the SOC_n-I carrier the lysine side chains (i, i+3) are located on one site of the helix. Consequently, it is likely that the adsorption of the $(PPGMRPP)_n$-SOC_n-II conjugate to the ELISA plate which should be realized mainly by the PPGMRPP peptides, is less strong due to their hydrophilic nature, compared to the $(PPGMRPP)_n$-SOC_n-I, which should be achieved by the rather hydrophobic oligopeptide backbone of the carrier. Thus, the limited coating of the $(PPGMRPP)_n$-SOC_n-II to the ELISA plate, as well as the fact that the adsorbed PPGMRPP antigenic peptides of the carrier are not disposed to the antibodies, may explain the rather low binding level of the $(PPGMRPP)_n$-SOC_n-II to the anti-Sm/U1RNP sera, compared to the $(PPGMRPP)_n$-SOC_n-I.

6.2. THE SOCs-CONJUGATES AS IMMUNOGENS

Immunized animals with the SOCs conjugates produced high titers of antibodies recognizing the immunogen peptide. Depending on the epitope anchored to SOCs, it was identified either an immune spreading covering various peptide sequences on the protein, as well as the intact protein, or a limited expansion of the B-cell repertoir [15,25,26]. Three examples are discussed in the following.

The gp63-SRYD-SOC_n-I (n=5,6) were used for rabbit immunizations [15] and the obtained antisera were tested for their specific reactivity in ELISA against gp63-SRYD peptide and purified gp63. The two SOC_n-I conjugates gave satisfactory antibody titers vs the gp63SRYD octapeptide, and they were found better immunogens, when compared to the same peptides conjugated to rabbit serum albumin. The produced antibodies recognized the purified Leishmania gp63 and the gp63 protein on intact parasites. It is concluded that the gp63-SRYD-SOC_n-I conjugates are effective immunogens, recognizing their cognate protein, and they could possibly induce protective immunity against leishmaniosis. Immunizations with the SOC_n-I carrier alone showed that it was not immunogenic [15].

Immunization of rabbits with the B-cell epitopes of the La/SSB autoantigen indentified by our group [13] resulted in the production of antibodies directed: (a) to particular immunizing epitope, (b) to other B-cell epitopes of the same autoantigen and (c) to rLa/SSB antigen as a whole [25]. These findings, as well as the cross inhibition experiments with the rabbit antisera and the B-cell epitopes of the La/SSB argue in favour of an immune spreading.

New Zealand white rabbits immunized with $(PPGMRPP)_5$-SOC_5 in complete Freund's adjuvant (CFA) developed anti-$(PPGMRPP)_5$-SOC_5 antibodies and renal disease compatible with SLE. Neither, anti-$(PPGMRPP)_5$-SOC_5 antibodies nor renal disease occured after immunization with SOC_5 alone, PPGMRPP in the free form or with La/SSB autoantigen B-cell epitopes. Although, the bulk of antibodies, produced in rabbits, reacted to the $(PPGMRPP)_5$-SOC_5, western blotting experiments, using HeLa nuclear extract as antigen source, indicated a reactivity to a whole 70kD polypeptide in a manner comparable with the reactivity of human anti-U1RNP sera to this extract. This reactivity, however, was associated with the presence of glomerulonephritis [26]. Two possible explanations are currently under investigation for the role of SOCs in the enhancement of the immune response agaist the attached B-cell epitopes:

(i) The SOCs-conjugate resembles a T-cell independent type 2 antigen and does not follow the hapten-carrier conjugate paradigme. It seems rather, that multiple cross-linking of the B-cell receptor by such a thymus-independent-type 2 like antigen can lead to autoantibody production either in the absence of, or augmented by, T-cell help.

(ii) The T-cell receptor repertoir is as extended as the B-cell receptor repertoir; therefore, there is a chance that at least the linear B-cell epitopes of some autoantigens can be recognized by specific T-cell receptors, in other words, some linear B-cell epitopes of autoantigens may also function as T-cell epitopes. In that case the construct (B-cell autoepitopes)$_n$-SOC_n provides a critical size and structure «molecule» which can be taken, processed and presented by B-cells.

7. Conclusions

Our concept in the design of the presented sequential oligopeptide carriers (SOCs) was to construct an artificial support with structural rigidity and regularity, so that peptide epitopes could be anchored without conformational restrictions and steric hindrances. 1HNMR studies and molecular modeling showed that SOC_n-I adopt a distorded 3_{10}-helical structure, which allows a favorable orientation of the lysine side chains and therefore of the attached peptides. Conformational analysis by 1HNMR spectroscopy and molecular modeling of the SOC_n-I conjugates pointed out that the peptides anchored to SOC_n-I retain their original «active» conformation. In addition, covalent coupling of the epitopes to the carrier imposes the prevalence of one conformer, enchancing specificity and sensitivity, confirming thus our initial design.

The SOC_n-I conjugates, when used as antigens, displayed significant biological reactivity, while the developed immunoassays were sensitive, convenient and reproducible in screening antibody specificities related to autoimmune diseases. It is very probable that the helicoid structure of SOC_n-I offers an optimal epitope presentation and helps the reconstruction and/or mimicing of the native epitopes.

Immunizations with the SOC_n-I conjugates generated in animals high titers of antibodies recognizing, in all cases, the immunogen peptide. Depending on the anchored to SOC_n-I peptide, it was identified either an immune spreading covering various peptide sequences on the protein, as well as the intact protein, or a limited expansion of the B-cell repertoir. Multiple cross-linking of the B-cell receptor by the B-cell epitopes coupled to SOC_n-I or uptake by B-cells and presentation to T-cell receptors which cross-react with these B cell epitopes, are two alternative explanations for the mode of action of $(epitope)_n$-SOC_n-I conjugates and the continous secretion of specific antibodies long time after immunization.

The conformational study, by 1HNMR, CD and FT-IR spectroscopy of SOC_n-II indicated that the carrier displays a pronounced 3_{10}-helix, compared to SOC_n-I. The sequence PPGMRPP of the Sm and U1RNP autoantigens, being the main target of anti-Sm/U1RNP autoantibodies in patients with SLE, was grafted to the Lys-$N^{\varepsilon}H_2$ groups of the carrier in order to use the $(PPGMRPP)_n$-SOC_n-II constructions in immunoassays. However, the binding capacity of the tested anti-Sm/U1RNP sera to the SOC_n-II conjugate was found rather limited compared to the SOC_n-I.

It is concluded that despite the well-defined structural motif (pronounced 3_{10}-helix) of SOC_n-II, the distribution of lysines (i, i+4) all around the axis of the 3_{10}-helix is less favorable, compared to the SOC_n-I lysines (i, i+3) located on one site of the helix, for a satisfactory coating of the $(PPGMRPP)_n$-SOC_n-II construction to the ELISA plates. These observations confirm the importance of the lysine positions along the oligopeptide backbone of the carriers.

Studies on further applications with novel SOCs-conjugares are in progress in our Laboratory with myelopeptides [27] and MUC1 peptides [28] for leukemia and breast cancer treatment. Further developments on the SOCs themselves are also in progress in our Laboratory for introducing lipoSOCs or/and adjuvant-conjugated-SOCs as a new generation of SOCs, as well as for anchoring several different, more than one kind, epitopes on the same SOC molecule by selective protection and deprotection of the lysine side chains.

References

1. Elkon, K.B. (1992) Use of synthetic peptides in the detection and quantification of autoantibodies, *Mol. Biol. Rep.* **16**, 207-212.
2. Kaumaya, P.T.P., Kobs-Conrad, S., Seo, Y.H., Lee, H., VanBuskirk, A.M., Feng, N., Sheridan, J.F., and Steven, V. (1993) Peptide Vaccines Incorporating a «Promiscuous» T-cell Epitope Bypass Certain Haplotype Restricted Immune Responses and Provide Broad Spectrum Immunogenicity, *J. Mol. Recogn.* **6**, 81-94.
3. Posnett, D.N., McGrath, H., and Tam, J.P. (1988) A Novel Method for Producing Anti-peptide Antibodies. Production of site-specific antibodies to the T-cell antigen receptor β-chain, *J. Biol. Chem.* **263**, 1719-1725.
4. Metzger, J., Wiesmuller, K.-H., Schaude, R., Bessler, W.G., and Jung, G. (1991) Synthesis of novel immunologically active tripalmitoyl-S-glycerylcysteinyl lipopeptides as useful intermediates for immunogen preparations, *Int. J. Peptide Protein Res* **37**, 46-57.
5. Tuchscherere, G., Domer, B., Sila, U., Kamber, B., and Mutter, M. (1993) The TASP concept: Mimetics of peptide ligands, protein surfaces and folding units, *Tetrahedron* **49**, 3559-3575.
6. Veprek, P. and Jezek, J. (1999) Peptide and Glycopeptide dendrimers. Part I, *J. Peptide Sci.* **5**, 5-23.

7. Veprek, P. and Jezek, J. (1999) Peptide and Glycopeptide dendrimers. Part II. *J. Peptide Sci.* **5**, 203-220.

8. Tsikaris, V., Sakarellos, C., Cung, M.T., Marraud, M., and Sakarellos-Daitsiotis, M. (1996) Concept and Design of a New Class of Sequential Oligopeptide Carriers (SOC) for Covalent Attachment of Multiple Antigenic Peptides, *Biopolymers* **381**, 291-293.

9. Sakarellos-Daitsiotis, M., Tsikaris, V., Vlachoyiannopoulos, P.G., Tzioufas, A.G., Moutsopoulos, H.M., and Sakarellos, C. (1999) Peptide Carriers: A Helicoid-Type Sequential Oligopeptide Carrier (SOCn) for multiple anchoring of antigenic/immunogenic peptides, *Methods: A companion to Methods in Enzymology*, **19**, 133-141.

10. Sakarellos-Daitsiotis, M., Tsikaris, V., Sakarellos, C., Vlachoyiannopoulos, P.G., Tzioufas, A.G., and Moutsopoulos, H.M. (1999) A new helicoid-type sequential oligopeptide carrier (SOCn) for developing potent antigens and immunogens, *Vaccine*, **18**, 302-310.

11. Tsikaris, V., Sakarellos, C., Sakarellos-Daitsiotis, M., Orlewski, P., Marraud, M., Cung, M.T., Vatzaki, E., and Tzartos, S. (1996) Construction and application of a new class of sequential oligopeptide carriers (SOC_n) for multiple anchoring of antigenic peptides-application to the acetylcholine receptor (AChR) main immunogenic region, *Int. J. Biol. Macromol.* **19**, 195-205.

12. Alexopoulos, Ch., Tsikaris, V., Sakarellos-Daitsiotis, M., Sakarellos, C., Cung, M.T., and Marraud, M. (1999) Studies on the synthesis and structure-conformation of artificial carriers for the development of Potent Antigens or/and immunogens,in S. Bajusz and F. Hudecz (eds.), *Peptides 1998*, Akadémiai Kiadó, Budapest, pp 340-341.

13. Alexopoulos, Ch., Tsikaris, V., Rizou, C., Sakarellos-Daitsiotis, M., Sakarellos, C., Cung, M.T., Marraud, M., Vlachoyiannopoulos, P.G., and Moutsopoulos, H.M. (1999) The position of the Lys-$N^{\varepsilon}H_2$-grafted antigens along the sequential oligopeptide carrier, Ac-(Aib-Lys-Aib-Gly)$_n$ (SOC_n-II) influences the antibody recognition: Application to the Sm main autoimmune epitope, *unpublished results, in preparation.*

14. Karle, I.L. and Balaram, P. (1990) Structural characteristics of α-Helical peptide molecules, containing Aib residues, *Biochemistry* **29**, 6747-6756.

15. Tsikaris, V., Sakarellos, C., Sakarellos-Daitsiotis, M., Cung, M,T., Marraud, M., Konidou, G., Tzinia, A., and Soteriadou, K.P. (1996) Use of Sequential Oligopeptide Carriers (SOC_n) in the Design of Potent *Leishmania* gp63 Immunogenic Peptides. *Peptide Res* **5**, 240-247.

16. Tzioufas, A.G., Yiannaki, E., Sakarellos-Daitsiotis, M., Routsias, J.G., Sakarellos, C., and Moutsopoulos, H.M. (1997) Fine specificity of autoantibodies to La/SSB: epitope mapping and characterization, *Clin. Exp. Immunol.* **108**, 191-198.

17. Tsikaris, V., Vlachoyiannopoulos, P.G., Panou-Pomonis, E., Marraud, M., Sakarellos, C., Moutsopoulos, H.M., and Sakarellos-Daitsiotis, M. (1996) Immunoreactivity and conformation of the P-P-G-M-R-P-P repetitive epitope of the Sm autoantigen, *Int. J. Peptide Protein Res* **48**, 319-327.

18. Sakarellos, C., Tsikaris, V., Panou-Pomonis, E., Alexopoulos, Ch., Sakarellos-Daitsiotis, M., Petrovas, C., Vlachoyiannopoulos, P.G., and Moutsopoulos, H.M. (1997) The PPGMRPP repetitive epitope of the Sm autoantigen: Antigenic specificity induced by conformational changes. Application of the Sequential Oligopeptide Carriers (SOCs), *Letters in peptide Science (LIPS)* **4**, 447-454.

19. Tsikaris, V., Detsikas, E., Sakarellos-Daitsiotis, M., Sakarellos, C., Vatzaki, E., Tzartos, S.J., Marraud, M., and Cung, M.T. (1993) Conformational requirements for molecular recognition of AChR MIR analogues by monoclonal anti-MIR antibody: A 2D-NMR and molecular Dynamics Approach, *Biopolymers* **33**, 1123-1134.

20. Orlewski, P., Marraud, M., Cung, M.T., Tsikaris, V., Sakarellos-Daitsiotis, M., Sakarellos, C., Vatzaki, E., and Tzartos, S.J. (1996) Compared structures of the free nicotinic acetylcholine receptor main immunogenic region (MIR) decapeptide and antibody-bound [A^{76}]MIR analogue: A molecular dynamics simulation from two-dimensional NMR data, *Biopolymers* **40**, 419-432.

21. Tsikaris, V., Cung, M.T., Sakarellos, C., Tzinia, A., Soteriadou, K.P., and Sakarellos-Daitsiotis, M. (1994) NMR study on an SRYD-containing Fibronektin-like sequence (250-257) of Leishmania gp63: Contribution of residual water in the dimethyl sulfoxide solution structure, *J. Chem. Soc., Perkin Trans.* **2**, 821-826.

22. Petrovas, C.J., Vlachoyiannopoulos, P.G., Tzioufas, A.G., Alexopoulos, Ch., Tsikaris, V., Sakarellos-Daitsiotis, M., Sakarellos, C., and Moutsopoulos, H.M. (1998) A major Sm epitope anchored to sequential oligopeptide carriers is a suitable antigenic substrate to detect anti-Sm antibodies, *J. Immunol. Methods* **220**, 59-68.

23. Papadouli, E., Potamiannos, S., Hadjidakis, I., Bairaktari, E., Tsikaris, V., Sakarellos, C., Cung, M.T., Marraud, M., and Tzartos, S.J. (1990) Antigenic role of single residues within the main immunogenic region of the nicotinic acetylcholine receptor, *Biochem. J.* **269**, 239-245.

24. Yiannaki, E.E., Tzioufas, A.G., Bachmann, M., Hantoumi, J., Tsikaris, V., Sakarellos-Daitsiotis, M., Sakarellos, C., and Moutsopoulos, H.M. (1998) The value of synthetic linear epitope analogues of La/SSB for the detection of autoantibodies to La/SSB specificity and comparison of methods, *Clin Exp. Immunol.* **112**, 152-158.

25. Yiannaki, E.E., Tzioufas, A.G., Manoussakis, M.N., Sakarellos-Daitsiotis, M., Sakarellos, C., and Moutsopoulos, H.M. (1998) Immunizations of synthetic epitope analogues of La/SSB autoantigen elicits intramolecular but not intermolecular spreading, *Clin. Exp. Rheumatol.* **16**, 208.

26. Sakarellos, C., Alexopoulos, Ch., Tsikaris, V., Sakarellos-Daitsiotis, M., Petrovas, C., Vlachoyiannopoulos, P.G., Tzioufas, A.G., and Moutsopoulos, H.M. (1998) Development of an experimental systemic lupus erythematosus (SLE) like disease after immunization with the Sm epitope PPGMRPP anchored to an artificial Carrier. XVIII European Workshop for Rheumatology Research, *Clin. Exp. Rheumatol.* **16**, 208.

27. Petrov, R.V., Mikhailova, A.A., and Forrina, L.A. (1997) Bone marrow immunoregulatory peptides (myelopeptides): Isolation, structure, and functional activity, *Biopolymers*, **43**, 139-146.

28. Apostolopoulos, V., Karanikas, V., Haurum, J.S., and McKenzie, I.F. (1997) Induction of HLA-A2-restricted CTLS to the mucin 1 human breast cancer antigen, *J. Immunol.* **159**, 5211-5218.

MOLECULAR RECOGNITION IN THE MEMBRANE: ROLE IN THE FOLDING OF MEMBRANE PROTEINS

YECHIEL SHAI
Department of Biological Chemistry, Weizmann Institute of Science
Rehovot, 76100 Israel. Email: Yechiel.Shai@weizmann.ac.il

1. INTRODUCTION

Elucidating the principles that govern the interaction between proteins and membranes, and understanding the molecular forces underlying peptide-peptide interaction within the lipid environment of the cell membrane are major goals in biochemistry. Both types of interactions play key roles in numerous biological events, some of which are vital for cell survival and others cause cell death. Membrane proteins can be divided into two major groups: (i) Water/membrane soluble proteins that are soluble in water but can undergo drastic conformational changes that can allow them to interact and insert into the cell membrane. The list includes bacterial toxins and envelope proteins of viruses. (ii) Integral membrane proteins such as ion channel, transporters and receptors. Membrane proteins are composed of extramembrane and membrane domains. It is believed that the membranous domains of various membrane proteins fold according to principles similar to those of soluble proteins (Gierasch & King, 1989; Glasser & Scheraga, 1991). However, the organization of the membrane-inserted domains is poorly known, mainly because of the difficulty to obtain high resolution structures. Three dimensional structures were obtained for only few membrane proteins as compared to numerous number of soluble proteins. The list includes bacteriorhodopsin, halorhodopsin, rhodopsin, the photosynthetic reaction center, the light harvesting complex, photosystem I, porin, the nicotinic acetylcholine receptor (reviewed in (White, 1992; Lemmon & Engelman, 1994; Shai, 1995; White & Wimley, 1999)) and the membrane inserted domain of a bacterial potassium channel (Doyle *et al.*, 1998). The structures of membrane proteins, are therefore hypothesized based of putative trans-membrane topology that was proposed on the basis of hydropathic analysis, analytical methods that predict local secondary structures for short segments, energy minimization calculations and functional studies combined with site directed mutagenesis. However, unambiguous structures cannot be predicted, and it is already clear that the hypothetical models do not always succeed in accounting for known functional features. Thus, direct experimental evidence as to the secondary structure of the proteins (i.e. of trans-membrane segments) is essential. Transmembrane segments were found in proteins with known high resolution structures to contain predominantly α-helical structures, but some were also ß-sheeted.

A central question is: What are the underlying forces that dictate a specific

A. Aggeli et al. (eds.), Self-assembling Peptide Systems in Biology, Medicine and Engineering, 273–294.

organization within the membrane to a certain polypeptide or protein, that in turn is responsible for its biological role? more specifically:

(a) Is there a common mechanism for the interaction of polypeptides, having similar overall conformations, with membranes?
(b) What are the driving forces that cause polypeptides to go from a predominantly unordered structure in solution to a well organized structure in the membrane?
(c) Are there specific interactions between polypeptides and various phospholipid membranes?
(d) How specific are peptide-peptide interactions within the hydrophobic interior of the bilayers?
(e) What are the underlying forces that govern peptide-peptide interactions within membranes?
(f) How are membrane-embedded segments organized?
(g) What is the precise mechanism that leads to membrane destabilization that is reflected by pore-formation, membrane disruption or membrane fusion?

It is reasonable to assume that the assembly of membrane proteins in the cell membrane requires a combination of recognition and interaction sites in different domains of the protein. Various studies showed that such sites can be intracellular, such as in the intracellular domains of potassium channels (Li *et al.*, 1992; Shen *et al.*, 1993; Babila *et al.*, 1994; Lee *et al.*, 1994). Other studies showed they can be extracellular such as the extracellular domains of the acetylcholine receptor subunits (Yu & Hall, 1991; Verrall & Hall, 1992), the platelet integrin GPIIb/IIIa (Frachet *et al.*, 1992) and the Na^{+}, K^{+}-ATPase α-subunit (Lemas *et al.*, 1992), or they can be in the hydrophobic transmembrane domains of integral membrane proteins, such as, glycophorin A (Bormann *et al.*, 1989; Lemmon *et al.*, 1992), bacteriorhodopsin (Kahn & Engelman, 1992), the T cell receptor complex (Bonifacino *et al.*, 1990; Manolios *et al.*, 1990), the tyrosine kinase receptor family (Sternberg & Gullick, 1990), the aspartate sensory receptor of *Escherichia coli* (Lynch & Koshland, 1991), the lactose permease of *Escherichia coli* .(Sahin-Toth *et al.*, 1992), potassium channels (Peled & Shai, 1993; Ben Efraim *et al.*, 1994; Pouny & Shai, 1995; Ben Efraim & Shai, 1996; Ben Efraim *et al.*, 1996; Peled *et al.*, 1996; Ben Efraim & Shai, 1997), and the insecticide δ-endotoxin (Gazit & Shai, 1993a; Gazit & Shai, 1995; Gazit *et al.*, 1998). In addition, short membrane permeating native polypeptide toxins were also shown to assemble within the membrane to form the active oligomer. Examples are; the shark repellent pardaxin (Rapaport & Shai, 1991; Rapaport & Shai, 1992) and the antimicrobial peptide dermaseptins (Strahilevitz *et al.*, 1994; Ghosh *et al.*, 1997), and the human antimicrobial peptide LL-37 (Oren *et al.*, 1999).

The present chapter describes the incorporation and organization within membranes of low molecular weight membrane permeating peptides and high molecular weight membrane proteins, based on studies with synthetic peptides. It is divided to two major parts;

1. The identification of two distinct mechanisms for membrane permeation by polypeptides; channel or pore formation via the "barrel-stave" mechanism, versus membrane disruption via the "carpet" mechanism.

2. The implication of the results obtained in the studies with native short polypeptides to the study of large proteins. In that we include the prediction of the structure and organization of pore forming domain of δ-endotoxins and a K^+ ion channels.

1.1. MEMBRANE PERMEATION BY MEMBRANOUS POLYPEPTIDES OCCURS VIA TWO DISTINCT MECHANISMS.

Membrane permeating peptides serve in the vertebrates and invertebrates world for both ofensive and defensive purposes. They have been isolated from insects, amphibians, mammals and recently from plants too. These peptides can be classified into three major groups; those which are active selectively against eukaryotic cells, prokaryotic cells or both eukaryotic and prokaryotic cells. Antimicrobial peptides are the largest group within these families. They were found initially in invertebrates (Boman, 1995), but later on also in vertebrates, including human (Agerberth *et al.*, 1995; Gudmundsson *et al.*, 1996). They are used as a cell-free defense mechanisms in addition to, or complementary to, the highly specific cell-mediated immune response (Hoffmann *et al.*, 1999). This secondary, chemical immune system provides organisms with a repertoire of small peptides that are promptly synthesized upon induction, and act against invasion by occasional and obligate pathogens (Saberwal & Nagaraj, 1994; Boman, 1995; Nicolas & Mor, 1995; Hancock & Lehrer, 1998). So far, more than 500 different antimicrobial peptides have been isolated and characterized. Most of the antimicrobial peptides are composed of L-amino acids, with a defined α-helix or β-sheet secondary structures. Some are linear, mostly helical, without cysteines, while others contain one or more disulfides bonds forming β-sheet or both β-sheet and α-helix structures (Boman, 1995). In most cases, the peptides' mode of action appears to be by direct lysis of the pathogenic cell membrane. A second native group of peptides antibiotics, although smaller, is composed of both L and D-amino acids. This group includes gramicidins, actinomycins, bacitracin, polymyxins, lantibiotics and bombinins H 3-5 (Kreil, 1994).

Membrane-Lytic peptides have been studied extensively in order to understand general aspects related to peptide protein interactions, peptide-peptide interaction within the membrane milieu, as well as, the relation of these interactions to the biological function of these peptides. The largest and most studied group out of those described above includes short linear polypeptides (40 a.a.) which are devoid of disulfide bridges (see reviews by (Segrest *et al.*, 1990; Nicolas & Mor, 1995)). These polypeptides vary considerably in chain length, hydrophobicity and overall distribution of charges, but share a common α-helical structure when associated with phospholipid membranes (Segrest et al., 1990). An amphipathic α-helical structure requires that polar amino acids are arranged along one side of the helix as a consequence of 1,3 and 1,4 periodicities and the hydrophobic amino acids along the other side. Some of these peptides are not cell selective (e.g. the bee venom melittin (Habermann & Jentsch, 1967; Blondelle & Houghten, 1991)), the Moses Sole fish lytic peptide pardaxin (Shai *et al.*, 1988; Shai *et al.*, 1990), and the human cathelicidin-like LL-37 (Johansson *et al.*, 1998), being able to lyse both bacterial and mammalian cells. Other are selective either to mammalian cells but not bacteria (e.g. δ- hemolysin from *Staphylococcus aureus*) (Dhople & Nagaraj, 1993) or visa versa, i.e. cytotoxic to various pathogenic microorganisms but not normal mammalian cells (e.g. cecropins, isolated from the cecropia moth (Steiner *et al.*, 1981) and from the hemolymph and cuticular extracts of other Lepidopteran and Dipterian insects (for a review, see: (Boman & Hultmark, 1987; Hultmark, 1993)), magainins (Zasloff, 1987) and dermaseptins (Mor *et al.*, 1991a; Mor *et al.*, 1991b; Mor *et al.*, 1994a; Mor *et al.*, 1994b; Charpentier *et al.*, 1998) both isolated from the skin of frogs. Each antimicrobial peptide has a broad but not identical spectrum of antimicrobial activity, providing the host maximum coverage against a rather broad spectrum of microbial organisms.

It is well accepted that one major reason for the cell-selective or non cell-selective activity of these peptides resides from their different abilities to bind cells of various types. One major difference between the outer surface of bacteria and normal mammalian cells in their net charge. The outer surface of Gram-negative bacteria contains lipopolysaccarides (LPS) and that of Gram-positive bacteria acidic polysaccharides (teichoic acids), giving the surface of both Gram-positive and Gram-negative bacteria a negative charge (Brock, 1974). Therefore, the net positive charge of the antibacterial peptides facilitates their initial binding to the bacterial surface. In contrast, the outer leaflet of human erythrocytes (representatives of normal mammalian cells) is composed predominantly of zwitterionic phosphatydilcholine (PC) and sphingomyelin phospholipids (Verkleij *et al.*, 1973).

Two major alternative mechanisms were proposed to explain the ability of amphipatic α-helical peptide to permeate membranes; (i) transmembrane pore formation via a "barrel-stave" mechanism (Ehrenstein & Lecar, 1977) (Figure 1, right panel). According to this model amphipatic α-helices insert into the hydrophobic core of the membrane and form transmembrane pores. (ii) membrane destruction/solubilization via a "carpet" mechanism (Figure 1, left panel). In this model, proposed for the first time in (Pouny *et al.*, 1992; Oren & Shai, 1998; Shai, 1999), the peptides which do not need necessarily to adopt amphipatic α-helical structure, are in contact with the lipid head group during the whole process of membrane permeation and do not insert into the hydrophobic core of the membrane. A major question is whether a particular mechanism can be assigned to a particular biological function (i.e., antibacterial activity versus cytotoxicity to mammalian cells).

1.1.1. Transmembrane Pore Formation via the "barrel-stave" mechanism

A transmembrane pore can be formed when several amphipatic α-helices insert into the membrane and form bundles, in which their hydrophobic surfaces interact with the lipid core of the membrane, and their hydrophilic surfaces point inward, producing a pore (Figure 1). The recruitment of several helices has been described as a "barrel-stave" mechanism (Ehrenstein & Lecar, 1977). Since these peptides can insert into the hydrophobic core of the membrane their interaction with the target membrane is predominantly driven by hydrophobic interaction. As a consequence, they can bind to both zwitterionic and charged phospholipid membranes, and therefore can lyse both bacteria and mammalian cells. In addition, since the hydrophilic surfaces of several amphipatic α-helical peptides facing each others while forming a transmembrane pore, it is unlikely that such helices will be highly homogeneously charged. Formation of a transmembrae pore via the "barrel-stave" mechanism should involve four major steps: (i) binding of the monomers to the membrane in a helical structure; (ii) membrane bound monomers should recognize each other, and therefore, assemble already at low surface density of bound peptide; (iii) insertion of at least two assembled monomers into the membrane to initiate the formation of a pore; and (iv) progressive recruitment of additional monomers to increase the pore size.

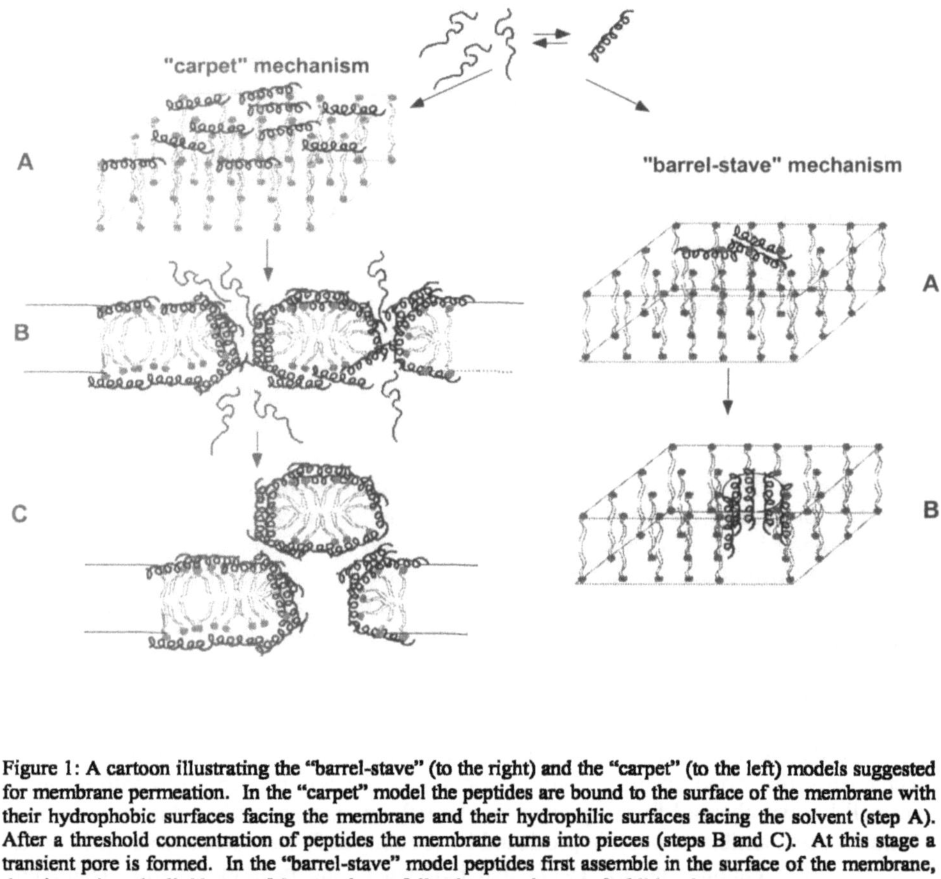

Figure 1: A cartoon illustrating the "barrel-stave" (to the right) and the "carpet" (to the left) models suggested for membrane permeation. In the "carpet" model the peptides are bound to the surface of the membrane with their hydrophobic surfaces facing the membrane and their hydrophilic surfaces facing the solvent (step A). After a threshold concentration of peptides the membrane turns into pieces (steps B and C). At this stage a transient pore is formed. In the "barrel-stave" model peptides first assemble in the surface of the membrane, then insert into the lipid core of the membrane following recruitment of additional monomers.

1.1.2 Membrane Disruption via the "Carpet" Mechanism

We proposed the "carpet" model for the first time to describe the mode of action of the antimicrobial peptide dermaseptin S (Pouny et al., 1992), and later on the same model was used to describe the mode of action of other antimicrobial peptides, such as, dermaseptin natural analogues (Pouny et al., 1992; Strahilevitz et al., 1994; Ghosh et al., 1997), cecropins (Gazit *et al.*, 1994b; Gazit *et al.*, 1995; Gazit *et al.*, 1996), the human antimicrobial peptide LL-37 (Oren et al., 1999), caerin 1.1 (Wong *et al.*, 1997) and diasteriomers of lytic peptides (Shai & Oren, 1996; Oren *et al.*, 1997; Oren & Shai,

1997; Hong *et al.*, 1999). According to this model membrane permeating peptides are in contact with the phospholipid head group through the entire process of membrane permeation. The peptides initially bind onto the surface of the target membrane and cover it in a carpet-like manner. Membrane permeation occures only if there is a high local concentration of membrane bound peptide. High local concentration on the surface of the membrane can occure either after all the surface of the membrane is covered with peptide monomers, or alternatively, there is a recognition between membrane bound peptides and therefore a local surface is covered with the peptide forming a local "carpet". In the "carpet" model, contrary to the "barrel-stave" model, peptides are not inserted into the hydrophobic core of the membrane (left panel in Figure 1), neither they assemble with their hydrophilic surfaces facing each other. Furhtermore, a peptide that permeat the membrane via this mechanism does not necessarily require to adopt a specific structure upon its binding to the membrane. A common feature of antimicrobial peptides is their high net positive charge. Initial interaction between these peptides and the negatively-charged target membrane is therefore electrostatically driven. Four steps were proposed to be involved in this process: (i) preferential binding of peptide monomers to the phospholipid head groups; (ii) laying of the peptide monomers on the surface of the membrane so that their hydrophilic surface is facing the phospholipid headgroups or water molecules, (iii) rotation of the molecule leading to reorientation of the hydrophobic residues toward the hydrophobic core of the membrane (Gazit et al., 1996; La Rocca *et al.*, 1999), and (iv) disintegrating the membrane by disrupting the bilayer curvature. An initial step before the collapse of the membrane packing may include transient holes in the membrane. Holes like these may enable the passage of low molecular weight molecules prior to complete membranal lysis. Such holes were proposed to describe the mode of action of dermaseptin (Mor et al., 1994a) and named a toroidal (or wormhole) model (Ludtke *et al.*, 1996) (Step B in left panel of Figure 1). As seen in Figure 1, these holes may allow the passage of peptide molecules from the outer membrane into the inner membrane of, for example, Gram-negative bacteria in a process which may be referred to as "self-promoting uptake" (Sawyer *et al.*, 1988; Falla *et al.*, 1996).

In order to allow a comparison between the "barrel-stave" and the "carpet" mechanisms, the properties of the antimicrobial peptides will be compared with those of two well known channel forming peptide, alamethicin and pardaxin. Alamethicin is a 20-amino acids antibiotic peptide produced by the fungus *Trichoderma viride* (Meyer & Reusser, 1967; Pandery *et al.*, 1977) that belongs to the peptaibol family (Sansom, 1993). The peptide assumes an amphipatic α-helical structure that is bent about Pro^{14}. The middle of the peptides and the C-terminus possess H-bonding patterns characteristic of 3_{10} helix (Fox & Richards, 1982). Alamethicin has been studied for over 20 years as a model for voltage-gated ion channels. The high concentration dependence of conduction, and the multistep conductances seen in single-channel recording were interpreted in terms of a barrel-stave model for the channel pore (for reviews see (Woolley & Wallace, 1992; Sansom, 1993; Cafiso, 1994)).

Pardaxin a 33-mer polypeptide, is an excitatory neurotoxin that has been purified from the Red Sea Moses Sole *Pardachirus marmoratus* (Lazarovici *et al.*, 1986; Shai et al., 1988) and from the Peacock Sole of the western Pacific *Pardachirus pavoninus*

(Thompson *et al.*, 1986). Pardaxin possesses a variety of biological activities depending upon its concentration (reviewed in (Shai, 1994)). Its biological roles have been attributed to its interference with the ionic transport of the osmoregulatory system in epithelium and to presynaptic activity by forming ion channels that are voltage dependent and slightly selective to cations. A "barrel-stave" mechanism for insertion of pardaxin into membranes was proposed on the basis of its structure and various biophysical studies (Rapaport & Shai, 1991; Rapaport & Shai, 1992) (reviewed in (Shai, 1994)). Pardaxin has a helix-hinge-helix structure; the N-helix includes residues 7-11 and the C-helix includes residues 14-26. The helices are separated by a proline residue situated at position 13 (Zagorski *et al.*, 1991).

The conclusions obtained from the studies on pardaxin and its analogues, as well as on selected segments from δ- endotoxin (Gazit & Shai, 1993a; Gazit & Shai, 1995), revealed several characteristics of pore forming polypeptides as follows:

(i) They bind to membranes in a cooperative manner.
(ii) They bind strongly to both charged and zwitterionic phospholipids, which implies that hydrophobic interactions play a major role in their mode of action.
(iii) They penetrate into the hydrophobic region of the membrane.
(iv) They specifically self-associate in their membrane-bound state.
(v) They induce single channel fluctuations in the current amplitude that increase greatly with time. (Shai et al., 1990; Gazit *et al.*, 1994a; Shai, 1994).

Studies on the interaction of antimicrobial peptides with model phospholipid membranes revealed low affinity to zwitterionic phospholipids compared to acidic phospholipids. This has been demonstrated with cecropins (Gazit et al., 1994b; Gazit et al., 1995), magainins (Matsuzaki *et al.*, 1989; Williams *et al.*, 1990; Gomes *et al.*, 1993) dermaseptins (Pouny et al., 1992; Strahilevitz et al., 1994) and others (Shai & Oren, 1996; Latal *et al.*, 1997; Oren et al., 1997; Oren & Shai, 1997). The low affinity of antimicrobial peptides to zwitterionic membranes might explain their inability to lyse erythrocytes. However, there are exceptions. Functional studies with dermaseptin S4 (Ghosh et al., 1997) and the human like cecropin, LL-37 (Oren et al., 1999) demonstrated that they bind and permeat efficiently both zwitterionic and negatively-charged phospholipid vesicles, compared to other native antibacterial peptides which permeate efficiently only negatively-charged membranes. The high affinity of these peptides towards zwitterionic PC phospholipid membranes is surprising in light of their high net positive charge. This may suggest the involvement of hydrophobic interactions between the peptides and the zwitterionic membranes. LL-37 and dermaeptin S4 reaches PC membranes as oligomers. Since the N-terminus of both peptides is hydrophobic, a bundle of N-termini regions can initiate binding to the membrane.

1.1.3 Why "Carpet" Mechanism May Be Preferred By Antibacterial Polypeptides?

The finding of a common mechanism of action for antibacterial polypeptides, i.e. non-cooperative binding to membranes and discrimination between charged and zwitterionic phospholipids, led to speculate why such peptides are efficient antibacterial agents. The most likely target of the membrane permeating antibacterial peptides is the inner membrane of bacteria, which typically contains the electron-transport chain and the enzymatic apparatus necessary for oxidative phosphorylation (Westerhoff *et al.*,

1989). To reach the target membrane, the peptides have to traverse the bacterial wall [composed of lipopolysaccarides (LPS) in the case of Gram-negative bacteria] and the peptidoglycan layer, both of which are negatively charged. The net positive charge of the antibacterial peptides should facilitate their binding to bacteria. A non-cooperative, as opposed to cooperative binding of the peptides to the outer surface of bacteria wall will make it easier for the peptide molecules to diffuse into the inner membrane and to permeate it, rather than to remain on the outer surface as large aggregates.

1.2. ELUCIDATING THE ORGANIZATION OF MEMBRANE PROTEINS THROUGH STUDIES WITH FRAGMENTS AND SYNTHETIC PEPTIDES DERIVED FROM THE PROTEINS

Significant number of studies were done on isolated fragments of membrane proteins to determine the oligomerization state and the organization of the transmembrane domains, because of the very limited success in studying the structure of intact membrane proteins. In line of this approach a "two-stage" model was proposed for membrane proteins folding and oligomerization (Popot *et al.*, 1987; Popot & Engelman, 1990). The model suggests that the final structure of a protein in membranes results from the packing of smaller elements, each of which reaches thermodynamic equilibrium with the lipid and aqueous phases before packing. The model excludes structures incorporating transmembrane segments that are not individually stable in the membrane. In support of this hypothesis, several studies have demonstrated that isolated transmembrane helices of membrane proteins can assemble in vitro within a bilayer environment into functional proteins. Examples are: bacteriorhodopsin (Liao *et al.*, 1983; Popot et al., 1987; Kahn & Engelman, 1992), lactose permease of *Escherichia coli* (Bibi & Kaback, 1990; Sahin-Toth et al., 1992) and the β_2-adrenergic receptor.(Kobilka *et al.*, 1988). These studies strengthen the notion that fragments of membrane proteins can adopt their native structure and position and therefore, support the two stage model. The examples described and others (reviewed in (Lemmon et al., 1992; Von Heijne, 1994; Shai, 1995)), suggest that transmembrane segments can contribute to specific recognition and assembly with other proteins as well.

Extensive studies were conducted with the single transmembrane domain of glycophorine A which dimerizes in detergents (Lemmon et al., 1992; Fleming *et al.*, 1997; MacKenzie *et al.*, 1997; MacKenzie & Engelman, 1998; Fisher *et al.*, 1999). Phospholamban, a 52-amino-acid protein that assembles into a pentamer in sarcoplasmic reticulum membranes, is another low molecular weight protein that contain a single transmembrane domain that was investigated thoroughly (Arkin *et al.*, 1997a; Arkin *et al.*, 1997b; Adams *et al.*, 1998). The protein has a role in the regulation of the resident calcium ATPase through an inhibitory association that can be reversed by phosphorylation. There is evidence that phospholamban may also function as a Ca^{2+}-selective ion channel. The structural properties of phospholamban have been studied by mutagenesis, modeling, and spectroscopy, resulting in a new view of the organization of this key molecule in membranes. However, glycophorine A and phospholamban represent membrane proteins with a single transmembrane domain, and hence only homo-oligomerization could be studied.

We have utilized a synthetic peptide approach to obtain information on the structure and organization within membranes of ion channels, pore forming toxins and envelope proteins of viruses. The principle underlying this approach is to characterize selected membrane segments from these membrane proteins, first as autonomous units, and then each in the presence of other membrane segments from the same protein, or from unrelated membrane proteins. The synthetic segments are labeled with fluorescent probes and subjected to various biophysical and functional studies to get information on their structure and organization within the membrane. The methodology includes the following steps: (i) Peptide synthesis and fluorescent labelling; (ii) Determination of the affinity of the synthetic segments to phospholipid membranes using spectroscopic methods; (iii) Determination of the structure of the segments in phospholipid membranes or membrane mimetic environments by means of Circular Dichroism (CD) spectroscopy and Attenuated Total Reflectance Fourier-transform infrared (ATR-FTIR) spectroscopy; (iv) Finding the localization of the segments in their membrane-bound state; (v) Determination of the orientation of the segments when membrane-bound; (vi) Detection of a molecular recognition between membrane-bound segments by means of fluorescence and ATR-FTIR spectroscopy; (vii) Observation of conformational/localization changes of a particular segment when it interacts with another segment. Other complemtary methods such as spin-labelled ESR and nmr were also used to investigate the structure and organization of synthetic segments of ion channels (Aggeli *et al.*, 1998) (reviewed by Marsh, 1996 (Marsh, 1996)).

A support to the segment analysis approach comes from several studies, which show a correlation between the structure and the organization of synthetic peptides as compared to their parent molecule. Such studies were performed for both soluble (Jaenicke, 1991; Dyson *et al.*, 1992; Kippen *et al.*, 1994) and membrane proteins (Kahn & Engelman, 1992; Lomize *et al.*, 1992; Pervushin & Arseniev, 1992). Other examples come from studies showing that various synthetic segments from bacteriorhodopsin (Barsukov *et al.*, 1992), the pore regions of δ-endotoxin (Gazit & Shai, 1993a), *Bacillus thuringiensis var. israelensis* cytolytic toxin (CytA) (Gazit & Shai, 1993b; Li *et al.*, 1996) and the influenza hemaglutanin (Carr & Kim, 1993), adopt conformations similar to those of their corresponding segments within the intact protein as determined by x-ray crystallography (reviewed in Shai, 1995 (Shai, 1995)). Further support for the segment analysis comes from the findings that synthetic peptides corresponding to the membrane-bound portion of envelope protein of viruses can specifically assemble with the corresponding parts in the intact proteins. This interaction interferes with the functional assembly of the protein and inhibit its function (Kliger *et al.*, 1997). Very interestingly, we found that peptide chirality does not play a role in peptide-peptide recognition within the membrane milieu. This is demonstrated by the findings that a synthetic all D-Amino Acid peptide corresponding the N-terminal sequence of HIV-1 gp41 (D-WT) of HIV-1 associates with its enantiomeric wild type fusion (WT) peptide in the membrane and inhibits cell fusion mediated by the HIV-1 envelope glycoprotein. D-WT does not inhibit cell fusion mediated by the HIV-2 envelope glycoprotein (Pritsker *et al.*, 1998).

Experiments done with synthetic peptides were also aimed at reconstructing an ion-selective pore in artificial bilayers whose properties match as closely as possible those of the native channel, in order to locate the pore-lining segments of different ion channels (reviewed in (Montal, 1995; Shai, 1995; Montal, 1996).

1.2.1 The Inward Rectifying ROMK1 K^+ Channel.

The ROMK1 channel is an inwardly rectifying K^+ channel that conduct an inward K^+ current at hyperpolarizing membrane potentials, and therefore play an important role in regulating the resting membrane potential and electrical excitability of cells in a variety of tissues, including the brain and heart (Ho *et al.*, 1993; Shuck *et al.*, 1994). Hydropathy plots have suggested only two potential transmembrane domains, M1 and M2 and a pore forming region, H5, which is involved in ion conductance (Figure 2, panel A) (Ho et al., 1993). For voltage-dependent K^+ channels, six transmembrane segments (S1 to S6) have been postulated (Papazian *et al.*, 1987). The H5 domain is highly homologous between inwardly rectifying K^+ channels and voltage gated K^+ channels. A small degree of similarity between M1 and S5 and between M2 and S6 also exists. Based on this homology, it was suggested that inwardly-rectifying channels are structurally analogous to the inner core of the *Shaker* superfamily of K^+ channels (Kubo *et al.*, 1993). However, hydropathy plots of the inwardly rectifying K^+ channels reveal areas of intermediate hydrophobicity in the extended C-terminal region (Taglialatela *et al.*, 1994), and the shorter N-terminal region (Ho et al., 1993), which are reminiscent of the degree of hydrophobicity of H5, a region that was initially believed to be extracellular (Catterall, 1988). Based on experimental results, it was suggested that in inwardly rectifying channels parts of the C-terminus are likely to be in the membrane and that the C-terminus makes a major contribution to the pore (Taglialatela et al., 1994; Pessia *et al.*, 1995).

We have utilized a detailed structural study using the synthetic segment approach to investigate of the ROMK1 K^+ channel (Ben Efraim & Shai, 1996; Ben Efraim & Shai, 1997). We investigated the structure, in a hydrophobic environment, localization, and the organization
within phospholipid membranes of six segments derived from the ROMK1 channel, namely Pre-M0, M0, M1, H5, M2 and post-M2 (Figure 2, panel A).

Three methods were utilized in order to evaluate the localization of the various peptides with respect to the membrane: (1) peptides were labeled with the enviromental sensitive fluorescent probe NBD (nitro benzo furazan) which has high quantum yield in hydrophobic enviroment. Based on the NBD shift we found that apart from Pre-M0, all other five peptides bind strongly to the membrane. (2) tryptophan quenching experiments using brominated phospholipids in the case when a peptide contained tryptophan within its sequence. (3) membrane bound peptides were subjected to proteolytic degradation to find out whether the peptides are localized on the surface of the membrane or inserted within the lipid environment. When the proteolytic enzyme proteinase K was applied to membrane-bound peptides only M0 and Post-M2 were cleaved, but M1, M2 and H5 were totally resistance to digestion.

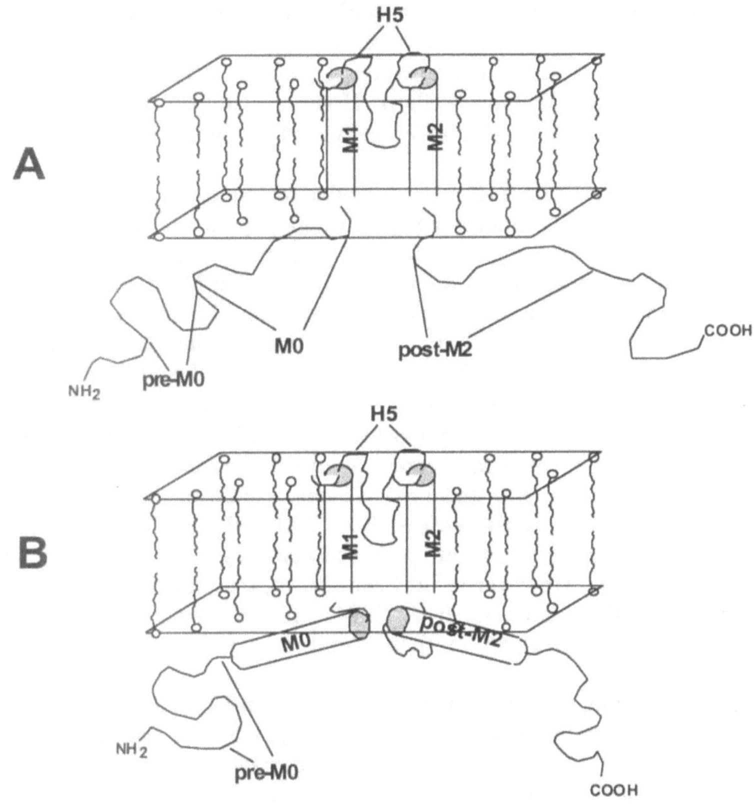

Figure 2: A schematic representation illustrating the proposed topology of the ROMK1 channel as predicted by Kyte and Doollitle Hydropathy plot (panel A) and based on the results obtained using the synthetic peptides (panel B).

Based on NBD shift, the NBD labeled N-terminal group of M2 and H5 seems to be surface localized, whereas the NBD labeled N-terminal group of M1 is membrane embedded. Based on these results we proposed that beside M1, M2 and H5, other domains from the N- and C-termini are also capable of interacting with membranes. A schematic view of our revised topology of the ROMK1 channel is shown in Figure 2 panel B.

These findings are in agreement with experimental evidence in which exchange of the C-terminus altered pore properties, which suggested that in inwardly rectifying channels parts of the C-terminus are likely to be in the membrane and that the C-terminus seems to make a major contribution to the pore (Taglialatela et al., 1994; Pessia et al., 1995).

Secondary structure determination using CD spectroscopy revealed that both M1 and M2 are predominantly α-helical when in a hydrophobic environment (40%

TFE). The structure of H5, however, could not be interpreted since the peptide did not yield any significant CD signal. Secondary structure determination of M0 and Post-M2 revealed that both are predominantly α-helices when in a membrane mimetic environment (40% TFE, and 1% SDS). However, compared to M1 and M2 their degree of α-helicity is lower. Pre-M0, which is located just before M0 did not yield significant CD signal.

We further demonstrated molecular recognition between pairs of membrane-bound segments using fluorescently labelled peptides with fluorescent donors and acceptors. Resonance energy transfer (RET) studies revealed that membrane bound H5 and M2 could self-associate in the membrane, but that M1 did not (Ben Efraim & Shai, 1997). Moreover, H5 was found to coassemble with M2, and M2 coassembled with M1. However, M1 did not coassemble with a membrane inserted oligomer of H5. These findings suggest that M2 and H5 form the inner part of the pore and that M1 participates in the formation of the outer ring of α-helices. We also suggest that other regions that participate in pore formation may exist in the C-and N-termini of ROMK1 (Ben Efraim & Shai, 1996). This was demonstrated by our finding that besides their ability to bind strongly to membranes M0 and Post-M2 coassemble in their membrane-bound state, although they do not self-associate. Both M0 and Post-M2 are positively charged, especially Post-M2, and it might be argued that they do not self-associate due to their net positive charge. However it has been shown, that positively charged peptides such as dermaseptin B (Strahilevitz et al., 1994) can self-assemble in membranes, therefore the lack of self-association in these peptides could not arise solely from charge repulsion.

In conclusion, based on membrane localization, RET and secondary structure studies of putative membranous peptide segments of the ROMK1 channel, we suggested a model for the overall putative organization of the peptides in the ROMK1 potassium channel as depicted in the scheme of Figure 3. If we assume that the channel is built up of tetramers, as suggested for the Shaker voltage activated potassium channel (MacKinnon, 1991), then the inner core is composed of a tetramer of H5 domains surrounded by M1 and M2 regions. Post-M2 and M0 are hypothesized to be surface localized and surround M1 and M2 forming interactions between each other. Most interestigly, the exact topology was found in a recently crystallized pore region of a homologous K^+ channel from Streptomyces lividans (Doyle et al., 1998). The agreement on the topology of the inner part of the channel obtained with the synthetic peptides compared to the x-ray data, suggests that the rest of the topology is correct too. Similar results on the structure and the membrane localization of the ROMK1 synthetic peptides were found by Haris et al., (Haris, 1998).

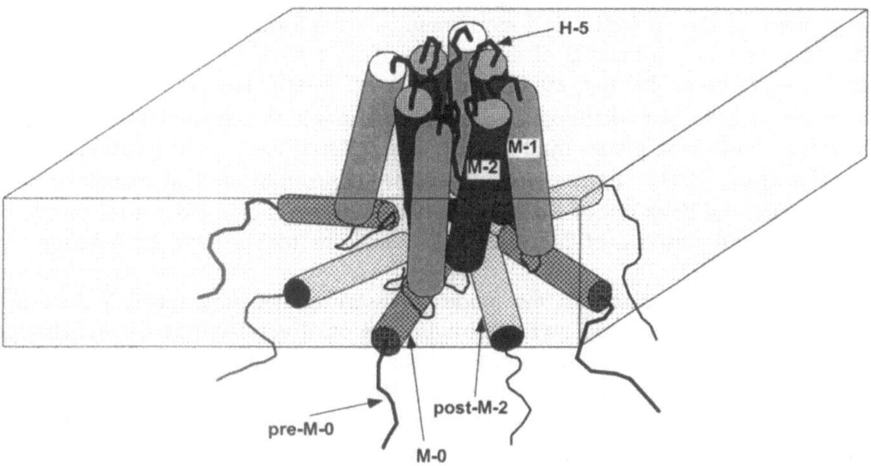

Figure 3: Upper view schematic illustration of a possible organization of the membrane domains of ROMK1 K^+ channel within the lipid environment based on the RET studies and assuming the channel is a tetramer.

1.3. FUNCTIONAL STRUCTURE OF MEMBRANE INSERTED δ-ENDOTOXIN.

The δ-endotoxins are highly potent insecticidal toxins produced by *Bacillus thuringiensis* bacteria. The use of δ- endotoxins rather than conventional chemical pesticides is preferable, both because of their high specificity and efficiency and because of their environmental safety and lack of harmful side effects. The δ-endotoxins are members of a larger group of membrane pore forming toxins of bacterial origin, such as colicin, α-toxin, and aerolysin (Li *et al.*, 1991; Parker & Pattus, 1993). All of these toxins are water soluble proteins that, in order to exert their activity, need to undergo a conformational change from hydrophilic to amphipatic structures that will allow them to insert into the hydrophobic core of the cell membrane. These toxins may therefore have similar mechanisms for their membrane pore forming activities.

Research over the past three decades has attempted to elucidate the structure and activity of δ-endotoxins. The three dimensional structure of several of these toxins has been solved and despite very different specificities, the structures are almost superimposable (Li et al., 1991; Grochulski *et al.*, 1995; Bravo, 1997). There are three domains whereas domain I which is comprised of a bundle of seven amphipathic α-helices is important for toxicity, and domain II for receptor binding (Figure 4, panel A). It is believed that one or more of these alpha helices inserts into the membrane to form ion channels. Either osmotic swelling of larval midgut cells or a more subtle perturbation of the ion flow across the cell membrane appears to account for the lethality. It is assumed that the trigger for the insertion of the pore-forming domain of the toxins into the epithelial cell membrane is a conformational change in the toxin, which occurs when domain II of the toxin binds to a receptor present on brush-border membranes (Ahmad & Ellar, 1990; Van Rie *et al.*, 1990). The pore-forming properties of the toxins have been demonstrated by studies in which activated δ-endotoxins form single ion channels in planar lipid bilayers and cultured insect cells (Slatin *et al.*, 1990; Schwartz *et al.*, 1991). Extensive mutagenesis studies indicate that mutations in α5, a highly conserved helix out of the seven helices of domain one, but not α2 or α6, result in a substantial number of inactive or low-activity toxins (Wu & Aronson, 1992; Walters *et al.*, 1993).

In order to elucidate the mode of insertion the pore forming domain into membranes, we synthesized and determined the relative affinities for membranes of peptides corresponding to the seven helices that compose the toxin pore-forming domain (Gazit & Shai, 1993a; Gazit et al., 1994a; Gazit & Shai, 1995; Gazit et al., 1998). In addition, their modes of membrane interaction, their structures within membranes, and their orientations relative to the membrane normal were investigated. The sequences of the peptides is given in Figure 4, panel B. Furthermore, we used resonance energy transfer measurements of all possible combinatorial pairs of membrane-bound helices to map the network of interactions between helices in their membrane-bound state. The interaction of the helices with the bilayer membrane was also probed by a Monte Carlo simulation protocol to determine lowest energy orientations. The results are consistent with a situation in which all the helices beside helix 1 interact with the membrane. Those helices may have an important role in the initial interaction of the pore-forming domain with the membrane surface. Furthermore, helices α4 and α5 insert into the membrane as a helical hairpin in an antiparallel manner, whilst the other helices lie on the membrane surface like the ribs of an umbrella (the "umbrella model"). Evidence suggesting that α4 and α5 have a structural role in the lining of the δ-endotoxins pores in an "umbrella-like" structure includes: (1) their ability to self- and coassemble within phospholipid membranes as determined by using resonance energy transfer; (2) their transmembrane orientation as determined by ATR-FTIR spectroscopy; (3) the membrane surface orientation of the rest of the membrane bound helices (helices α2, α3, α6 and α7,), and (4) the network of interactions between the membrane-bound helices. The data reveal that α4 and α5 insert into the membrane in an antiparallel manner as a helical hairpin in agreement with the hydrophobic hairpin hypothesis (Engelman & Steitz, 1981), suggested for the insertion of proteins into membranes. The other helices are spread on the surface of the membrane (Figure 4, panel B). The insertion the of α4-α5 hairpin into the membrane is also expected from theoretical considerations since (i) the C-terminal of α4, the loop between α4 and α5, and the N-terminal of α5 forms an hairpin that contains the least polar segment of domain I (8), and (ii) the helices are joined on the side of the pore-forming domain proximal to the membrane.

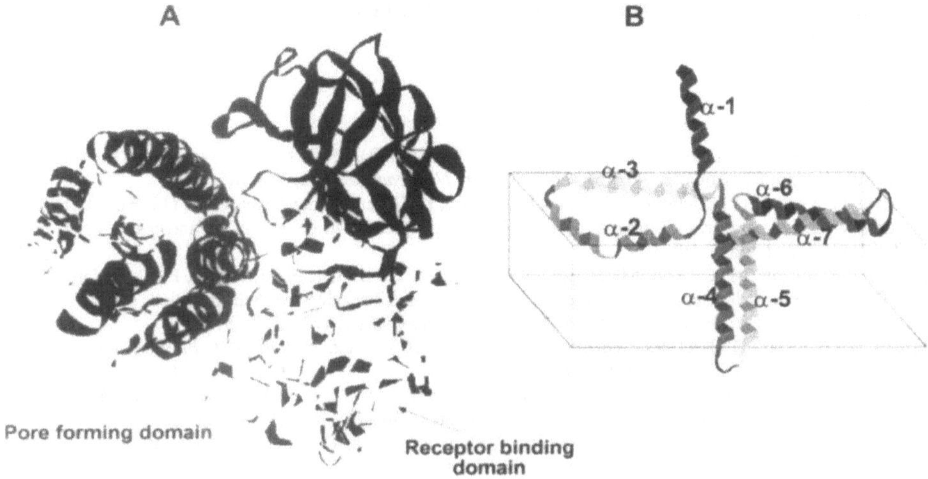

Figure 4: A. The structure of CryIIIA *B. thuringiensis* d- endotoxin as determined by Li et al., 1991). B. schematic presentation of a proposed model for the interaction of δ-endotoxin with phospholipid membranes based on the synthetic peptide approach.. The loop connecting a4 and a5 may be either in an intracellular localization or may interact with the inner leaflet of the membrane due to its hydrophobicity.

The "umbrella" model is also consistent with (i) mutational analysis demonstrating that mutations in α5, but not in α2 or α6, affect the toxicity of the δ-endotoxin (Aronson *et al.*, 1995); (ii) the observation that the introduction of a negative charge in the loop between α4 and α5, but not between α5 and α6 or α3 and α4, drastically reduced the toxicity of the toxin (Hussain *et al.*, 1996); and (iii) the introduction of a negative charge in the loop between α4 and α5 reduces irreversible binding, but not the Kd, to BBMV (Chen *et al.*, 1995). Following receptor binding, the network of contacts between α7, the helix in the interface between the pore-forming domain and the receptor-binding domain, and α5, α6, and presumably α4 helices, may assist the insertion of the α4-α5 hairpin into the membrane by the unpacking of the helical bundle that exists in the non-membrane bound form of the toxin. This hypothesis could account for the observation that α7 mutants are susceptible to proteolysis by either trypsin or midgut juice (Dean *et al.*, 1996). Presumably the mutations abolish the pre-packing of the non-membrane bound toxin. Further support for the ability of α5 and α4-α5 loop to insert into the membrane comes from studies with synthetic α5 (Cummings *et al.*, 1994) and mutation in the intact toxin. Recently, Schwartz and colleagues suggested a more detailed structure according to the 'Umbrella model' (Masson *et al.*, 1999). According to the revised model, domain I drifts away from domain II and III after initial binding. A hairpin inserts into the membrane with α4 facing the lumen of the channel and α5 facing the lipid interface, while the other helices spread on the surface of the membrane. Using cysteine labeling, they showed that α4 is accessible to the lumen of the channel. Their conclusion was that α4-loop-α5 is inserted into the membrane like in the original 'Umbrella model', and that α4 faces the lumen of the channel while α5 is at an outer circle, facing the hydrophobic environment of the membrane milieu.

REFERENCES

Adams, P. D., Lee, A. S., Brunger, A. T. & Engelman, D. M. (1998). Models for the transmembrane region of the phospholamban pentamer: which is correct? *Ann. NY Acad. Sci.* **853**, 178-185.

Agerberth, B., Gunne, H., Odeberg, J., Kogner, P., Boman, H. G. & Gudmundsson, G. H. (1995). FALL-39, a putative human peptide antibiotic, is cysteine-free and expressed in bone marrow and testis. *Proc. Natl. Acad. Sci. U.S.A.* **92**(1), 195-199.

Aggeli, A., Bannister, M. L., Bell, M., Boden, N., Findlay, J. B., Hunter, M., Knowles, P. F. & Yang, J. C. (1998). Conformation and ion-channeling activity of a 27-residue peptide modeled on the single-transmembrane segment of the IsK (minK) protein. *Biochemistry* **37**(22), 8121-8131.

Ahmad, W. & Ellar, D. J. (1990). Directed mutagenesis of selected regions of a *Bacillus thuringiensis* entomocidal protein. *FEMS Microbiol. Lett.* **68**, 97-104.

Arkin, I. T., Adams, P. D., Brunger, A. T., Aimoto, S., Engelman, D. M. & Smith, S. O. (1997a). Structure of the transmembrane cysteine residues in phospholamban. *J. Membr. Biol.* **155**(3), 199-206.

Arkin, I. T., Adams, P. D., Brunger, A. T., Smith, S. O. & Engelman, D. M. (1997b). Structural perspectives of phospholamban, a helical transmembrane pentamer. *Annu. Rev. Biophys. Biomol. Struct.* **26**, 157-179.

Aronson, A. I., Wu, D. & Zhang, C. (1995). Mutagenesis of specificity and toxicity regions of a *Bacillus thuringiensis* protoxin gene. *J. Bacteriol.* **177**(14), 4059-4065.

Babila, T., Moscucci, A., Wang, H., Weaver, F. E. & Koren, G. (1994). Assembly of mammalian voltage-gated potassium channels: evidence for an important role of the first transmembrane segment. *Neuron* **12**(3), 615-626.

Barsukov, I. L., Nolde, D. E., Lomize, A. L. & Arseniev, A. S. (1992). Three-dimensional structure of proteolytic fragment 163-231 of bacterioopsin determined from nuclear magnetic resonance data in solution. *Eur J Biochem* **206**(3), 665-672.

Ben Efraim, I. & Shai, Y. (1996). Secondary structure, membrane localization, and coassembly within phospholipid membranes of synthetic segments derived from the N- and C- termini regions of the ROMK1 K+ channel. *Protein Sci.* **5**(11), 2287-2297.

Ben Efraim, I. & Shai, Y. (1997). The structure and organization of synthetic putative membranous segments of ROMK1 channel in phospholipid membranes. *Biophys. J.* **72**(1), 85-96.

Ben Efraim, I., Shai, Y. & Attali, B. (1996). Cytoplasmic and extracellular IsK peptides activate endogenous K+ and Cl- channels in Xenopus oocytes. Evidence for regulatory function. *J. Biol. Chem.* **271**(15), 8768-8771.

Ben Efraim, I., Strahilevitz, J., Bach, D. & Shai, Y. (1994). Secondary structure and membrane localization of synthetic segments and a truncated form of the IsK (minK) protein. *Biochemistry* **33**(22), 6966-6973.

Bibi, E. & Kaback, H. R. (1990). In vivo expression of the lacY gene in two segments leads to functional lac permease. *Proc. Natl. Acad. Sci. USA* **87**, 4325-4329.

Blondelle, S. E. & Houghten, R. A. (1991). Hemolytic and antimicrobial activities of the twenty-four individual omission analogues of melittin. *Biochemistry* **30**(19), 4671-4678.

Boman, H. G. (1995). Peptide antibiotics and their role in innate immunity. *Annu. Rev. Immun.* **13**, 61-92.

Boman, H. G. & Hultmark, D. (1987). Cell-free immunity in insects. *Annu. Rev. Microbiol.*

Bonifacino, J. S., Suzuki, C. K. & Klausner, R. D. (1990). A Peptide Sequence Confers Retention and Rapid Degredation in the Endoplasmic Reticulum. *Science* **247**, 79-82.

Bormann, B. J., Knowles, W. J. & Marchesi, V. T. (1989). Synthetic Peptides Mimic the Assembly of Transmembrane Glycoproteins. *J. Biol. Chem* **264**(7), 4033-4037.

Bravo, A. (1997). Phylogenetic relationships of Bacillus thuringiensis delta-endotoxin family proteins and their functional domains. *J. Bacteriol.* **179**(9), 2793-2801.

Brock, T. D. (1974). *Biology of Microorganisms*. 2nd edition edit, Prentice-Hall Inc., Englewood Cliffs, N.J.

Cafiso, D. S. (1994). Alamethicin: a peptide model for voltage gating and protein-membrane interactions. *Annu. Rev. Biophys. Biomol. Struct.* **23**, 141-165.

Carr, C. M. & Kim, P. S. (1993). A spring-loaded mechanism for the conformational change of influenza hemagglutinin. *Cell* **73**(4), 823-832.

Catterall, W. A. (1988). Structure and function of voltage-sensitive ion channels. *Science* **242**, 50-61.

Charpentier, S., Amiche, M., Mester, J., Vouille, V., Le Caer, J. P., Nicolas, P. & Delfour, A. (1998). Structure, synthesis, and molecular cloning of dermaseptins B, a family of skin peptide antibiotics. *J.*

Biol. Chem. **273**(24), 14690-14697.

Chen, X. J., Curtiss, A., Alcantara, E. & Dean, D. H. (1995). Mutations in domain I of *Bacillus thuringiensis* δ-endotoxin CryIAb reduce the irreversible binding of toxin to *Manduca sexta* brush border membrane vesicles. *J. Biol. Chem.* **270**, 6412-6419.

Cummings, C. E., Armstrong, G., Hodgmann, T. C. & Ellar, D. J. (1994). Structural and functional studies of a synthetic peptide mimicking a proposed membrane inserting region of a *Bacillus thuringiensis* δ-endotoxin. *Mol. Memb. Biol.* **11**, 87-92.

Dean, D. H., Rajamohan, F., Lee, M. K., Wu, S.-J., Chen, X. J., Alcantara, E. & Hussain, S. R. (1996). Probing the mechanism of action of *Bacillus thuringiensis* insecticidal proteins by site directed mutagenesis - a minireview. *Gene* **179**, 111-117.

Dhople, V. M. & Nagaraj, R. (1993). δ-toxin, unlike melittin, has only hemolytic activity and no antimicrobial activity: rationalization of this specific biological activity. *Biosci. Rep.* **13**(4), 245-250.

Doyle, D. A., Morais Cabral, J., Pfuetzner, R. A., Kuo, A., Gulbis, J. M., Cohen, S. L., Chait, B. T. & MacKinnon, R. (1998). The structure of the potassium channel: molecular basis of K+ conduction and selectivity. *Science* **280**(5360), 69-77.

Dyson, H. J., Sayre, J. R., Merutka, G., Shin, H. C., Lerner, R. A. & Wright, P. E. (1992). Folding of peptide fragments comprising the complete sequence of proteins. Models for initiation of protein folding. II. Plastocyanin. *J. Mol. Biol.* **226**(3), 819-835.

Ehrenstein, G. & Lecar, H. (1977). Electrically gated ionic channels in lipid bilayers. *Q. Rev. Biophys.* **10**, 1-34.

Engelman, D. M. & Steitz, T. A. (1981). The spontaneous insertion of proteins into and across membranes: the helical hairpin hypothesis. *Cell* **23**(2), 411-422.

Falla, T. J., Karunaratne, D. N. & Hancock, R. E. W. (1996). Mode of action of the antimicrobial peptide indolicidin. *J. Biol. Chem.* **271**(32), 19298-19303.

Fisher, L. E., Engelman, D. M. & Sturgis, J. N. (1999). Detergents modulate dimerization, but not helicity, of the glycophorin A transmembrane domain. *J. Mol. Biol.* **293**(3), 639-651.

Fleming, K. G., Ackerman, A. L. & Engelman, D. M. (1997). The effect of point mutations on the free energy of transmembrane alpha- helix dimerization. *J. Mol. Biol.* **272**(2), 266-275.

Fox, R. O., Jr. & Richards, F. M. (1982). A voltage-gated ion channel model inferred from the crystal structure of alamethicin at 1.5-A resolution. *Nature* **300**(5890), 325-330.

Frachet, P., Duperray, A., Delachanal, E. & Marguerie, G. (1992). Role of transmembrane and cytoplasmic domains in the assembly andsurface exposure of the platelet integrin GPIIb/IIIa. *Biochemistry* **31**, 2408-2415.

Gazit, E., Bach, D., Kerr, I. D., Sansom, M. S., Chejanovsky, N. & Shai, Y. (1994a). The alpha-5 segment of Bacillus thuringiensis delta-endotoxin: in vitro activity, ion channel formation and molecular modelling. *Biochem. J.* **304**(Pt 3)), 895-902.

Gazit, E., Boman, A., Boman, H. G. & Shai, Y. (1995). Interaction of the mammalian antibacterial peptide cecropin P1 with phospholipid vesicles. *Biochemistry* **34**(36), 11479-11488.

Gazit, E., La Rocca, P., Sansom, M. S. & Shai, Y. (1998). The structure and organization within the membrane of the helices composing the pore-forming domain of Bacillus thuringiensis delta- endotoxin are consistent with an "umbrella-like" structure of the pore. *Proc. Natl. Acad. Sci. USA* **95**(21), 12289-12294.

Gazit, E., Lee, W. J., Brey, P. T. & Shai, Y. (1994b). Mode of action of the antibacterial cecropin B2: a spectrofluorometric study. *Biochemistry* **33**(35), 10681-10692.

Gazit, E., Miller, I. R., Biggin, P. C., Sansom, M. S. P. & Shai, Y. (1996). Structure and orientation of the mammalian antibacterial peptide cecropin P1 within phospholipid membranes. *J. Mol. Biol.* **258**(5), 860-870.

Gazit, E. & Shai, Y. (1993a). Structural and functional characterization of the α-5 segment of Bacillus thuringiensis δ-endotoxin. *Biochemistry* **32**(13), 3429-3436.

Gazit, E. & Shai, Y. (1993b). Structural characterization, membrane interaction, and specific assembly within phospholipid membranes of hydrophobic segments from *Bacillus thuringiensis* var. *israelensis* cytolytic toxin. *Biochemistry* **32**(46), 12363-12371.

Gazit, E. & Shai, Y. (1995). The assembly and organization of the alpha 5 and alpha 7 helices from the pore-forming domain of Bacillus thuringiensis delta-endotoxin. Relevance to a functional model. *J. Biol. Chem.* **270**(6), 2571-2578.

Ghosh, J. K., Shaool, D., Guillaud, P., Ciceron, L., Mazier, D., Kustanovich, I., Shai, Y. & Mor, A. (1997). Selective cytotoxicity of dermaseptin S3 toward intraerythrocytic Plasmodium falciparum and the underlying molecular basis. *J. Biol. Chem.* **272**(50), 31609-31616.

Gierasch, L. M. & King, J. (1989). *Protein Folding: Deciphering the second half of the genetic code*, The American Association for the Advancement of Science, Washington, D.C.

Glasser, L. & Scheraga, H. A. (1991). Investigation of a physical basis for conformational similarity in proteins. *J Protein Chem* **10**(3), 273-85.

Gomes, A. V., de Waal, A., Berden, J. A. & Westerhoff, H. V. (1993). Electric potentiation, cooperativity, and synergism of magainin peptides in protein-free liposomes. *Biochemistry* **32**(20), 5365-5372.

Grochulski, P., Masson, L., Borisova, S., Pusztai-Carey, M., Schwartz, J. L., Brousseau, R. & Cygler, M. (1995). Bacillus thuringiensis CryIA(a) insecticidal toxin: crystal structure and channel formation. *J. Mol. Biol.* **254**(3), 447-464.

Gudmundsson, G. H., Agerberth, B., Odeberg, J., Bergman, T., Olsson, B. & Salcedo, R. (1996). The human gene FALL39 and processing of the cathelin precursor to the antibacterial peptide LL-37 in granulocytes. *Eur. J. Biochem.* **238**, 325-332.

Habermann, E. & Jentsch, J. (1967). *Hoppe Seyler's Z. Physiol. Chem.* **348**, 37-50.

Hancock, R. E. & Lehrer, R. (1998). Cationic peptides: a new source of antibiotics. *Trends Biotechnol.* **16**(2), 82-88.

Haris, P. I. (1998). Synthetic peptide fragments as probes for structure determination of potassium ion-channel proteins. *Biosci. Rep.* **18**(6), 299-312.

Ho, K., Nichols, C. G., Lederer, W. J., Lytton, J., Vassilev, P. M., Kanazirska, M. V. & Hebert, S. C. (1993). Cloning and expression of an inwardly rectifying ATP-regulated potassium channel. *Nature* **362**(6415), 31-38.

Hoffmann, J. A., Kafatos, F. C., Janeway, C. A. & Ezekowitz, R. A. (1999). Phylogenetic perspectives in innate immunity. *Science* **284**(5418), 1313-1318.

Hong, J., Oren, Z. & Shai, Y. (1999). Structure and Organization of Hemolytic and Nonhemolytic Diastereomers of Antimicrobial Peptides in Membranes. *Biochemistry* **38**(51), 16963-16973.

Hultmark, D. (1993). Immune reactions in Drosophila and other insects: a model for innate immunity. *Trends Genet.* **9**(5), 178-183.

Hussain, S. A., Aronson, A. I. & Dean, D. H. (1996). Substitution of residues on the proximal side of CryIA *Bacillus thuringiensis* δ-endotoxins affects irreversible binding to *Manduca sexta* midgut membrane. *Biochem. Biophys. Res. Commun.* **226**, 8-14.

Jaenicke, R. (1991). Protein folding: local structures, domains, subunits, and assemblies. *Biochemistry* **30**(13), 3147-3161.

Johansson, J., Gudmundsson, G. H., Rottenberg, M. E., Berndt, K. D. & Agerberth, B. (1998). Conformation-dependent Antibacterial Activity of the Naturally-Occuring Human Peptide LL-37. *J. Biol. Chem.* **273**, 3718-3724.

Kahn, T. W. & Engelman, D. M. (1992). Bacteriorhodopsin can be refolded from two independently stable transmembrane helices and the complementary Five-helix fragment. *Biochemistry* **31**, 6144-6151.

Kippen, A. D., Arcus, V. L. & Fersht, A. R. (1994). Structural studies on peptides corresponding to mutants of the major alpha- helix of barnase. *Biochemistry* **33**(33), 10013-10021.

Kliger, Y., Aharoni, A., Rapaport, D., Jones, P., Blumenthal, R. & Shai, Y. (1997). Fusion peptides derived from the HIV type 1 glycoprotein 41 associate within phospholipid membranes and inhibit cell-cell Fusion. Structure- function study. *J. Biol. Chem.* **272**(21), 13496-13505.

Kobilka, B. K., Kobilka, T. S., Daniel, K., Regan, J. W., Caron, M. G. & Lefkowitz, R. J. (1988). Chimeric &2-ß2-adrenergic receptors: Delineation of domains involved in Effector coupling and ligand binding specificity. *Science* **240**, 1310-1316.

Kreil, G. (1994). Peptides containing a D-amino acid from frogs and Molluscs. *J. Biol. Chem.* **266**, 6264-6270.

Kubo, Y., Baldwin, T. J., Jan, Y. N. & Jan, L. Y. (1993). Primary structure and functional expression of a mouse inward rectifier potassium channel. *Nature* **362**(6416), 127-133.

La Rocca, P., Shai, Y. & Sansom, M. S. (1999). Peptide-bilayer interactions: simulations of dermaseptin B, an antimicrobial peptide. *Biophys. Chem.* **76**(2), 145-159.

Latal, A., Degovics, G., Epand, R. F., Epand, R. M. & Lohner, K. (1997). Structural aspects of the interaction of peptidyl-glycylleucine- carboxyamide, a highly potent antimicrobial peptide from frog skin, with lipids. *Eur. J. Biochem.* **248**(3), 938-946.

Lazarovici, P., Primor, N. & Loew, L. M. (1986). Purification and pore-forming activity of two hydrophobic polypeptides from the secretion of the Red Sea Moses sole (Pardachirus marmoratus). *J. Biol. Chem.* **261**(35), 16704-16713.

Lee, T. H., Philipson, L. H., Kuznetsov, A. & Nelson, D. J. (1994). Structural determinant for assembly of mammalian potassium channels. *Biophys. J.* **66**, 667-673.

Lemas, M. V., Takeyasu, K. & Fambrough, D. M. (1992). The carboxyl-terminal 161 amino acids of the Na,K-ATPase α–subunit are sufficient for assembly with the β-subunit. *J. BIol. Chem.* **267**(29), 20987-20991.

Lemmon, M. A. & Engelman, D. M. (1994). Specificity and promiscuity in membrane helix interactions. *Quater. Rev. Biophys.* **27**(2), 157-218.

Lemmon, M. A., Flanagan, J. M., Hunt, J. F., Adair, B. D., Bormann, B.-J., Dempsey, C. E. & Engelman, D. M. (1992). Glycophorin A dimerization is driven by specific interactions between transmembrane α–helices. *J. Biol. Chem.* **267**(11), 7683-7689.

Li, J. D., Carroll, J. & Ellar, D. J. (1991). Crystal structure of insecticidal δ-endotoxin from *Bacillus thuringiensis* at 2.5 A resolution. *Nature* **353**(6347), 815-821.

Li, J. D., Koni, P. A. & Ellar, D. J. (1996). Structure of the Mosquitocidal δ-endotoxin CytB from *Bacillus thuringiensis ssp.* kyushuensis and implications for membrane pore formation. *J. Mol. Biol.*

Li, M., Jan, Y. N. & Jan, L. Y. (1992). Specification of subunit assembly by the hydrophilic amino-terminal domain of the Shaker potassium channel. *Science* **257**(5074), 1225-1230.

Liao, M. J., London, E. & Khorana, H. G. (1983). Regeneration of the native bacteriorhodopsin structure from two chymotryptic fragments. *J Biol Chem* **258**(16), 9949-9955.

Lomize, A. L., Pervushin, K. V. & Arseniev, A. S. (1992). Spatial structure of (34-65)bacterioopsin polypeptide in SDS micelles determined from nuclear magnetic resonance data. *J. Biomol. NMR.* **2**(4), 361-372.

Ludtke, S. J., He, K., Heller, W. T., Harroun, T. A., Yang, L. & Huang, H. W. (1996). Membrane pores induced by magainin. *Biochemistry* **35**(43), 13723-13728.

Lynch, B. A. & Koshland, D. E., Jr. (1991). Disulfide cross-linking studies of the transmembrane regions of the aspartate sensory receptor of Eshrichia coli. *Proc. Natl. Acad. Sci. USA* **88**, 10402-10406.

MacKenzie, K. R. & Engelman, D. M. (1998). Structure-based prediction of the stability of transmembrane helix- helix interactions: the sequence dependence of glycophorin A dimerization. *Proc. Natl. Acad. Sci. USA* **95**(7), 3583-3590.

MacKenzie, K. R., Prestegard, J. H. & Engelman, D. M. (1997). A transmembrane helix dimer: structure and implications. *Science* **276**(5309), 131-133.

MacKinnon, R. (1991). Determination of the subunit stoichiometry of a voltage-activated potassium channel. *Nature* **350**(6315), 232-235.

Manolios, N., Bonifacino, J. S. & Klausner, R. D. (1990). Transmembrane Helical Interactions and the Assembly of the T Cell Receptor Complex. *Science* **249**, 274-277.

Marsh, D. (1996). Peptide models for membrane channels. *Biochem. J.* **315**, 345-361.

Masson, L., Tabashnik, B. E., Liu, Y. B., Brousseau, R. & Schwartz, J. L. (1999). Helix 4 of the Bacillus thuringiensis Cry1Aa toxin lines the lumen of the ion channel. *J. Biol. Chem.* **274**(45), 31996-32000.

Matsuzaki, K., Harada, M., Handa, T., Funakoshi, S., Fujii, N., Yajima, H. & Miyajima, K. (1989). Magainin 1-induced leakage of entrapped calcein out of negatively- charged lipid vesicles. *Biochim. Biophys. Acta* **981**(1), 130-134.

Meyer, P. & Reusser, F. (1967). A polypeptide antibacterial agent isolated from Trichoderma viride. *Experientia* **23**, 85-86.

Montal, M. (1995). Design of molecular function: channels of communication. *Annu. Rev. Biophys. Biomol. Struct.* **24**, 31-57.

Montal, M. (1996). Protein folds in channel structure. *Curr. Opin. Struct. Biol.* **6**(4), 499-510.

Mor, A., Amiche, M. & Nicolas, P. (1994a). Structure, synthesis, and activity of dermaseptin b, a novel vertebrate defensive peptide from frog skin: relationship with adenoregulin. *Biochemistry* **33**(21), 6642-6650.

Mor, A., Hani, K. & Nicolas, P. (1994b). The vertebrate peptide antibiotics dermaseptins have overlapping structural features but target specific microorganisms. *J Biol Chem* **269**(50), 31635-31641.

Mor, A., Nguyen, V. H., Delfour, A., Migliore-Samour, D. & Nicolas, P. (1991a). Isolation, amino acid sequence, and synthesis of dermaseptin, a novel antimicrobial peptide of amphibian skin. *Biochemistry* **30**(36), 8824-8830.

Mor, A., Nguyen, V. H. & Nicolas, P. (1991b). *J. Mycol. Med.* **1**, 216-220.

Nicolas, P. & Mor, A. (1995). Peptides as weapons against microorganisms in the chemical defense system of vertebrates. *Annu. Rev. Microbiol.*

Oren, Z., Hong, J. & Shai, Y. (1997). A repertoire of novel antibacterial diastereomeric peptides with selective cytolytic activity. *J. Biol. Chem.* **272**(23), 14643-14649.

Oren, Z., Lerman, J. C., Gudmundsson, G. H., Agerberth, B. & Shai, Y. (1999). Structure and organization of the human antimicrobial peptide LL-37 in phospholipid membranes: relevance to the molecular basis for its non- cell-selective activity. *Biochem J* **341**(Pt 3)), 501-513.

Oren, Z. & Shai, Y. (1997). Selective lysis of bacteria but not mammalian cells by diastereomers of melittin: structure-function study. *Biochemistry* **36**(7), 1826-1835.

Oren, Z. & Shai, Y. (1998). Mode of action of linear amphipathic alpha-helical antimicrobial peptides. *Biopolymers* **47**(6), 451-463.

Pandery, R. C., Cook, J. C. & Rinehart, K. L. (1977). High resolution and field desorption mass spectrometry studies and revised structure of alamethicin I and II. *J. Am. Chem. Soc.* **99**, 8469-8483.

Papazian, D. M., Schwarz, T. L., Tempel, B. L., Jan, Y. N. & Jan, L. Y. (1987). Cloning of genomic and complementary DNA from Shaker, a putative potassium channel gene from Drosophila. *Science* **237**(4816), 749-753.

Parker, M. W. & Pattus, F. (1993). Rendering a membrane protein soluble in water: a common packing motif in bacterial protein toxins. *Trends. Biochem. Sci.* **18**(10), 391-395.

Peled, H. & Shai, Y. (1993). Membrane interaction and self-assembly within phospholipid membranes of synthetic segments corresponding to the H-5 region of the shaker K+ channel. *Biochemistry* **32**(31), 7879-7885.

Peled, Z. H., Arkin, I. T., Engelman, D. M. & Shai, Y. (1996). Coassembly of Synthetic Segments of Shaker K^+ Channel within Phospholipid Membranes. *Biochemistry*(35), 6828-6838.

Pervushin, K. V. & Arseniev, A. S. (1992). Three-dimensional structure of (1-36)bacterioopsin in methanol-chloroform mixture and SDS micelles determined by 2D 1H-NMR spectroscopy. *FEBS. Lett.* **308**(2), 190-196.

Pessia, M., Bond, C. T., Kavanaugh, M. P. & Adelman, J. P. (1995). Contribution of the C-terminal domain to gating properties of inward rectifier potassium channels. *Neuron* **14**, 1039-1045.

Popot, J.-L. & Engelman, D. M. (1990). Membrane Protein Folding and Oligomerization: The Two Stage Model. *Biochemistry* **29**(17), 4031-4037.

Popot, J.-L., Gerchman, S.-E. & Engelman, D. M. (1987). Refolding of Bacteriorhodopsin in Lipid Bilayers A Thermodynamically Controlled Two-stage Model. *J. Mol. Biol.* **198**, 655-676.

Pouny, Y., Rapaport, D., Mor, A., Nicolas, P. & Shai, Y. (1992). Interaction of antimicrobial dermaseptin and its fluorescently labeled analogues with phospholipid membranes. *Biochemistry* **31**(49), 12416-12423.

Pouny, Y. & Shai, Y. (1995). Synthetic Peptides Corresponding to the Four P-Regions of Electrophorus electricus Na+ Channel: Interaction with and Organization in Model Phospholipid Membranes. *Biochemstry* **34**, 7712-7721.

Pritsker, M., Jones, P., Blumenthal, R. & Shai, Y. (1998). A synthetic all D-amino acid peptide corresponding to the N-terminal sequence of HIV-1 gp41 recognizes the wild-type fusion peptide in the membrane and inhibits HIV-1 envelope glycoprotein-mediated cell fusion. *Proc. Natl. Acad. Sci. USA* **95**(13), 7287-7292.

Rapaport, D. & Shai, Y. (1991). Interaction of fluorescently labeled pardaxin and its analogues with lipid bilayers. *J. Biol. Chem.* **266**(35), 23769-23775.

Rapaport, D. & Shai, Y. (1992). Aggregation and organization of pardaxin in phospholipid membranes. A fluorescence energy transfer study. *J. Biol. Chem.* **267**(10), 6502-6509.

Saberwal, G. & Nagaraj, R. (1994). Cell-lytic and antibacterial peptides that act by perturbing the barrier function of membranes: facets of their conformational features, structure- function correlations and membrane-perturbing abilities. *Biochim. Biophys. Acta* **1197**(2), 109-131.

Sahin-Toth, M., Dunten, R. L., Gonzalez, A. & Kaback, H. R. (1992). Functional interactions between putative intramembrane charged residues in the lactose permease of Escherichia coli. *Proc. Natl. Acad. Sci. USA* **89**(21), 10547-10551.

Sansom, M. S. (1993). Alamethicin and related peptaibols--model ion channels. *Eur. Biophys. J.* **22**(2), 105-124.

Sawyer, J. G., Martin, N. L. & Hancock, R. E. (1988). Interaction of macrophage cationic proteins with the outer membrane of Pseudomonas aeruginosa. *Infect. Immun.* **56**(3), 693-698.

Schwartz, J. L., Garneau, L., Masson, L. & Brousseau, R. (1991). Early response of cultured lepidopteran cells to exposure to δ-endotoxin from *Bacillus thuringiensis*: involvement of calcium and anionic channels. *Biochim. Biphys. Acta* **1065**, 250-260.

Segrest, J. P., De, L. H., Dohlman, J. G., Brouillette, C. G. & Anantharamaiah, G. M. (1990). Amphipathic helix motif: classes and properties. *Proteins* **8**(2), 103-117.

Shai, Y. (1994). Pardaxin: channel formation by a shark repellant peptide from fish. *Toxicology* **87**(1-3), 109-129.

Shai, Y. (1995). Molecular recognition between membrane-spanning polypeptides. *Trends. Biochem. Sci.* **20**(11), 460-464.

Shai, Y. (1999). Mechanism of the binding, insertion and destabilization of phospholipid bilayer membranes by alpha-helical antimicrobial and cell non-selective membrane-lytic peptides. *Biochim. Biophys. Acta.* **1462**(1-2), 55-70.

Shai, Y., Bach, D. & Yanovsky, A. (1990). Channel formation properties of synthetic pardaxin and analogues. *J. Biol. Chem.* **265**(33), 20202-20209.

Shai, Y., Fox, J., Caratsch, C., Shih, Y. L., Edwards, C. & Lazarovici, P. (1988). Sequencing and synthesis of pardaxin, a polypeptide from the Red Sea Moses sole with ionophore activity. *FEBS Lett.* **242**(1), 161-166.

Shai, Y. & Oren, Z. (1996). Diastereoisomers of cytolysins, a novel class of potent antibacterial peptides. *J. Biol. Chem.* **271**(13), 7305-7308.

Shen, N. V., Chen, X., Boyer, M. M. & Pfaffinger, P. J. (1993). Deletion analysis of K+ channel assembly. *Neuron* **11**(1), 67-76.

Shuck, M. E., Bock, J. H., Benjamin, C. W., Tsai, T. D., Lee, K. S., Slightom, J. L. & Bienkowski, M. J. (1994). Cloning and characterization of multiple forms of the human kidney ROM-K potassium channel. *J Biol Chem* **269**(39), 24261-70.

Slatin, S. L., Abrams, C. K. & English, L. (1990). δ-endotoxins form cation-selective channels in planar lipid bilayers. *Biochem. Biophys. Res. Commun.* **169**, 765-772.

Steiner, H., Hultmark, D., Engstrom, A., Bennich, H. & Boman, H. G. (1981). Sequence and specificity of two antibacterial proteins involved in insect immunity. *Nature* **292**(5820), 246-248.

Sternberg, M. J. E. & Gullick, W. J. (1990). A Sequence Motif in the Transmembrane Region of Growth Factor Receptors with Tyrosine Kinase Activity mediates Dimerization. *Protein Eng* **3**(4), 245-248.

Strahilevitz, J., Mor, A., Nicolas, P. & Shai, Y. (1994). Spectrum of Antimicrobial Activity and Assembly of Dermaseptin-b and its Precursor Form in Phospholipid Membranes. *Biochemistry* **33**, 10951-10960.

Taglialatela, M., Champagne, M. S., Drewe, J. A. & Brown, A. M. (1994). Comparison of H5, S6, and H5-S6 exchanges on pore properties of voltage- dependent K+ channels. *J Biol Chem* **269**(19), 13867-13873.

Thompson, S. A., Tachibana, K., Nakanishi, K. & Kubota, I. (1986). Melittin-Like Peptides from the Shark-Repelling Defense Secretion of the Sole *Pardachirus pavoninus*. *Science* **233**, 341-343.

Van Rie, J., McGaughey, W. H., Johnson, D. E., Barnett, B. D. & Van Mellaert, H. (1990). Mechanism of insect resistance to the microbial insecticide *Bacillus thuringiensis*. *Science* **247**, 72-74.

Verkleij, A. J., Zwaal, R. F., Roelofsen, B., Comfurius, P., Kastelijn, D. & Deenen, L. V. (1973). The asymmetric distribution of phospholipids in the human red cell m embrane. A combined study using phospholipases and freeze-etch electron microscopy. *Biochim. Biophys. Acta* **323**(2), 178-193.

Verrall, S. & Hall, Z. W. (1992). The N-terminal domains of acetylcholine receptor subunits contain recognition signals for the initial steps of receptor assembly. *Cell* **68**, 23-31.

Von Heijne, G. (1994). Membrane proteins: From sequence to Structure. *Annu. Rev. Biophys. Biomol. Struct.* **23**, 167-192.

Walters, F. S., Slatin, S. L., Kulesza, C. A. & English, L. H. (1993). Ion channel activity of N-terminal fragments from CryIA(c) δ-endotoxin. *Biochem. Biophys. Res. Commun.* **196**(2), 921-926.

Westerhoff, H. V., Juretic, D., Hendler, R. W. & Zasloff, M. (1989). Magainins and the disruption of membrane-linked free-energy transduction. *Proc. Natl. Acad. Sci. USA* **86**(17), 6597-6601.

White, J. M. (1992). *Science* **258**, 917-924.

White, S. H. & Wimley, W. C. (1999). Membrane protein folding and stability: physical principles. *Annu. Rev. Biophys. Biomol. Struct.* **28**, 319-365.

Williams, R. W., Starman, R., Taylor, K. M., Gable, K., Beeler, T., Zasloff, M. & Covell, D. (1990). Raman spectroscopy of synthetic antimicrobial frog peptides magainin 2a and PGLa. *Biochemistry* **29**(18), 4490-4496.

Wong, H., Bowie, J. H. & Carver, J. A. (1997). The solution structure and activity of caerin 1.1, an antimicrobial peptide from the Australian green tree frog, Litoria splendida. *Eur. J. Biochem.* **247**(2), 545-557.

Woolley, G. A. & Wallace, B. A. (1992). Model ion channels: gramicidin and alamethicin. *J. Membr. Biol.* **129**(2), 109-136.

Wu, D. & Aronson, A. I. (1992). Localized mutagenesis defines regions of the *Bacillus thuringiensis* δ-endotoxin involved in toxicity and specificity. *J. Biol. Chem.* **267**(4), 2311-2317.

Yu, X.-M. & Hall, Z. W. (1991). Extracellular domains mediating ε–subunit interactions of muscle acetylcholine receptor. *Nature* **352**, 64-67.

Zagorski, M. G., Norman, D. G., Barrow, C. J., Iwashita, T., Tachibana, K. & Patel, D. J. (1991). Solution structure of pardaxin P-2. *Biochemistry* **30**(32), 8009-8017.

Zasloff, M. (1987). Magainins, a class of antimicrobial peptides from Xenopus skin: isolation, characterization of two active forms, and partial cDNA sequence of a precursor. *Proc. Natl. Acad. Sci. USA* **84**(15), 5449-5453.

NOVEL PEPTIDE NUCLEIC ACIDS WITH IMPROVED SOLUBILITY AND DNA-BINDING ABILITY

M. SISIDO and M. KUWAHARA

Department of Bioscience and Biotechnology,
Faculty of Engineering, Okayama University, 3-1-1
Tsushimanaka, Okayama 700-8530, Japan

1. Introduction

Synthesis of artificial polymers that posses nucleobase sequences along the main chain has been a target of many synthetic polymer chemists. The approach was first started as pure scientific study to construct synthetic analogs of DNA, but later accelerated by their potential application as antisense drugs [1]. Some 30 years ago, polymers from vinyl monomers with pendant nucleobases have been synthesized [2]. In those vinyl polymers however, specific base sequences cannot be obtained. From mid 80s, when the importance of the base sequences widely recognized in the pharmaceutical and medicinal fields , attention has been focused on the synthesis polymers that carry specific base sequences [3]. Since only stepwise polymerization, such as the peptide synthesis can afford the special base sequence, it was very natural for many researchers to start synthesis of peptides with nucleobases on the side chain. The latter types of peptides will be called hereafter as peptide nucleic acids (PNAs), although they contain no acid groups.

In 1991, Nielsen and coworkers discovered that PNAs of δ-amino acids with a $[NH-CH_2-CH_2-N(CO-CH_2-B)-CH_2-CO]_n$ backbone structure hybridize to the complementary DNAs with higher affinity than the DNA-DNA counterpart [4]. The Nielsen's PNAs may have a flexible backbone and this may allow them to hybridize to DNAs. A drawback of the Nielsen's PNA is its limited water solubility that restricts their applications.

In this paper, we show several attempts to find another types of PNAs that can hybridize to DNAs. The attempts includes several types of PNAs consisting of α-amino acids and a PNA of $[NH-C^*H(CH_2-CH_2-(Base)-CH_2-O-CH_2-CO]_n$ backbone (oxy-PNA = OPNA) [5]. Through these approaches we show that detailed molecular structures play a crucial role to determine the hybridization ability of peptides carrying nucleobases.

2. Peptides that Contain α-Amino Acids with Nucleobases on the Side Chain

A. Aggeli et al. (eds.), Self-assembling Peptide Systems in Biology, Medicine and Engineering, 295–309.

2.1. Peptides Containing L-Willardiine

Willardiine (I, Wil) is a naturally-occurring amino acid that contains an uracil group [6].

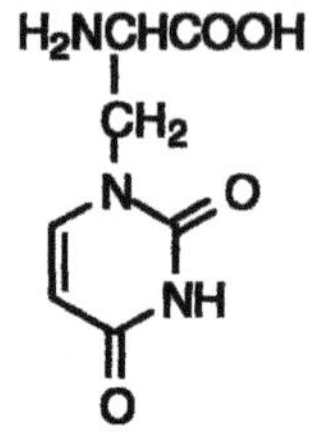

(I) L-Wil

N$^{\alpha}$-acetylWil was synthesized in the racemic form according to Dewar and the free L-amino acid was obtained by an acylase treatment. The molar ellipticity at 260 nm agreed with that of the authentic sample (-2.0 $M^{-1}cm^{-1}$). Peptides containing Wil were synthesized through Boc/Bop strategy on an oxim resin. The structures of the peptides containing three (Wil-Gly) units is shown below.

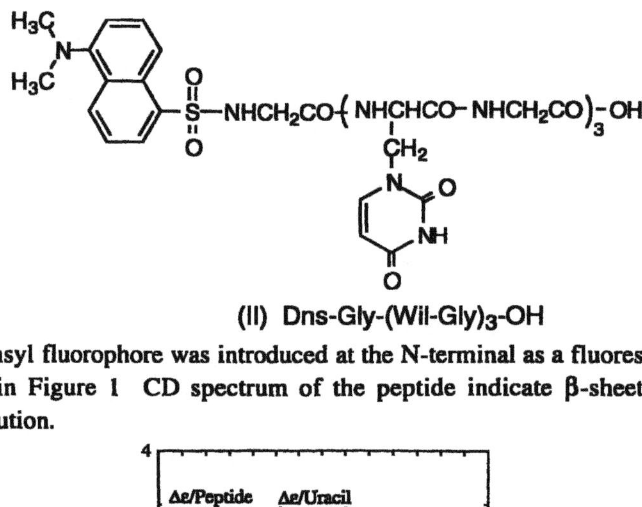

(II) Dns-Gly-(Wil-Gly)$_3$-OH

The dansyl fluorophore was introduced at the N-terminal as a fluorescence probe. As shown in Figure 1 CD spectrum of the peptide indicate β-sheet structure in aqueous solution.

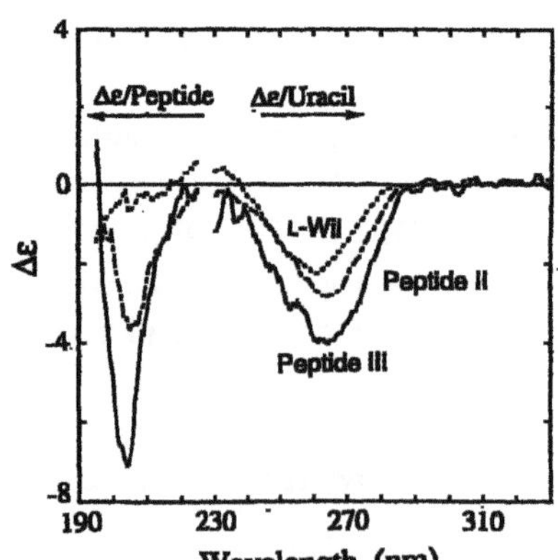

Figure 1. CD spectra of L-Wil and peptide II and III in water.

Interaction of the peptide II with $(dA)_{12}$ was studied by using absorption and fluorescence spectroscopy in aqueous solution at pH=7.0, 5°C. The concentration of the uracil group was $3.8x10^{-5}$ M, and the concentration of the adenine unit was varied from 0 to 3 fold to the uracil group. Each measurement was made after incubating the mixture for 3h at 5°C. Neither the fluorescence intensity nor the absorption intensity, however, changed by the addition of $(dA)_{12}$, indicating absence of the peptide-DNA interaction. The absence of the interaction may be explained in terms of insufficient flexibility of the peptide chain due to the intra- and/or inter-chain hydrogen bondings.

To increase the flexibility by reducing number of hydrogen bonds, the peptide III in which the amide hydrogen atom of glycine is replaced by a methyl group was prepared. The sarcosine (Sar)-containing peptide was synthesized by the stepwise technique on the solid support and purified to a single peak on an HPLC (ODS column). As shown in Figure 1, the peptide also favors β-sheet conformation in aqueous solution. The mixture of the peptide III with $(dA)_{12}$, however, did not show any sign of hybridization, suggesting the chain flexibility is not enough.

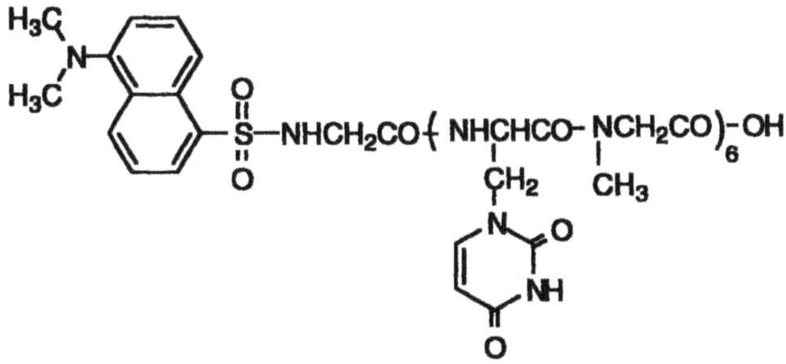

(III) Dns-Gly-(Wil-Sar)$_6$-OH

2.2. Depsipeptide Containing L-Willardiine or L-Homowillardiine

To increase the flexibility of the peptide, ester groups were introduced in the main chain. The octamer of depsipeptide [IV, H-(Wil-glc)$_8$-OH] was prepared by a liquid-phase method through a Boc/Bop strategy. First, Boc-Wil-glc-OBzl was synthesized by the coupling of Boc-Wil-OH with benzyl bromoacetate in DMF containing diisopropylethylamine. The depsipeptide was divided into two portions and one was treated with hydrochloric acid to remove Boc group and the other was hydrogenated to remove benzyl ester. Then the two components were coupled together in DMF in the presence of the Bop reagent. The product was purified by HPLC to give Boc-(Wil-glc)$_2$-OBzl. This procedure was repeated further two times to give 3 mg of Boc-(Wil-glc)$_8$-OBzl. (Scheme 1).

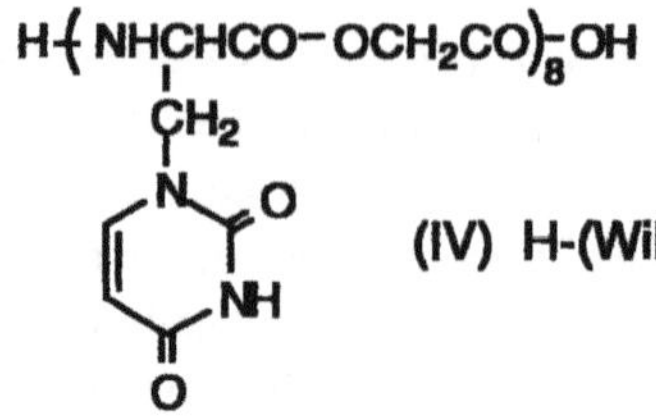

(IV) H-(Wil-glc)$_8$-OH

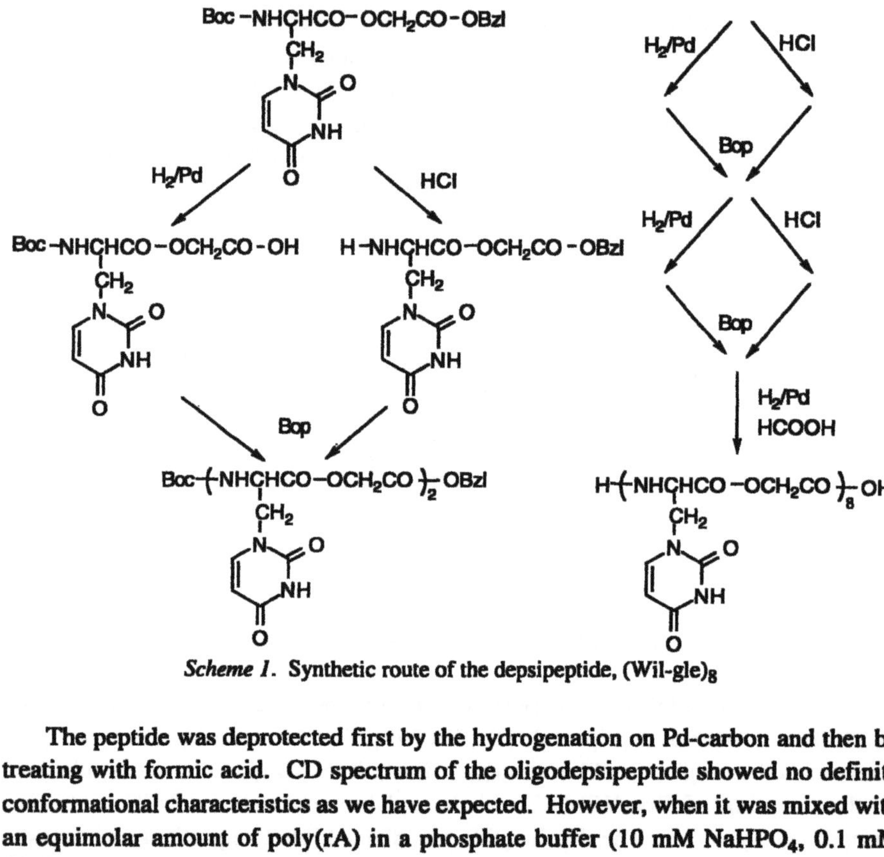

Scheme 1. Synthetic route of the depsipeptide, (Wil-gle)$_8$

The peptide was deprotected first by the hydrogenation on Pd-carbon and then by treating with formic acid. CD spectrum of the oligodepsipeptide showed no definite conformational characteristics as we have expected. However, when it was mixed with an equimolar amount of poly(rA) in a phosphate buffer (10 mM $NaHPO_4$, 0.1 mM EDTA, 1000 mM NaCl, pH=7.0), no clear hybridization was observed even after the mixture was stored at 5°C for 24 h. The depsipeptide may be still too rigid to hybridize with the nucleic acid.

To increase the flexibility of the peptide nucleic acid further, another depsipeptide that contains homowillardiine (V) was synthesized.

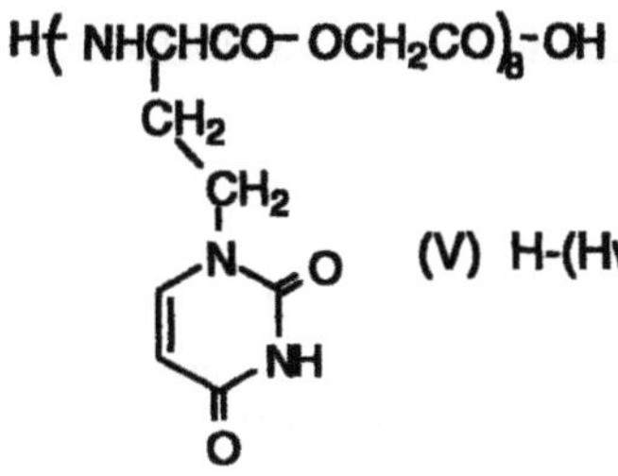

The amino acid was prepared from uracil through 4 steps as shown in Scheme 2.

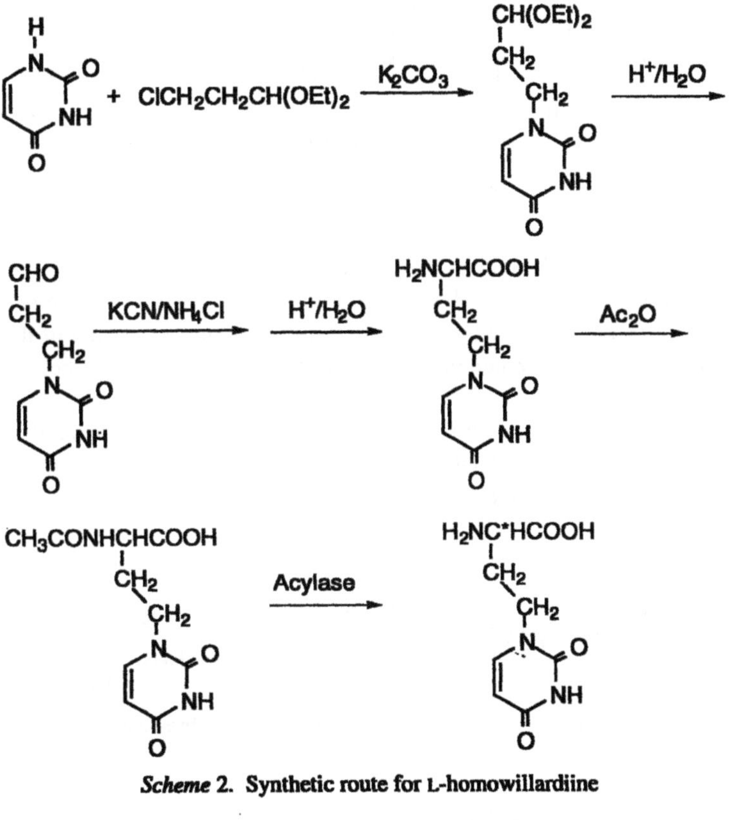

Scheme 2. Synthetic route for L-homowillardiine

The depsipeptide (V) was synthesized by the liquid-phase synthesis in a similar manner as in the case of octamer of Wil-glc. The final product, Boc-(Hwl-glc)$_8$-OBzl was purified by HPLC and deprotected as described above.

Very unfortunately again, the depsipeptide showed any sign of hybridization neither with (dA)$_8$ nor with (rA)$_8$. Furthermore, the depsipeptide was found to be hydrolyzed in the aqueous solution at pH 7 with a half lifetime of 105 h. These undesirable properties enforced us to discard the exploration of peptide nucleic acids that consist of α-amino acids carrying nucleobases on the side chain.

3. Peptides that Contain δ-Amino Acids with Nucleobases on the Side Chain

3.1. Design and Synthesis of δ-Amino Acids with Nucleobases on the Side Chain

To design a flexible peptide chain with nucleobases on the side chain, we have introduced an ether linkage in the main chain of δ-amino acid and the novel peptides of the δ-amino acids were named as oxy-PNA (OPNA). The chemical structure of OPNA (IV) is compared with that of the Nielsen's type PNA (VII) and with DNA (VIII).

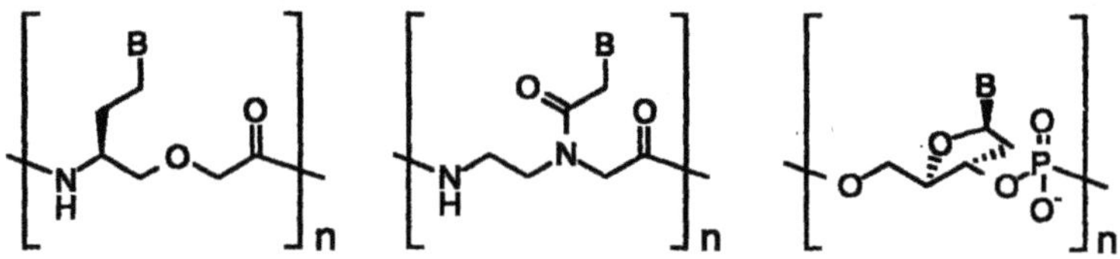

OPNA (VI) Nielsen's PNA (VII) DNA (VIII)

B = A, T, G, C, U

The ether linkages in the OPNA may afford sufficient flexibility in the main chain and improved water solubility. The limited solubility of the Nielsen type PNA for some sequences has been known to be one of the drawbacks. Another feature of the OPNA is the presence of a chiral center in the main chain that may extend their structural diversity.

The Fmoc-δ-amino acids with four types of nucleobases were prepared through a common intermediate [7], Boc-NHCH(CH_2CH_2OH)$CH_2OCH_2COO^tBu$, (IX) that was prepared from L-homoserine in 6 steps as shown in the Scheme 3.

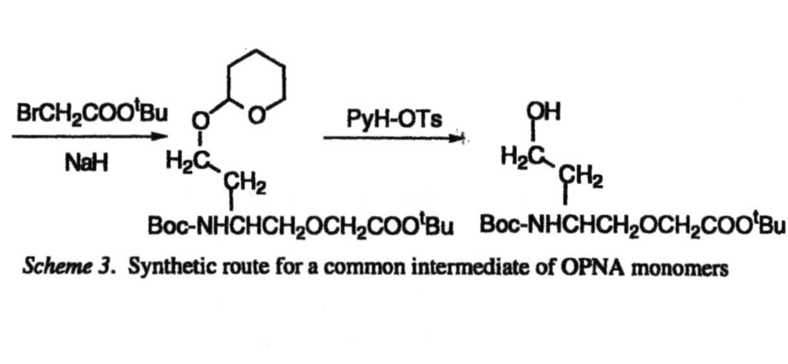

Scheme 3. Synthetic route for a common intermediate of OPNA monomers

Fmoc δ-amino acids with N[6]-benzoyladenine (Bz-A), (X) thymine (T) (XI) , and uracil (U) (XII) as a side-chain nucleobases were synthesized as shown in Scheme 4

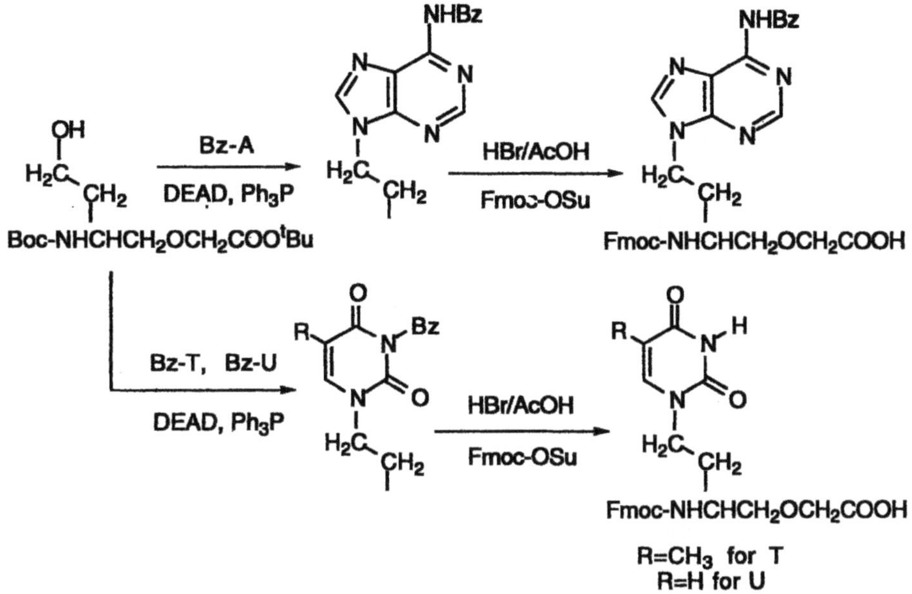

Scheme 4. Synthetic route to the OPNA monomers with Bz-A, T, and U side groups.

In the case of cytosine derivative, no product was obtained in a direct substitution of the hydroxy group under Mitsunobu conditions. Therefore, N[4]-benzoylcytosine (Bz-C) was introduced through a bromide derivative as shown in Scheme 5.

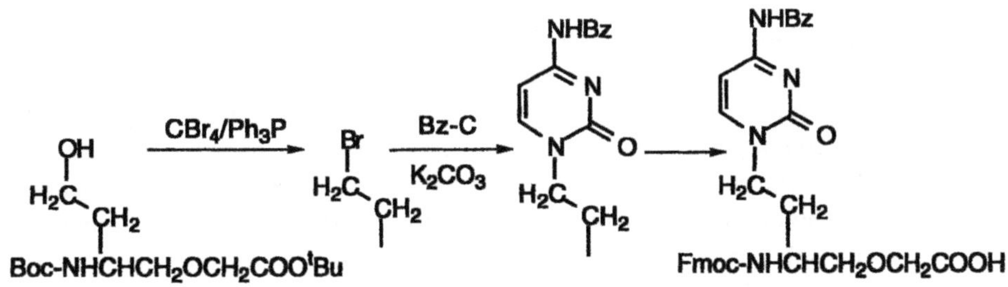

Scheme 5. Synthetic route to the OPNA monomer with Bz-C group.

Fmoc-δ-amino acid with a N[2]-isobutyrylguanine (Ibu-G) group in the side chain (XIV) was synthesized in somewhat complex manner as shown in Scheme 6

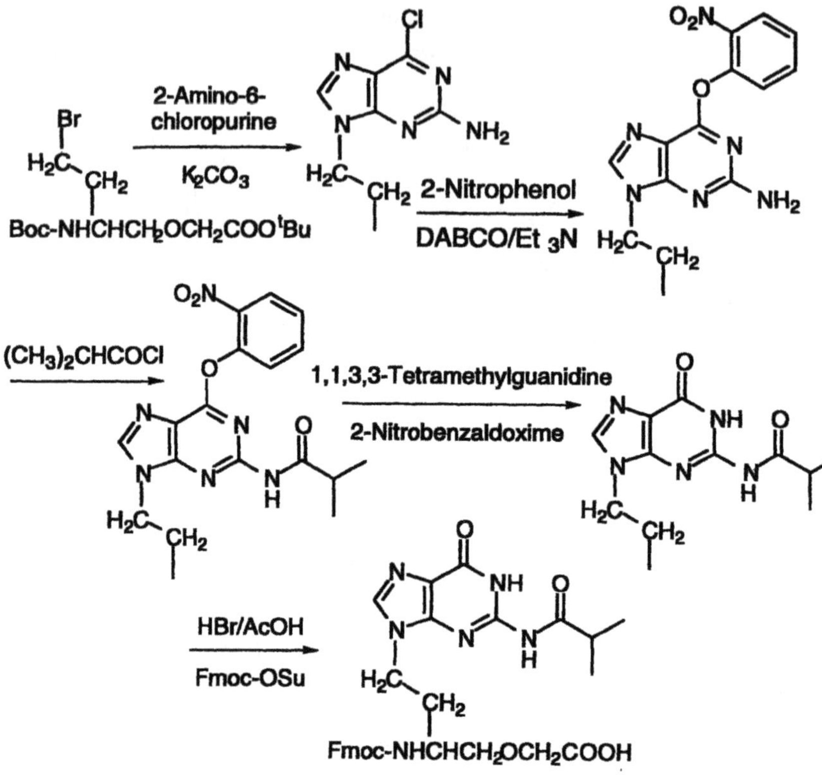

Scheme 6. Synthetic route to the OPNA monomer with Ibu-G side group.

3.2. Preparation of OPNA through Solid-Phase Peptide Synthesis

The Fmoc δ-amino acids with nucleobases on the side chain were inked together through solid-phase method on an Fmoc-NH-SAL-PEG resin (super acid-labile polyethyleneglycol resin from Watanabe Chemicals, Hiroshima, Japan) with benzotriazole-1-yl-oxy-tripyrrolidinephosphonium hexaflurophosphate (PyBop)/HOBt as the coupling reagents. First, $^{\varepsilon}N$-Boc-lysine was linked to the resin and the coupling of the Fmoc-δ-amino acid was repeated for appropriate times The benzoyl groups of Bz-A and the isobutyryl group of Ibu-G were removed by treating the resin with 1:1 mixture of ethylenediamine and ethanol overnight. The benzoyl group of Bz-C was removed by treating the resin with a 1:1 mixture of aqueous ammonia and dioxane at 60 °C overnight. The unprotected OPNA was cleaved off from the resin by the treatment with TFA containing 20% *m*-cresol. The crude OPNA was purified by a RP HPLC (C18 column) with a linear gradient of 7-32% of eluent A (0.1% TFA in acetonitrile) in eluent B (0.1% TFA in water) over 50 min at a flow rate 5 mL/min.

3.3. Hybridization of OPNA with the Complementary DNA

As the simple example, OPNA(A_{12}) was synthesized and mixed with DNA(T_{12}) at different molar ratios. Figure 2 shows CD spectra for the mixtures. At the 1/1 ratio the CD pattern is different either from that of the each component or from their mixtures. As shown in Figure 3, the Job plot shows a minimum at 1/1 molar ratio. In contrast, no minimum or maximum was observed for the mixtures of OPNA(A_{12})-DNA(C_{12}), indicating sequence specific interactions between OPNA and DNA. Similar behavior was also observed in absorption spectroscopy. A large hypochromicity was observed for the mixtures of OPNA(A_{12}) and DNA(T_{12}) and the hypochromicity was largest at the 1/1 molar ratio. No hypochromicity was detected for the mixture of OPNA(A_{12}) and DNA(C_{12}).

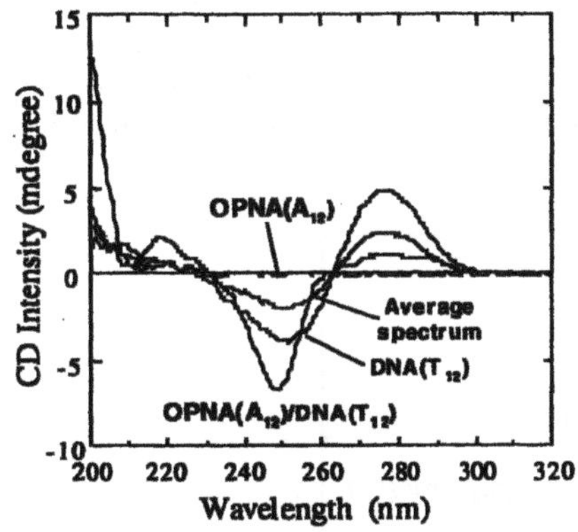

Figure 2 CD spectra of OPNA(A_{12}), DNA(T_{12}), and their 1/1 mixture. The average spectrum of the two component is also shown.

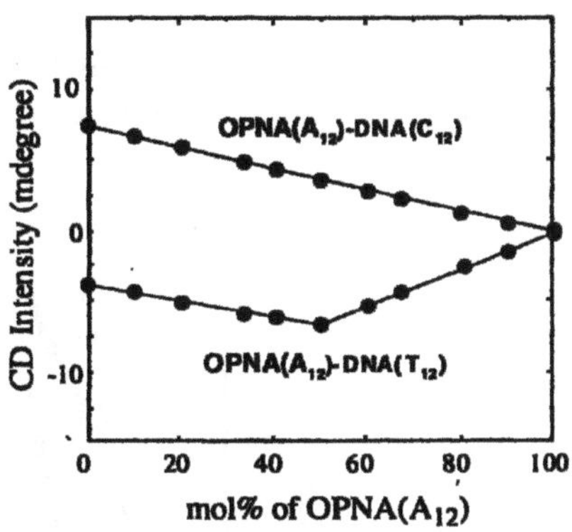

Figure 3. Job plot for the OPNA(A_{12})-DNA(T_{12}) mixture and for the OPNA(A_{12})-DNA(C_{12}) mixture.

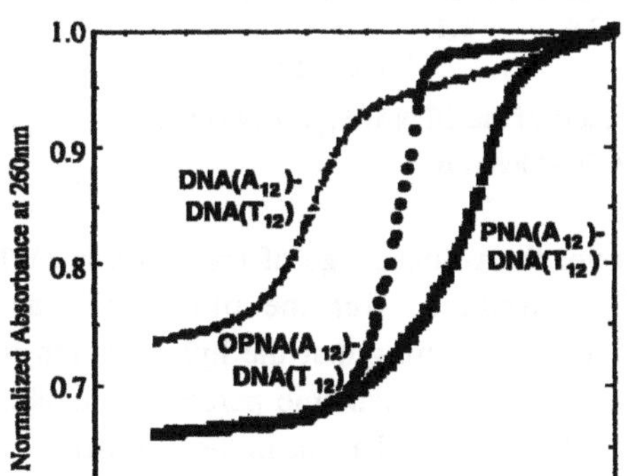

Figure 4. Melting curves for the OPNA(A_{12}),-DNA(T_{12}), PNA(A_{12}),-DNA(T_{12}), and DNA(A_{12}),-DNA(T_{12}) hybrids.

Temperature dependence of the absorption intensity of the OPNA(A_{12})/ DNA(T_{12}) (1/1) mixture at 260 nm is shown in Figure 4. At high temperature, the hypochromicity disappeared, due to the dissociation of the hybrid. When the temperature was lowered down to 43°C, the absorption intensity dropped sharply, indicating the hybridization began to take place. The temperature dependence was reversible. Similar melting curves were also observed for the Nielsen's PNA(A_{12})/DNA(T_{12}) (1/1) mixture and for the DNA(A_{12})/DNA(T_{12}) (1/1) mixture. A remarkable feature of the OPNA(A_{12})-DNA(T_{12}) hybrid is its very sharp transition between the hybridized and dissociated forms as compared with the DNA/DNA and DNA/PNA cases. The temperature range for the 5-95% transition was 15° for the OPNA, 29.5° for the PNA, and 18.5° for the DNA, respectively. The transition temperature (T_m) for the OPNA(A_{12})-DNA(T_{12}) complex (43 °C) is higher than the DNA(A_{12})-DNA(T_{12}) complex (30 °C) but lower than the PNA(A_{12})-DNA(T_{12}) complex (55 °C).

The details of the transition may be discussed in terms of the thermodynamics of the melting curve The van't Hoff plots of the melting curves are shown in Figure 5.

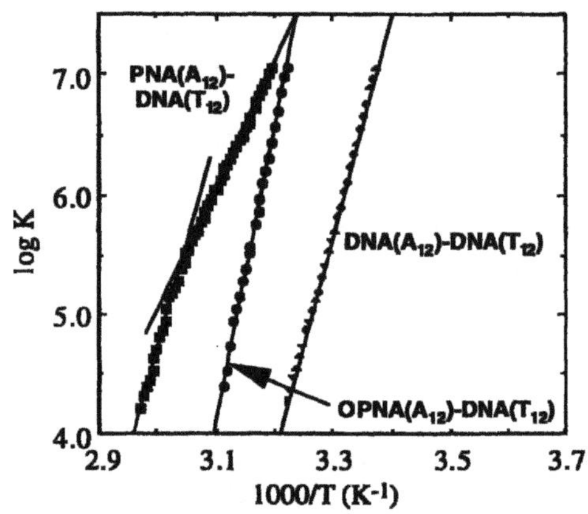

Figure 5. van't Hoff plots of the OPNA(A_{12}),-DNA(T_{12}), PNA(A_{12}),-DNA(T_{12}), and DNA(A_{12}),-DNA(T_{12}) hybrids.

The plot is straight over the whole range of transition in OPNA(A_{12})/DNA(T_{12}) and DNA(A_{12})-DNA(T_{12}) mixture. But the plot shows a curvature for the PNA(A_{12})/DNA(T_{12}) mixture, indicating that the hybridization in the latter mixture proceeds through two steps. Thermodynamic parameters evaluated from the van't Hoff plot are listed in Table I. Both the changes in the stabilizing energy ($-\Delta H$) and the conformational constraints ($-\Delta S$) are largest in the hybridization of the OPNA(A_{12})-DNA(T_{12}) pair, suggesting that, by decreasing temperature, the randomly-coiled OPNA with large conformational entropy is folded into a regular and rigid double-stranded helix with nearly optimized stabilization energy.

Table I Thermodynamic parameters for the hybridizations of the mixtures of $DNA(A_{12})$, $OPNA(A_{12})$, and $PNA(A_{12})$ with $DNA(T_{12})$.

Hybrid Pair	T_m (°C)	ΔH (kcal mol^{-1})	ΔS (cal mol^{-1} K^{-1})
$DNA(A_{12})$-$DNA(T_{12})$	30	-84	-251
$OPNA(A_{12})$-$DNA(T_{12})$	43	-112	-329
$PNA(A_{12})$-$DNA(T_{12})$	55	-48	-120[a]
		-80	-217[b]

[a]Taken from the low temperature side of the van't Hoff plot.
[b]Taken from the high temperature side of the van't Hoff plot.

The randomly coiled conformation of the free OPNA chain is supported from the absence of CD signals, even though the presence of a chiral center in each monomer unit. The helical conformation of the hybrid is also supported from the large CD signals as compared with the single-stranded DNA. The large enthalpy change may be explained in terms of the hybrid structure that are nearly optimized for the inter-base hydrogen bondings. The difference of the thermodynamics of the DNA/DNA or DNA/OPNA hybridizations are illustrated in Figure 6.

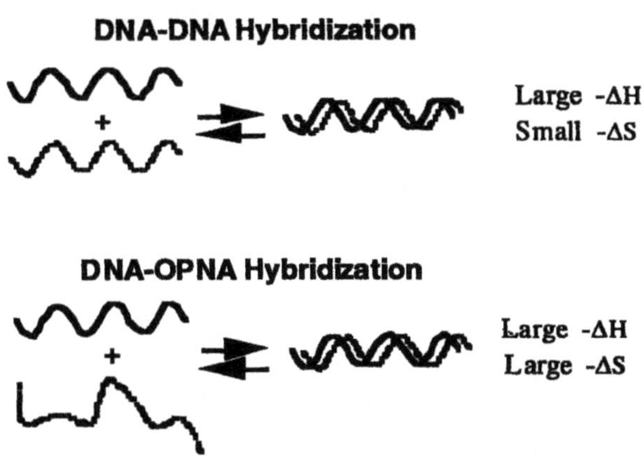

Figure 6. Illustration to explain large enthalpy loss and large entropy loss for the OPNA-DNA hybridization.

The sequence specificity of the OPNA was examined by using $DNA(T_6CT_5)$, instead of $DNA(T_{12})$ as the counterpart. The melting curves for the $OPNA(A_{12})$, $DNA(A_{12})$, and $PNA(A_{12})$, respectively, mixed with $DNA(T_6CT_5)$ were measured. In both OPNA and PNA cases, the transition temperatures were lowered by 13° from those of the complementary pairs. In the case of OPNA, the decrease of T_m is on the same order as the temperature range of the 5-95% transition (15 °), indicating that OPNA can detect even a single mismatching. In the case of PNA, however, the decrease of T_m is about half of the 0-95% transition range, suggesting that the detection of single mismatch is difficult in this case.

4. Sequence-Specific Hybridization between Two Artificial Nucleic Acid Analogs

4.1. Hybridization between PNA(T_{12}) and OPNA(A_{12})

As described above, the OPNA hybridizes with the complementary DNA. This finding is remarkable, since a number of peptide nucleic acid analogs have been reported, but only a very few of them successfully hybridize with the complementary DNA. Since the hybridization property of OPNA is comparable with that of the Nielsen's PNA, it will be interesting to see if the hybridization between the complementary pair of PNA and OPNA also takes place or not [8].

Hybridization between PNA(T_{12}) and OPNA(A_{12}) was examined. The Job plot for the PNA(T_{12})-OPNA(A_{12}) mixtures is shown in Figure 7. The minimum at the 1:1 molar ratio indicates duplex formation between the two types of nucleic acid analogs, despite the fact that the PNA(T_{12})-DNA(A_{12}) and PNA(T_{12})-PNA(A_{12}) mixtures form a triplex with a 2:1 ratio. The preference of the duplex rather than the triplex is compatible with the finding that the OPNA(A_{12}) forms only duplex with DNA(T_{12}). Absorption intensities of PNA(A_{12})-OPNA(A_{12}) mixtures are also shown in Figure 7. The absence of hypochromicity in the non-complementary mixture indicates that the hybridization is sequence-specific.

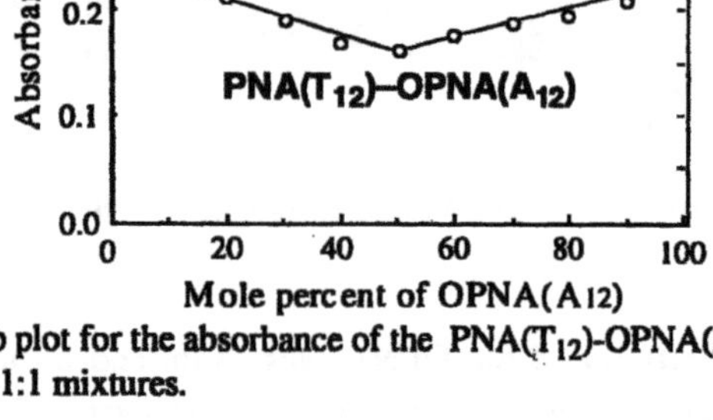

Figure 7. Job plot for the absorbance of the PNA(T_{12})-OPNA(A_{12}) and PNA(A_{12})-OPNA(A_{12}) 1:1 mixtures.

Hybridization between the two nucleic acid analogs, PNA(T_{12}) and OPNA(A_{12}) is the first example for artificial nucleic acid analogues to hybridize with sequence specificity. The hybridization indicates either that the duplexes from DNA-DNA, DNA-PNA, DNA-OPNA, PNA-PNA, and PNA-OPNA pairs take similar conformations or that the OPNA is so flexible as to adapt to different duplex structures. No CD signal was observed for of the OPNA(A_{12})-PNA(T_{12}) mixture indicating no regular helical structure for the duplex. Presumably, OPNA(A_{12}) has a very flexible chain and can adapt to form duplex with DNA and PNA. Since DNA main chain is relatively rigid, the duplex has a definite helical conformation as indicated from strong CD (Figure 2), but the duplex with achiral PNA has a non-helical structure.

A marked hysteresis was observed in the melting curve of the PNA(T_{12})-OPNA(A_{12}) mixture (Figure 8). On heating, the duplex dissociates at higher temperature than 80 °C, but on cooling it associates at lower temperature than about 35 °C. The marked hysteresis indicates that the duplex once formed is difficult to dissociate and the dissociated pair is difficult to form the duplex.

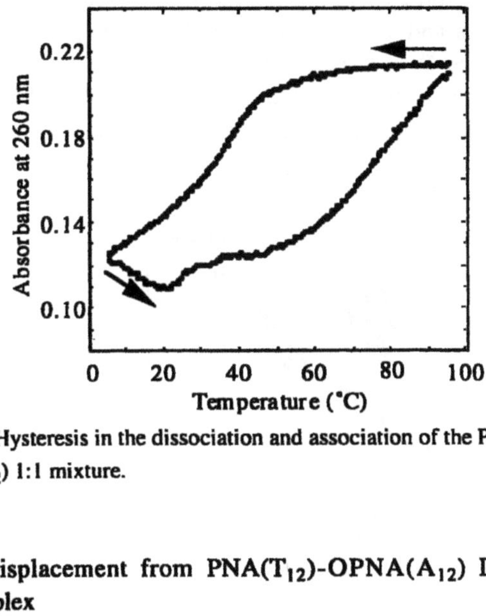

Figure 8. Hysteresis in the dissociation and association of the PNA(T_{12})-OPNA(A_{12}) 1:1 mixture.

4.2. Strand Displacement from PNA(T_{12})-OPNA(A_{12}) Duplex to DNA(T_{12})-OPNA(A_{12}) Duplex

To compare relative stabilities of the duplexes, melting curves of the DNA(T_{12})-PNA(T_{12})-OPNA(A_{12}) 1:1:1 mixtures were measured with different orders of mixing. First, OPNA(A_{12}) was added to the mixture of DNA(T_{12}) and PNA(T_{12}). Since no interaction is expected between the latter two components, this mixing process corresponds to a concomitant mixing of the three components. The CD spectrum of the three-component mixture was coincident to that of the DNA(T_{12})-OPNA(A_{12}) duplex. The melting curve also agreed with that of the latter duplex, indicating only DNA(T_{12})-OPNA(A_{12}) duplex is formed in the three-component mixture. This result is in consistent with the melting curves of DNA(T_{12})-OPNA(A_{12}) (Figure 4) and of PNA(T_{12})-OPNA(A_{12}) (Figure 8). Since the melting temperature is 41 °C for DNA(T_{12})-OPNA(A_{12}) duplex and >80 °C for PNA(T_{12})-OPNA(A_{12}), the OPNA can associate only with DNA near room temperature.

In the second experiment, DNA(T_{12})-OPNA(A_{12}) duplex was first formed and after incubating the solution for 1h at 25 °C, an equivalent amount of PNA(T_{12}) was added. CD spectrum after storing the three-component mixture at 25 °C for 1 h, was virtually identical to that of the original mixture, indicating neither the strand displacement nor the triplex formation took place.

In the third experiment, a PNA(T_{12})-OPNA(A_{12}) duplex was preformed and the mixture was incubated for 30 min at 25 °C. Then an equivalent amount of DNA(T_{12}) was added. The melting curve of the ternary mixture was measured as shown in Figure 9 (Trace A to B). During the first heating process, the melting curve showed a gentle slope with the melting temperature around 60 °C, suggesting that the dissociation of PNA(T_{12})-OPNA(A_{12}) duplex is the major event. After dissociation at elevated temperature, the three-component mixture behaved similarly as the above three-component mixtures and showed the reversible melting profile of DNA(T_{12})-OPNA(A_{12}) duplex (Trace B to C and C to B). The melting behavior obviously indicate a strand displacement from a PNA(T_{12})-OPNA(A_{12}) duplex to a DNA(T_{12})-OPNA(A_{12}) duplex. CD spectrum also indicated the strand displacement.

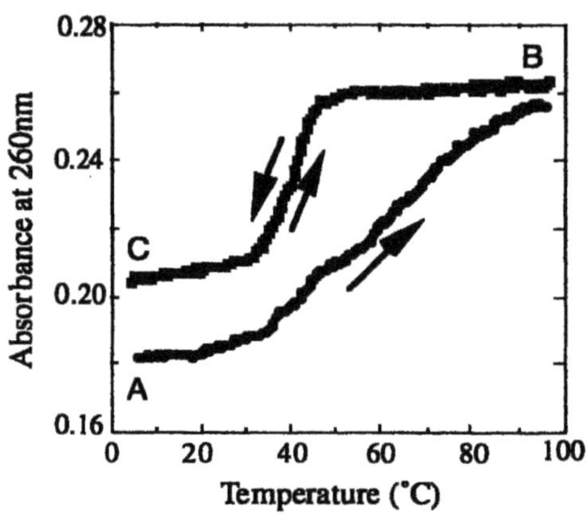

Figure 9. Melting behavior of the DNA(T_{12})-PNA(T_{12})-OPNA(A_{12}) ternary mixture starting from the PNA(T_{12})-OPNA(A_{12}) duplex added with ANA(T_{12}) (Point A).

The sequence-specific interaction between the two artificial nucleic acid analogs will open a new field of artificial molecular recognition.

5. Conclusions

A novel peptide nucleic analog (OPNA) was presented. The OPNA contains ether linkages in the main chain that provides sufficient flexibility to achieve successful hybridization with DNA and with Nielsen-type PNA. OPNA will be used as an impvoved version of PNA or find unique applications that have not been reached by the PNA.

6. References

1. Uhlmann, E., and Peyman, A.: Antisense oligonucleotides: A new therapeutic principle, *Chem. Rev.* **90** (1990), 543-584.
2. Ueda, N., Kondo, K., Kondo, M., Takemoto, K., and Imoto, K.: Vinyl compounds of nucleic acid bases. I. Synthesis of N-vinyluracil, N-vinyladenine, *Makro. Chem.* **120** (1968), 13-20.
3. (a) Garner, P., and Yoo, J. U.: Peptide-based nucleic acid surrogates incorporating Ser[CH_2B]-Gly subunits, *Tetrahedron Lett.* **34** (1993), 1275-1278. (b) Lewis, I.: Peptide analogues of DNA incorporating nucleobase-Ala-Pro subunits, *Tetrahedron Lett.* **34** (1993), 5697-5700. (c) van der Laan, A. C., Strömberg, R., van Boom, J. H., Kuyl-Yeheskiely, E., Efimov, V. A., and Chakhmakhcheva, O. G.: An approach towards the synthesis of oligomers containing a N-2-hydroxyethyl-aminomethylphosphonate backbone: A novel PNA analogue, *Tetrahedron Lett.* **37** (1996) 7857-7860. (d) Gangamani, B. P., Kumar, V. A., and Ganesh, K. N.: Synthesis of N^{α}-(purinyl/pyrimidinyl acetyl)-4-aminoproline diastereomers with potential use in PNA synthesis, *Tetrahedron* **52** (1996) 15017-15030. (e) Tsantrizos, Y. S., Lunetta, J. F., Boyd, M., Fader, L. D., and Wilson, M. -C.: Stereoselective synthesis of a thymine derivative of (S)-2-hydroxy-4-(2-aminophenyl)butanoic acid. A novel building block for the synthesis of aromatic peptidic nucleic acid oligomers, *J. Org. Chem.* **62** (1997) 5451-5457. (f) Lowe, G., and Vilaivan, T.: Solid-phase synthesis of novel peptide nucleic acids, *J. Chem. Soc. Perkin Trans. 1* **4** (1997), 555-560. (g) Bergmeier, S. C., and Fundy, S. L.: Synthesis of oligo(5-aminopentanoic acid)-nucleobases (APN): potential antisense agents, *Bioorg. Med. Chem. Lett.* **7** (1997) 3135-3138. (h) Fujii, M., Yoshida, K., Hidaka, J., and Ohtsu, T.: Hybridization properties of nucleic acid analogs β-aminoalanine modified with nucleobases, *Chem. Commun.* **6** (1998) 717-718. (i) van der Laan, A. C., van Amsterdam, I., Tesser, G. I., van Boom, J. H., and Kuyl-Yeheskiely, E.: Synthesis of chirally pure ornithine based PNA analogues, *Nucleosides Nucleotides* **17** (1998), 219-231.
4. Nielsen, P. E., Egholm, M., Berg, R. H., and Buchardt, O.: Sequence-selective recognition of DNA by strand displacement with a thymine-substituted polyamide, *Science* **254** (1991), 1497-1500.
5. Kuwahara, M., Arimitsu, M., and Sisido, M.: Novel peptide nucleic acid that shows high sequence specificity and all-or-none-type hybridization with the complementary DNA, *J. Am. Chem. Soc.* **121** (1999), 256-257.
6. Dewar, J. H., and Shaw, G.: A Synthesis of Willardiine, *J. Chem. Soc.* (1962), 583-585.
7. Kuwahara, M., Arimitsu, and M., Sisido, M.: Synthesis of δ-amino acids with an ether linkage in the main chain and nucleobases on the side chain as monomer units for oxy-peptide nucleic acids, *Tetrahedron* **55** (1999), 10067-10078.
8. Kuwahara, M., Arimitsu, M., and Sisido, M.: Sequence-specific hybridization between two different types of peptide nucleic acids, *Bull. Chem. Soc. Jpn.* **72** (1999), 1747-1752.

CHIRAL LIPID TUBULES

MARK S. SPECTOR, RONALD R. PRICE, AND JOEL M. SCHNUR
Center for Bio/Molecular Science and Engineering, Naval Research Laboratory, Washington, D.C. 20375-5348, U.S.A. *

Recent fundamental research has revealed the possibility to rationally form supramolecular aggregates. These self-assembled structures range in size from several nanometers (containing a few molecules) to hundreds of microns (containing billions of molecules). The ability to control both the morphology and size of these aggregates through modifications of both molecular structure and solution conditions suggests the possibility that they might be exploited to solve important problems in the development of advanced material applications. It is also possible that the lesson learned for the study of chiral behavior in lipid may have some relevance to the study of the ways in which polypeptide tapes self assemble in to fibers. Tubules formed from phospholipids constitute a unique self-assembled microstructure and show promise in a number of biotechnological applications [1]. This article will provide an outline of the talk that was given in June 1999 at the International Meeting on Polypeptide self-assembly held in Crete. It will discuss advances that have led to the understanding of chiral behavior and the subsequent ability to control the structure of lipid-derived tubules and the resulting impact of this on materials applications.

Modifications to the molecular structure of lipids have been made to increase the versatility of liposomes as delivery agents for pharmaceutical agents [2]. Phospholipids with photopolymerizable diacetylenic moieties in the acyl chains were synthesized to increase the durability of liposomes [3-5]. However, Yager and Schoen found that some polymerizable lipids self-assemble to form hollow, cylindrical structures known as tubules [6]. Remarkably, tens of billions of molecules self-organize to form these

* Portions of this paper were previously published in *Advanced Materials* **11**, 337-340 (1999).

A. Aggeli et al. (eds.), Self-assembling Peptide Systems in Biology, Medicine and Engineering, 311–321.

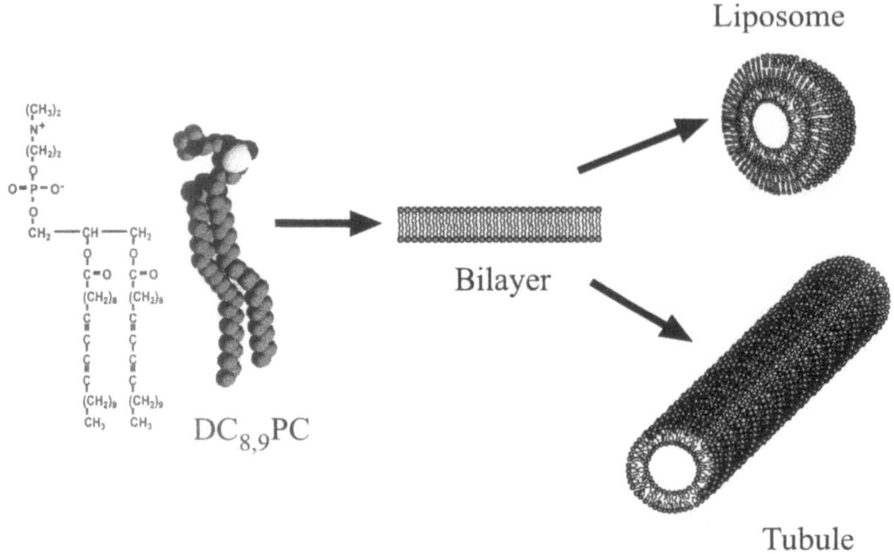

Figure 1. Schematic representation of lipid tubule formation. On the left is shown the chemical structure of the diacetylenic lipid $DC_{8,9}PC$ along with a model representing its three-dimensional structure. In the chain disordered L_α phase, $DC_{8,9}PC$ bilayers form spherical liposomes depicted on the top right. Upon cooling into the $L_{\beta'}$ gel phase, the bilayers assemble into helical tubules shown on the bottom right.

aggregates with a well-defined diameter and very little bending along their axis. Unlike spherical liposomes, lipid tubules reflect the chiral nature of these lipids. This chirality in molecular packing is reflected in helical markings often visible in electron micrographs of tubules and in large peaks observed in their circular dichroism spectra, both of which change handedness when the opposite enantiomer lipid is used [7,8].

The tubules observed by Yager and Schoen assembled from liposomes of 1,2-bis(tricosa-10,12-diynoyl)-*sn*-glycero-3-phosphocholine (abbreviated $DC_{8,9}PC$) in water [6]. Tubules formed upon cooling from the lipid bilayer from the chain disordered phase, known as the L_α phase, into an ordered phase, called the $L_{\beta'}$ gel phase. Infrared and Raman spectroscopy, as well as x-ray diffraction, show that the acyl chains are highly ordered and tilted in tubules [1,9]. The chemical structure of $DC_{8,9}PC$ along with a schematic representation of a liposome and a tubule are shown in Figure 1. Tubules formed from $DC_{8,9}PC$ have an average diameter of 0.5 μm and lengths which range from 50-200 μm. The size and stability of tubules formed in water are sensitive to preparation. Longer and more robust tubules can be formed by first dissolving the lipids in alcohol and then mixing with water above their melting temperature (T_m ~ 37 °C) and slowly cooling the mixture [10]. Several studies have found that the length distribution and thickness of tubules depend on the alcohol used, while the diameter appears independent of preparation conditions. Tubules formed in mixtures of methanol and water at low lipid concentration consist of single bilayer walls and are of maximal length at 85% methanol, while those in ethanol/water contain more than five bilayers and are longest at 70% ethanol [11]. Decreasing the cooling rate through T_m has been shown to greatly increase the average length of tubules and decrease the wall thickness in ethanol/water solutions [9]. Results from spectroscopic studies have led to the production of tubules with an average length in excess of 100 μm and an optimum thickness for applications of two bilayers (based on cost and fragility issues) [8].

One technique that has proved particularly useful in studying chiral lipid architectures is circular dichroism spectroscopy. Circular dichroism (CD), the difference in the absorption of right and left circularly polarized light, arises from the chirality of the molecular architecture. This chirality can arise from either the structure of individual molecules or from the chiral packing of molecules into larger aggregates. Kunitake and co-workers found that chiral bilayers can show greatly enhanced CD spectra [12]. Lipid tubules were also found to have intense peaks in their CD spectra indicating large chiral correlations in the molecular packing [7]. This chiral order can no longer be maintained when the chains disorder above T_m, and the tubules melt, leading to a decrease in CD peak intensity by four orders of magnitude. These observations are consistent with theoretical models which suggest that tubules formation is driven by a twisting of the bilayer due to symmetry breaking in the packing of chiral molecules which can only occur when the lipids are in the ordered $L_{\beta'}$ phase [13].

Recent CD experiments revealed that, in addition to chiral packing within the bilayer, chiral correlations exist between bilayers in multi-bilayer tubules [14]. Representative CD spectra from tubules of varying thickness are shown in Figure 2. For the single bilayer tubule formed in methanol/water solution, a sharp peak at 195 nm associated with chiral packing of the diacetylene group is the predominant spectral feature, while a broader peak near 205 nm due to chiral ordering of headgroups is seen in the many bilayer tubule formed in ethanol/water. The loss of chiral order in the L_α phase is reflected in the CD spectrum shown in Fig. 2 for $DC_{8,9}PC$ liposomes above the lipid melting temperature, where the peak magnitude is 10^4 times smaller that from the tubule ($L_{\beta'}$) phase. Such observations showed the usefulness of CD as a tool to investigate tubule morphology and allowed for optimization of $DC_{8,9}PC$ tubules for technological applications [8]. For example, by monitoring the CD spectra as a function of lipid concentration in methanol/water tubules, we

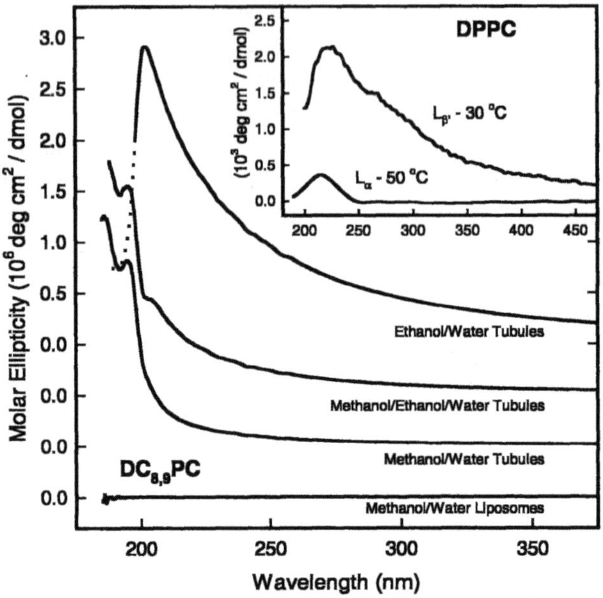

Figure 2. Circular dichroism spectra of lipid bilayer aggregates. The main figure shows the spectra of $DC_{8,9}PC$ tubules prepared in ethanol/water (7:3), methanol/ethanol/water (64:16:20), and methanol water (85:15), and $DC_{8,9}PC$ liposomes above the melting temperature. All samples were prepared at a lipid concentration of 5.0 mg/ml and the spectra for the tubules were recorded at 25 °C. The liposome spectra was recorded at 40 °C and the peak intensity is about 10^4 smaller than that from tubules. The inset shows CD spectra from DPPC bilayers where an eight-fold enhancement is seen in the $L_{\beta'}$ phase (30 °C) relative to the L_α phase (50 °C).

found a crossover from single-bilayer to multiple-bilayer tubules with increasing concentration which was subsequently verified using electron microscopy. Figure 3 shows the concentration dependence of the CD spectra from $DC_{8,9}PC$ tubules in 80:20 (v/v) methanol/water at three different lipid concentrations. The peak at 205 nm is prominent in the 5 mg/ml sample, whereas it only appears as a small elbow in the spectra at lower concentrations. The size of the 195 nm peak is similar at all concentrations. These results suggest a transformation from single-bilayer to multiple-bilayer tubules near a lipid concentration of 5 mg/ml in 80:20 methanol/water. The crossover concentration depends on the methanol/water ratio, with the crossover occurring at higher concentrations as this ratio increases [15].

Tubules with an optimal thickness of two bilayers can be formed by either varying the lipid concentration or using mixed alcohol solutions. Examining the tubule morphology at the crossover concentration in methanol/water

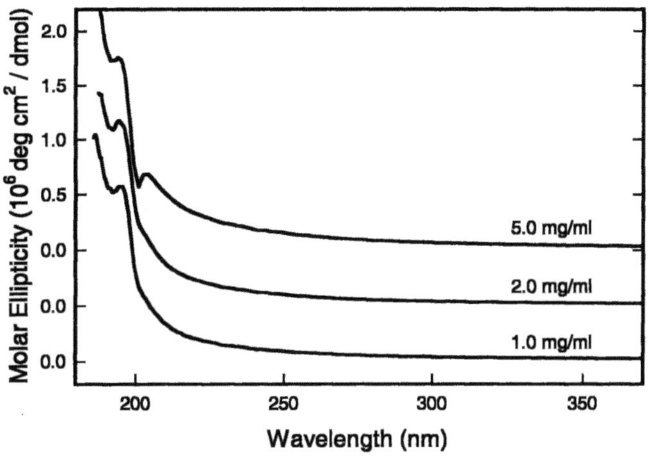

Figure 3. Circular dichroism spectra of $DC_{8,9}PC$ tubules prepared in 80:20 methanol/water at lipid concentrations of 1.0, 2.0, and 5.0 mg/ml taken at 25 °C. The curves have been offset vertically for display.

85:15, we found that most (> 90%) of the tubules have two-bilayer thick walls at a concentration of 5 mg/ml [8]. A transmission-mode electron micrograph of such a tubule is shown in Fig. 4. A complete cross section of this tubule is shown in the top panel, with an enlargement of the edge shown in the bottom panel. The wall thickness for the two-bilayer tubule in this micrograph is 16 ± 2 nm, implying a single bilayer thickness of about 8 nm which is consistent x-ray diffraction results [9,16]. These tubules also exhibit a very high aspect ratio. We have been able to use electroless plating to metallize these two-bilayer tubules, which we were unable to do with single-bilayer tubules. This provides a significant cost savings over metallizing ethanol/water tubules which typically have ten bilayers [11]. Unfortunately, this process results in highly thixotropic suspensions which are difficult to process for large quantity applications. Diluting the sample leads to single bilayer tubules which we are unable to coat with metal. This problem has been overcome through the use of mixed alcohol solvents. The CD spectrum from two-bilayer tubules made in a solution of methanol, ethanol, and water is also shown in Fig. 2. It has both a large peak at 195 nm and a broad shoulder indicating an additional peak near 205 nm.

We note that the two-bilayer tubule in Fig. 4 shows clear helical markings. Such markings are characteristic of multiple-bilayer ethanol/water tubules, but are not seen in single-bilayer methanol/water tubules. These markings may be associated with defects in the tilt direction of the lipid molecules on the tubules [13] or they may be the edges of helical ribbons wrapped around the

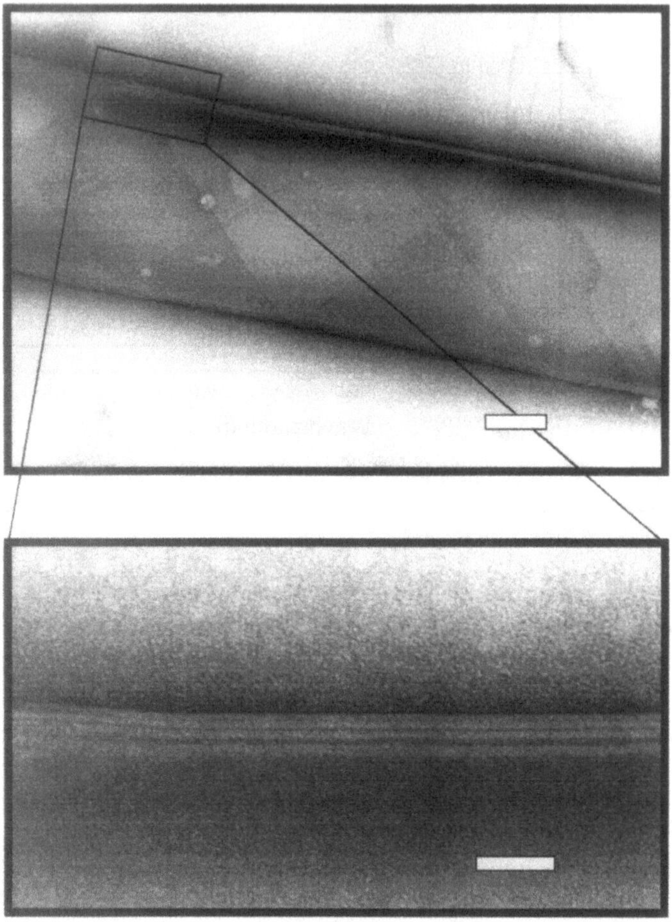

Figure 4. Negative stained electron micrograph of a $DC_{8,9}PC$ tubule from a 5 mg/ml sample in methanol/water (85:15) . This sample predominantly contains two-bilayer tubules evidenced in the enlargement of the tubule edge shown in the bottom panel. Bar = 200 nm (top); 50 nm (bottom).

inner tubule core. Although it is sometimes difficult to differentiate the helical markings from the top and bottom bilayers, observation of the tubule ends or taking stereo pairs of micrographs occasionally allows unambiguously the determination of the handedness of the helical markings. Only right-handed helices are observed for the L-enantiomer in our single component systems [8,17]. However, a small proportion of helices and ribbons with opposite handedness have recently been observed in enantiomerically pure systems by two groups [18,19]. In one case, a model bile system containing three chiral components was studied [18]. The presence of multiple components may lead to the formation of microstructures with different chirality.

The question then arises as to why tubules do not form in the $L_{\beta'}$ phase in saturated lipid bilayers. Indeed, an enhancement of the CD signal is seen in the gel phase of bilayers containing the saturated phospholipid DPPC (dipalmitoyl phosphatidylcholine) in methanol/water (7:3), as shown in the inset of Fig. 2. Here an eight-fold increase in the peak magnitude is seen in the $L_{\beta'}$ phase relative to the L_{α} phase. However, this is significantly smaller than the 10^4 increase seen for the diacetylenic lipid. Similar results have been reported in pure water, where an increase in peak CD intensity is found upon cooling DPPC liposomes below the main chain transition temperature [20]. This suggests that in the $L_{\beta'}$ phase the chiral ordering of DPPC bilayers is much shorter range than for $DC_{8,9}PC$ and this ordering is not enough to allow for tubule formation.

While $DC_{8,9}PC$ remains the lipid of choice for applications due to its commercial availability, other work has focused on structural modifications of $DC_{8,9}PC$ to improve tubule morphology. Key goals of this research include increasing the aspect ratio of tubules to improve function and decreasing the number of bilayers to reduce cost. Much of this work has been discussed in earlier reviews [1]. Recently, Thomas and coworkers have synthesized an analogue of $DC_{8,9}PC$ where the phosphate linkage between the choline headgroup and glycerol backbone is replaced by a phosphonate [21]. They find tubules formed under similar conditions to have twice the diameter of $DC_{8,9}PC$ tubules, but also to be significantly more fragile.

The sub-micron diameter, hollow cylindrical morphology presented by tubules offers attractive possibilities for applications in controlled release and electroactive composites. However, since the constituent lipid membranes are held together by weak hydrophobic forces, their technological utility is limited by a lack of mechanical strength and temperature stability. An electroless metallization technique, which allows the tubules to be clad with such metals as copper, gold, iron, and nickel, has been developed to increase the mechanical strength of lipid tubules [22]. Palladium-tin colloidal particles are first adsorbed on the lipid's polar head groups under acidic conditions. This dispersion is then added to commercially available metal plating bath. The colloidal palladium acts as a catalyst for the metallization. The thickness of the metal layer can be varied from 20 to 200 nm. The necessity to subject the lipid tubules to rapid changes in pH, osmolarity, and alcohol concentrations, and the generation of large quantities of hydrogen gas during the plating procedure, results in significant breakage leading to a reduction of the tubule length by at least 50%. Micrographs of metallic coated microtubules are shown in Figure 5. In the top of the figure, a scanning

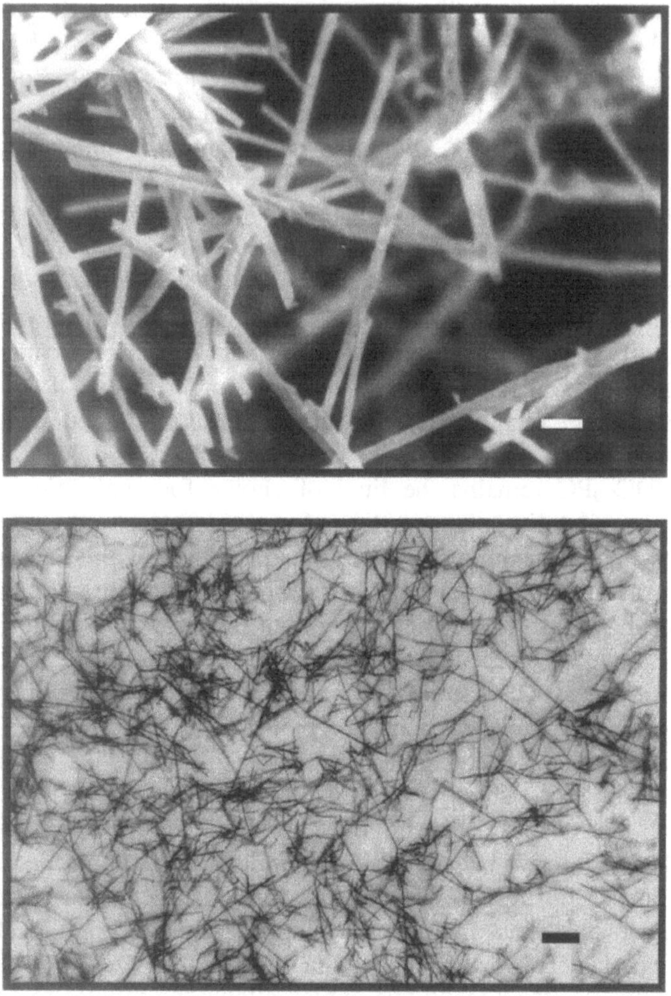

Figure 5. Micrographs of metal coated $DC_{8,9}PC$ tubules. The top panel shows a SEM micrograph of copper plated microtubules (Bar = 2.0 μm). Note the uniformity in diameter and the relatively straight morphology of the tubules. The bottom panel shows an optical micrograph of iron coated microtubules embedded in an acrylic/urethane clear coating (Bar = 25 μm).

electron microscope image of copper plated microtubules illustrates the uniformity in diameter and the relatively straight morphology of the tubules. The lower optical micrograph illustrates iron coated microtubules which have an average length in excess of 50 um embedded in an acrylic/urethane clear coating.

Several prototypes for applications using metal clad tubules have been demonstrated. Metallized tubules are easily incorporated into composites and coatings. They can serve as submicron sized microcapillaries which are suitable for entrapping a range of materials within their hollow core making them useful for controlled release applications [23]. The entrapment is controlled by both absorption and capillary forces and the fixed diffusional areas of the tubule ends aid in controlling the release rate of entrapped materials. Release characteristics may be further altered by the addition of materials with differing permeability to the active ingredient used and further by covering or capping the ends with a membrane of known permeability for the active ingredient. Further modification may be made by covering the ends with materials that will dissolve when the conditions are appropriate for the initiation of release or by the application of physical or heat stress. Tubules are being evaluated for pharmaceutical applications and in antifouling paints. Tubules containing biocides have shown enhanced antifouling properties for periods exceeding six months [23].

The high aspect ratio of tubules (> 100:1), coupled with their small size, also affords the possibilities for electroactive applications. Recently, Browning *et al.* have shown that percolation can be reached at very low loading using metal clad tubules in epoxy coatings [24]. This result suggests that very light weight paints can be fabricated having the high dielectric constants and properties associated with materials showing or being near to conductive percolation. In addition, the aspect ratio can be used to tune these properties by aligning the metal clad tubules. This may lead to the fabrication of coatings with controlled anisotropic dielectric properties.

While the utility of tubules in several applications has been clearly shown, their cost limits the realistic field of use. Reducing the number of layers has cut the lipid material cost by at least 80% [8], but the lipid is still too expensive for many applications. The likely successful commercial applications will be those with significant value added. Leading candidates for commercialization lie in the pharmaceutical and electronics industries.

This article has briefly discussed the supramolecular morphology of self-assembled lipid tubules and the consequences of this for material applications. The rational control of the morphologies requires the ability to correlate the modifications in structure and conditions with observations of structural change. Recent advances in spectroscopy, electron microscopy, and synthesis, coupled with a more fundamental understanding of the basis for intermolecular interactions in self-assembled systems have enabled the advances discussed above. Progress in submicron and nanoscale characterization techniques continues at a rapid rate. The recent new characterizations of the self-assembly of polypeptides to form fibers has been quite interesting to the authors. It seems possible that some of the lessons learned in the study of the structure and formation of lipid tubules might be of use in the understanding of the role that chirality might have in the formation of these polypeptide fibers. The future should be one of new understanding of the processes involved in controlling the fabrication of biological materials and the subsequent application of these principles for material development as we learn more about the way in which the molecular architecture can be manipulated to control the properties of self-assembled materials.

We acknowledge the support of the Office of Naval Research for the research reported in this paper.

References

1. Schnur, J. M. (1993) Lipid tubules: A paradigm for molecularly engineered structures, *Science* **262**, 1669-1676.
2. Lasic, D. D. (1993) *Liposomes: From Physics to Applications*; Plenum: New York.
3. Johnston, D. S., Sanghera, S., Pons, M., and Chapman, D. (1980) Phospholipid polymers – Synthesis and spectral characteristics, *Biotic. Biophys. Acta* **602**, 57-69.
4. Hub, H. H., Hupfer, B., Koch, H., and Ringsdorf, H. (1980) Polyreactions in ordered systems 20. Polymerizable phospholipid analogs – New stable biomembrane and cell models, *Angew. Chem. Int. Ed. Engl.* **19**, 938-940.
5. Kusumi A., *et al.* (1983) Dynamic and structural properties of polymerized phosphatidylcholine vesicle membranes, *J. Am. Chem. Soc.* **105**, 2975-2980.
6. Yager, P. and Schoen, P. E. (1984) Formation of tubules by a polymerizable surfactant, *Mol. Cryst. Liq. Cryst.* **106**, 371-381.
7. Schnur, J. M., Ratna, B. R., Selinger, J. V., Singh, A., Jyothi, G., and Easwaran, K. R. K. (1994) Diacetylenic lipid tubules: Experimental evidence for a chiral molecular architecture, *Science* **264**, 945-947.
8. Spector, M. S., Selinger, J. V., Singh, A., Rodriguez, J. M., Price, R. R., and Schnur, J. M. (1998) Controlling the morphology of chiral lipid tubules, *Langmuir* **14**, 3493-3500.
9. Thomas, B. N., Safinya, C. R., Plano, R. J., and Clark, N. A. (1995) Lipid tubule self-assembly: Length dependence on cooling rate through a first-order phase transition, *Science* **267**, 1635-1638.
10. Georger, J. H., Singh, A., Price, R. R., Schnur, J. M., Yager, P., and Schoen, P. E. (1987) Helical and tubular microstructures formed by polymerizable phosphatidylcholines, *J. Am. Chem. Soc.* **109**, 6169-6175.
11. Ratna, B. R., Baral-Tosh, S., Kahn, B., Schnur, J. M., and Rudolph, A. S. (1992) Effect of alcohol chain length on tubule formation in 1,2-bis(tricosa-10,12-diynoyl)-*sn*-glycero-3-phosphocholine, *Chem. Phys. Lipids* **63**, 47-53.

12. Kunitake, T., Nakashima, N., Shimomura, M., Okahata, Y., Kano, K., and Ogawa, T. (1980) Unique properties of chromophore containing bilayer aggregates: Enhanced chirality and photochemically induced morphological change, *J. Am. Chem. Soc.* **102**, 6642-6644.
13. Selinger, J. V., MacKintosh, F., and Schnur, J. M. (1996) Theory of cylindrical tubules and helical ribbons of chiral lipid membranes, *Phys. Rev. E* **53**, 3804-3818 and references therein.
14. Spector, M. S., Easwaran, K. R. K., Jyothi, G., Selinger, J. V., Singh, A., and Schnur, J. M. (1996) Chiral molecular self-assembly of phospholipid tubules: A circular dichroism study, *Proc. Nat. Acad. Sci. USA* **93**, 12943-12946.
15. Spector, M. S., Selinger, J. V., and Schnur, J. M. (1997) Thermodynamics of phospholipid tubules in alcohol/water solutions, *J. Am. Chem. Soc.* **193**, 8533-8539.
16. Caffrey, M., Hogan, J., and Rudolph, A. S. (1991) Diacetylenic lipid microstructures: Structural characterization by x-ray diffraction, *Biochem.* **30**, 2134-2146.
17. Singh, A., Burke, T. G., Calvert, J. M., Georger, J. H., Herendeen, B., Price, R. R., Schoen, P. E., and Yager, P. (1988) Lateral phase separation based on chirality in a polymerizable lipid and its influence on formation of tubular microstructures, *Chem. Phys. Lipids* **47**, 135-148.
18. Zastavker, Y. V., Asherie, N., Lomakin, A., Pande, J., Donovan, J. M., Schnur, J. M., and Benedek, G. B. (1999) Self-assembly of helical ribbons, *Proc. Nat. Acad. Sci. USA* **96**, 7883-7887.
19. Thomas, B. N., Lindemann, C. M., and Clark, N. A. (1999) Left- and right-handed helical tubule intermediates from a pure chiral phospholipid, *Phys. Rev. E* **59**, 3040-3047.
20. Walde, P. and Blöchliger, E. (1997) Circular dichroic properties of phosphatidylcholine liposomes, *Langmuir* **13**, 1668-1671.
21. Thomas, B. N., Corcoran, R. C., Cotant, C. L., Lindemann, C. M., Kirsch, J. E., and Persichini, P. J. (1998) Phosphonate lipid tubules, *J. Am. Chem. Soc.* **120**, 12178-12186.
22. Schnur, J. M., Price, R., Schoen, P., Yager, P., Calvert, J. M., Georger, J., and Singh, A. (1987) Lipid-based tubule microstructures, *Thin Solid Films* **152**, 181-206.
23. Price, R. R., Patchan, M., Clare, A., Rittschof, D., and Bonaventura, J. (1992) Performance enhancement of natural antifouling compounds and their analogs through microencapsulation and controlled release, *Biofouling* **6**, 207-216.
24. Browning, S. L., Lodge, J., Price, R. R., Schelleng, J., Schoen, P. E., and Zabetakis, D. (1998) Fabrication and radio frequency characterization of high dielectric loss tubule-based composites near percolation, *J. Appl. Phys.* **11**, 6109-6113.

ΔT_t-MECHANISM IN THE DESIGN OF SELF-ASSEMBLING STRUCTURES

Dan W. Urry,[*#] Larry Hayes,[*] Chixiang Luan,[*] D. Channe Gowda,[*] David McPherson,[*] Jie Xu,[*] and Timothy Parker[*]

[*]Bioelastics Research, Ltd., OADI Technology Center, 2800 Milan Court, Suite 386, Birmingham AL 35211-6918
[#]University of Minnesota, Biological Process Technology Institute, 1479 Gortner Ave., St Paul MN 55108-6106

ABSTRACT: Protein-based polymers can be designed in which self-assembly occurs as the temperature is raised above the onset temperature, T_t, of an inverse temperature transition for hydrophobic folding and assembly. Instead of changing the temperature, however, by many means the value of T_t can be lowered from above to below an operating temperature to drive hydrophobic folding and assembly. This is the ΔT_t-mechanism. Modulation of charges on the polymer provides the most dramatic means of controlling T_t and therefore becomes the most effective means for controlling self-assembly. The formation of charge raises the value of T_t and causes disassembly, whereas neutralization of charge by lowering degree of ionization or by increasing ion-pairing drives self-assembly.

For example, a polymer with one Asp(COO^-) or Glu(COO^-) per 30 residues can be in solution with its T_t above 100°C. Titration with a cationic drug lowers T_t to below 25°C and results in self-assembly into a drug delivery vehicle capable of a constant release profile for a constant surface area, and the vehicle simply disperses as the drug is released. Similarly an anionic drug can induce self-assemby of a cationic, e.g., Lys(NH_3^+)-containing, protein-based polymer.

Two solutions of protein-based polymers, one polymer with negative charges, e.g., COO^-, and the other with positive charges, e.g., NH_3^+ and both with hydrophobic residues sufficient to shift the pKa values of their respective functional groups, can exhibit their individual inverse temperature transitions at temperatures much higher than body temperature, even greater than 100°C. On combining the two solutions, the polymers self-assemble with a T_t below room temperature. Each polymer by ion pairing with the other dramatically lowers the temperature of the inverse temperature transition for polymer self-assembly. The effectiveness of this self-assembly of two

A. Aggeli et al. (eds.), Self-assembling Peptide Systems in Biology, Medicine and Engineering, 323–342.

oppositely charged protein-based polymers increases as the individual hydrophobic-induced pKa shifts are larger and as steric matching occurs in the ion-pairing between the pair of polymers. Furthermore, the same protein-based polymer can contain both positive and negative charges with hydrophobically shifted pKa values to become locked in self-assembled structures. This bears analogy to certain globular proteins, which on unfolding would similarly self-assemble to form the neurofibrillary tangles and amyloid plaques of Alzheimers and related diseases.

In summary, ion-pairing within properly designed protein-based polymers results in self-assembling materials and structures by means of the ΔT_t-mechanism.

1. Introduction

Our understanding of self-assembling protein-based materials begins with the polymer, poly(Gly-Val-Gly-Val-Pro) or more simply poly(GVGVP) where the amino acid residues are glycine (Gly, G), valine (Val, V), and proline (Pro, P). Even in the absence of any charge, this polymer is miscible in all proportions with water as long as the temperature is below 25°C. On raising the temperature from 25 to 30°C, phase separation begins; the solution becomes cloudy, and on standing a more-dense phase develops that is approximately 40% polymer and 60% water by weight when the initial concentration is less than 400 mg/ml [1,2].

Making the polymer more hydrophobic, simply by adding an oil-like CH_2 moiety to one of the two Val residues resulting in a more hydrophobic isoleucine (Ile, I) residue, has two inter-related consequences; 1) it lowers the temperature interval over which the phase transition occurs [3] and 2) it raises the free energy of charged species attached to the polymer. The latter translates into an increase of the pKa of a carboxyl moiety and a decrease in the pKa of an amino moiety [4,5,6].

Adding a small amount of charge, as in poly[0.8(GVGVP),0.2(GEGVP)] where E is glutamic acid (Glu) giving four carboxylates (COO^-) in 100 residues at neutral pH, raises the temperature interval for the phase transition beyond reach in an aqueous system. Addition of 0.15 N NaCl partially neutralizes the effect of the charge and lowers the onset temperature, T_t, for the phase transition, when there are four Glu COO^- in 100 residues, to approximately 70°C.

1.1 INVERSE TEMPERATURE TRANSITIONS AND THE ΔT_t -MECHANISM

The phase transition, which has been extensively correlated with hydrophobicity, results in self-assembling of protein-based polymers, that is, results in an increase in order of the model protein on increasing the temperature. Normally for a phase transition, of course, the low temperature side of the transition is the more ordered state and the high temperature side of the transition is the disordered state. The situation for this hydrophobically dominated transition is exactly the inverse; the disordered state is on the low temperature side of the transition, and the more ordered state is on the high temperature side of the transition. For this reason the observed phase transition is called an ***inverse temperature transition***. The onset temperature for the inverse temperature transition, as the temperature is raised, is designated as T_t [7].

The effect of increasing the hydrophobicity is to lower the value of T_t, and the effect of increasing the degree of ionization is to raise the value of T_t. Thus, when the value of T_t is above an operating temperature, the protein-based materials are disassembled, but when the value of T_t is lowered from above to below an operating temperature, i.e., there is a negative ΔT_t as for example resulting from a decrease in the number, or from neutralization, of charged species in the polymer, then the protein-based materials self-assemble. This process whereby self-assembling structures are achieved is called the ***ΔT_t-mechanism***.

1.2 TEMPERATURE OF THE INVERSE TEMPERATURE TRANSITION: THE FUNDAMENTAL OPERATING PROPERTY OF THE SYSTEM

When operating at constant temperature, the magnitude of the small heat of an inverse temperature transition matters little. On the other hand, it matters greatly whether the value of T_t is above or below the operating temperature. If T_t is 10 to 15°C below the operating temperature, the protein-based material will be completely self-assembled, whereas if T_t is immediately above the operating temperature, the protein-based material will be disassembled.

Should the variable be concentration of proton, or of a counterion, or of an organic solute, or of another protein-based polymer each of which interacts with the original protein-based polymer with its particular affinity, as a practical matter for self-assembling materials under the control of an inverse temperature transition, it matters only where T_t is in relation to the operating temperature. It matters only what concentration of the independent variable is required to shift T_t above or below the operating temperature in order to disassemble or assemble, respectively, the structure.

1.3 THE CENTRAL ROLE OF WATER OF HYDROPHOBIC HYDRATION

In our interpretation of the underlying physical basis for these effects, the temperature interval, over which the phase transition occurs, depends inversely on the amount of water of hydrophobic hydration [8]; an increase in the amount of structured water at the surface of hydrophobic groups lowers the temperature interval, and a decrease in the amount of structured water surrounding hydrophobic groups raises the temperature range over which the phase transition occurs. Increased numbers of hydrophobic residues in the protein-based polymer increase the amount of water of hydrophobic hydration and, thereby, lower the temperature interval. On the other hand, charged species of the polymer, in order to achieve their required hydration, destructure significant amounts of structured water at the surface of hydrophobic residues [9,10], and, thereby, raise the temperature interval over which the phase transition occurs. Furthermore, as charged species necessarily destructure water of hydrophobic hydration in order to improve their own limited hydration, the free energy of the charged species increases, thereby causing a shift in the pKa. This competition for hydration has been termed an apolar-polar repulsive free energy of hydration, ΔGap [2,7,11].

1.4 EFFECT OF CHARGED SIDE CHAINS ON THE TEMPERATURE OF INVERSE TEMPERATURE TRANSITIONS

Using the model protein-based polymer, poly[f_V(GVGVP),f_X(GXGVP)] where f_V and f_X are mole fractions with $f_V + f_X = 1$ and X is, in general, any one of the naturally occurring amino acid residues, but in the present concern the relative effects of each substitution of an ionizable side chain is made by comparison to a reference state of f_X = 1, i.e., for poly(GXGVP). The four principle ionizable side chains of proteins are those of the amino acid residues, histidine (His, H), lysine (Lys, K), aspartate (Asp, D), and glutamate (Glu, E). Each of these are commonly charged near physiological pH of 7.4.

The shift in the temperature at which the inverse temperature transition begins on raising the temperature, i.e., the change in the value of T_t, on conversion to the charged state varies significantly among the series from H^+: (ΔT_t = 40°C), K^+: (ΔT_t = 85°C), D^- : (ΔT_t = 125°C), and E^-: (ΔT_t = 220°C) [11]. Thus, it is the carboxylate of glutamate that most profoundly destructures the waters of hydrophobic hydration as shown by most dramatically increasing the value of T_t.

There is a biologically accessible charged species that has an even more profound effect, which makes the case most emphatically for the effect of charge on T_t; it is the phosphorylation of serine (Ser, S), or of threonine (Thr, T), or of tyrosine (Tyr, T) by specific protein kinases. Using the model system poly[30(GVGIP),(RGYSLG)], the enzymatic phosphorylation of the **S** residue results in a calculated value of T_t for poly(GS{$PO_4^=$}GIP) of greater than 1000°C [12]. Thus, for phosphorylation of a serine residue, the comparable change in T_t would be a ΔT_t > 1000°C. Phosphorylation, therefore, is found to be the most profound means of driving the unfolding of hydrophobically folded domains of proteins and protein-based polymers, and dephosphorylation, or ion-pairing of phosphate with a multivalent cation, would be the most dramatic way to drive hydrophobic folding and assembly.

Phosphorylation is included in the present discussion of self-assembling structures because it presents such a dramatic demonstration of the role of charge in shifting T_t and because of the striking use of this feature in the assembly and disassembly of cytoskeletal structures and in protein structure-function in general. The unfolding of hydrophobically folded globular G-actin to form hydrophobically assembled fibrous F-actin would be, in our view, a primary example of the role of high energy phosphate in controlling assembly of biological structures by controlling the value of T_t.

1.5 EFFECT OF HYDROPHOBIC SIDE CHAINS ON pKa IN PROTEIN-BASED POLYMERS

One means of quantifying the competition for hydration that exists between charged side chains and hydrophobic side chains of a protein-based polymer is to study the conditions that alter the affinity of a carboxylate, COO^-, or of an amino, NH_2, group for a proton, H^+. This amounts to the characterization of hydrophobic-induced pKa shifts by comparison of pKa values for a series of designed protein-based polymers where the Val (V) residue with the $-CH(CH_3)_2$ side chain is replaced with the more hydrophobic Ile (I) residue with the $-CH(CH_3)CH_2CH_3$ side chain ($\Delta T_t = -14°C$), and with the even more hydrophobic Phe (F) residue with the $-CH_2$-phenyl side chain ($\Delta T_t = -54°C$).

Proton affinity, of course, is measured by a binding or association constant, e.g., $Ka = [COO^-][H+]/[COOH]$ and $Ka = [NH_2][H^+]/[NH_3^+]$, and $pKa = -logKa$. Accordingly, from equilibrium theory, $\Delta G = -RTlnKa = -2.3RTlogKa = 2.3RTpKa$, that is, the change in Gibbs free energy due to protonation is directly proportional to the pKa.

1.5.1 *Hydrophobic-induced pKa Shifts*

Hydrophobic-induced pKa shifts have been observed and systematically evaluated in three different protein-based polymer constructs. The first construct utilizes poly[f_V(GVGIP),f_X(GXGIP)] where f_X varies from 1.0 to 0.06 in approximately ten steps with X in three different series being E, D, or K [4,5,6]. In this example, the more polar X residue is systematically replaced by the more hydrophobic V residue, and the pKa is determined for each composition. As shown in Figure 1A for data obtained in 0.15 N NaCl, it is not an increasing concentration of charge (as would be the case for charge-charge repulsion) that gives rise to the large pKa shifts. It is an increase in more hydrophobic residues as the charged species decrease in number that results in the largest pKa shifts.

In the next series of constructs, the ionizable function is kept constant at one per 30 residues (one per 6 pentamers), and the hydrophobicity is increased by replacing Val residues by more hydrophobic Phe (F) residues, the number of Phe residues being 0, 2, 3, 4, and 5 per 30 mer. Again, the pKa is determined for each composition. As the number of more hydrophobic Phe residues increases, as plotted in Figure 1B, the pKa increases in a strikingly non-linear manner [13].

1.5.2 *Stretch-induced pKa Shifts*

Finally, in the third construct, poly[0.82(GVGIP),0.18(GEGIP)], the composition is kept constant; the polymer is cross-linked to form an elastic band; the hydrophobically folded elastic band is stretched, exposing progressively more hydrophobic groups, and the pKa is determined at each load required to achieve the different extensions. Again, the pKa increases are strikingly non-linear with linear increase in mechanical work [14], that is, with what are expected to be approximately linear increases in hydrophobicity, as seen in Figure 1C.

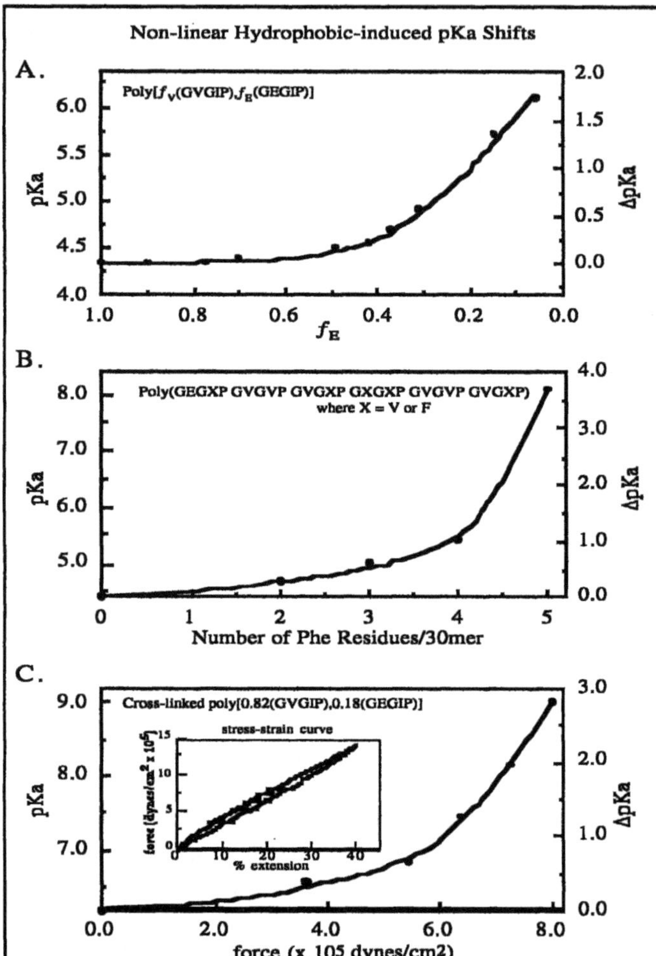

Figure 1: Plots of pKa (left ordinate) and of the ΔpKa (right ordinate) resulting from a linear increase in hydrophobicity as the independent variable. See text for discussion.
Adapted with permission from references 4,13 and 14

In each of the three experimental means of changing hydrophobicity, the general forms of the curves of linear increases in hydrophobicity versus ΔpKa are quite similar as apparent in comparing Figures 2A, B, and C for the three examples above using the Glu (E) functional group. To the extent examined, the other two functional groups, Asp (D) and Lys (K), give qualitatively similar results, though of the three Asp exhibited the larger shifts and Lys gave the smaller pKa shifts.

1.6 IMPLICATIONS OF FREE ENERGY CHANGES REPRESENTED BY pKa SHIFTS

So unfavorable is the formation of a negatively charged carboxylate, $-COO^-$, groups in high molecular weight and chemically synthesized poly(GDGFP GVGVP GVGFP GFGFP GVGVP GVGFP) with a 0.167 mole fraction of Asp(D)-containing pentamers that it can extract protons from a solution that is 10^{-10} M in proton, whereas for poly[0.83(GVGVP),0.17(GDGVP)], the less hydrophobic random polymer, a concentration of 10^{-4} M protons is required before 50% protonation occurs. Since by definition $\Delta\Delta G = 2.3RT\Delta pKa$, this means that the increase in free energy of the carboxylate at 25°C can be greater than 8kcal/mole. Our perspective is that this measures the competition for hydration between the increasing number of hydrophobic Phe residues and the unchanging carboxylate, as shown in Figure 1B for the Glu residue. This is a direct measure of what we have termed the apolar-polar repulsive free energy of hydration, i.e., $\Delta G_{ap} = \Delta\Delta G$.

An analogous situation exists for the amino functional group of Lys, except, of course, that it is the free energy of the positively charged, $-NH_3^+$, moiety that is increased.

While protonation of a negatively charged carboxylate or deprotonation of a positively charged amino function become a means of relieving the unfavorable increase in free energy of charged species as the hydrophobicity is increased, other means of doing so, such as ion-pairing, can be used to achieve self-assembly of protein-based structures. As discussed below, the increase in free energy of a charged species, which predisposes the moiety to some means of neutralization to lower free energy is called "poising".

2. Materials and Methods

The materials and methods will include both the chemical synthesis and the microbial biosynthesis of the protein-based polymers and the determination of T_t.

2.1 PREPARATION OF POLYMERS

While there are specific advantages to chemical synthesis, such as the specific radio-labeling of one particular residue, the comparison of the two approaches, apparent here, demonstrates the dramatic advantage of using recombinant DNA technology. A single fermentation run, once the gene has been made and the microbe transformed, can achieve in a day what would take chemical synthesis many many months to produce a less perfect product.

2.1.1 *Chemical Syntheses of Protein-based Polymers*

Synthesis of Pentamers. The peptides Boc-Ile-Pro-OBzl, Boc-Gly-Lys(2Clz)Gly-Ile-Pro-OH, Boc-Gly-Val-Gly-Val-Pro-OH, Boc-Gly-Val-Gly-Ile-Pro-OBzl, Boc-Gly-Phe-Gly-Val-Pro-OH, Boc-Gly-Glu(OCHx)-Gly-Val-Pro-OH, and Boc-Gly-Glu(OCHx)-Gly-Ile-Pro-OH were all prepared as previously described [6,15-18].

Boc-Ile-Gly-OBzl (1). Boc-Ile-OH·1/2 H_2O (Boc: tert-butyloxycarbonyl; 12.02 g, 0.0500 mol) was dissolved in 100 ml of dimethylformamide (DMF) and cooled to -15°C, 1-hydroxybenzotriazole (HOBt; 7.44 g, 0.055 mol) and 1-ethyl-3-(3-dimethylaminopropyl) carbodiimide (EDCI; 10.54 g, 0.055 mol) were added. After 20 min, a pre-cooled solution of Tos·H·Gly·OBzl (Tos; p-toluenesulfonyl; OBzl; benzyl ester; 16.87 g, 0.05 mol) and N-methylmorpholine (NMM; 5.50 ml, 0.05 mol) was added. The reaction mixture was stirred overnight at room temperature. The DMF was removed under reduced pressure and the residue was extracted into $CHCl_3$. The $CHCl_3$ extract was washed with water, 10% citric acid, 5% $NaHCO_3$, water and dried over Na_2SO_4. The solvent was removed under reduced pressure and dried to obtain 14.2 g (yield 75.05%) of I.

Boc-Gly-Ile-Gly-OBzl (II). Compound I (14.08 g, 0.0372 mol) was deprotected with 4.0 N HCl/dioxane for 1.5 h. Excess HCl and dioxane were removed under reduced pressure, triturated with ether, filtered, washed with ether and petroleum ether and dried (yield 100%). This was coupled to Boc-Gly-OH(6.51 g, 0.0372 mol) using EDCI with HOBt. The reaction was worked up by acid and base extractions to obtain 14.1 g (yield 87.14%) of II.

Boc-Gly-Ile-Gly-OH (III). Compound II (14.1 g, 0.0324 mol) in glacial acetic acid (140 ml) was hydrogenated in presence of 10% palladized charcoal catalyst at 40 psi. The catalyst was filtered off and the solvent was removed under reduced pressure. The residue was triturated with ether, filtered, washed with ether and petroleum ether and dried to obtain 11.18 g (yield 100%) of III.

Boc-Gly-Ile-Gly-Ile-Pro-OBzl (IV). Boc-Ile-Pro-OBzl (13.56 g, 0.0324 mol) was deprotected with HCl/dioxane and coupled to compound III (11.18 g, 0.0324 mol) using EDCI with HOBt. The reaction was worked up by acid and base extraction to obtain 17.98 g (yield 85.93%)

Boc-Gly-Glu(OCHx)-Gly-Val-Pro-Gly-Phe-Gly-Val-Pro-OBzl (V). Boc-Gly-Phe-Gly-Val-Pro-OBzl 3.95 g, 0.0059 mol) was deprotected with trifluoroacetic acid (TFA) for 45 min. Excess TFA was removed under reduced pressure, triturated with ether and dried (yield 100%). This was coupled to Boc-Gly-Glu-(OCHx)-Gly-Val-Pro-OH (3.8 g, 0.0059 mol) using EDCI with HOBt. The reaction was worked up by acid and base extractions to obtain 6.1 g (yield 85.67%) of V.

Boc-GFGVP GE(OCHx) GVP GFGVP-OBzl (VI). Compound V (5.69 g, 0.0048 mol) was deprotected with TFA and coupled to Boc-Gly-Phe-Gly-Val-Pro-OH

(2.76 g, 0.0048 mol) using EDCI with HOBt. The reaction was worked up by acid and base extractions to obtain 7.4 g (yield 93.72%) of VI.

Random Incorporation of Pentamers to form Protein-based Polymer. The pentamers, Boc-GVGIP-OBzl and Boc-GK(2-ClZ)GIP-OBzl were hydrogenated in the presence of 10% palladized charcoal catalyst at 40 psi and converted into p-nitrophenyl esters by reacting with bis-p-nitrophenylcarbonate. The ONP esters were mixed at appropriate ratios and deprotected with TFA for 45 minutes. A one molar solution of the TFA salt in DMSO was polymerized in the presence of 1.6 equiv. of NMM base. After 15 days, the polymers were dissolved in water, dialyzed using 3500 mol. wt. cutoff tubing and lyophilized. The polymers were then deprotected with HF:DMS:p-cresol (35:55:10, v/v) at 0°C for 2 hrs. They were triturated with ether and then dissolved in water, dialyzed using 50 kD mol. wt. cutoff tubing and lyophilized to obtain poly[f_k(GKGIP, f_v(GVGIP)]. Amino acid analysis and ^{13}C-NMR verified the ratios of pentamers incorporated.

Protein-based Polymers of Fixed Pentamer Sequences. Compound VI (5.92 g, 0.0036 mol) was hydrogenated and deprotected with TFA. A one molar solution of the TFA salt in DMSO was polymerized for 15 days using EDCI with HOBt and 1.6 equiv. of NMM as base. The polymer was dissolved in water, dialyzed using 3500 mol. wt. cutoff tubing and lyophilized. The material was then deprotected with HF: p-cresol (90:10, v/v) at 0° C for 1 h. It was triturated with ether and then dissolved in water, dialyzed, using 50 kD mol. wt. cutoff tubing and lyophilized to obtain 2.73 g (yield 55.96%) of poly [GFGVP GEGVP GFGVP] (VII).

The pentamers above were then used to construct two polytricosapeptides, poly(GEGIP GVGVP GVGIP GIGIP GVGVP GVGIP), designated E/5I, and poly(GKGIP GVGVP GVGIP GIGIP GVGVP GVGIP), designated as K/5I. These tricosapeptides were synthesized by the [(5+5+5)+(5+5+5)] fragment coupling strategy in solution phase. In the synthesis, the Boc group removal was effected by TFA. All coupling reactions were carried out by using EDCI with HOBt. The final deprotection and purification by dialysis were carried out as for polymer I above with the exception that the deprotection of polymer K/5I was carried out with HF:DMS:p-cresol (35: 55:10,v/v) at 0°C for 2 h. These peptides were checked at every step using thin layer chromatography. The final product was verified by means of carbon-13-NMR spectra.

2.1.2 *Microbial Biosyntheses*

Gene Construction of Polytricosapeptides. Glu-, Lys-, His-, and Asp-containing polytricosapeptides have been prepared by microbial biosynthesis. Only one, the Asp-polytricosapeptide, will be given as an example here. The others either have been or will be described elsewhere.

The Asp-Polytricosamer Example of Gene Construction. The construction of the basic monomer gene of Asp-polytricosamer, GVGVP GVGFP GDGFP GVGVP GVGFP GFGFP (designated as D/5F), utilized the corresponding oligonucleotide sequences of upper and lower strands. The two opposing oligonucleotides had a 20 base pair overlapping region, which was extended to form a double stranded full length DNA using PCR protocol. This basic gene sequence was flanked by the BamHI cloning site for the insertion into pUC118 and the PflM1 site for polymer gene construction [19]. Using this methodology, which has been a routine in our laboratory, we have successfully prepared some 60 basic monomer genes. After the sequence verification of the monomer clones, the amount of the plasmid was amplified by transformation and plasmid purification. The large amount of the catemer (monomer) gene fragment was prepared by releasing the gene from the plasmid, pUC118, on cleavage with PflM1, separation by electrophoresis through a polyacrylamide gel and followed by electroelution from the gel. The gene fragments were then concatenated by ligation through the PflM1 ends in the presence of cloning adaptors and inserted into pUC118. Concatemer clones were recovered and analyzed by restriction endonuclease digestion and electrophoresis versus sized standards, including a concatemer ligation ladder. Genes encoding for the Glu- and Lys-containing polymers were constructed in similar fashion.

E coli Transformation. The polymer production from the concatemer genes was achieved by using an expression system from Novagen Inc., the vector pET-11d. After subcloning the gene into pET-11d as a BamHI-NcoI fragment, this expression vector was used to transform the host *E coli* strain BL21(DE3).

Fermentation. Fermentation was done in a 16-liter New Brunswick fermentor with a working volume of 12 L. Cultures were grown in Luria Broth containing 100 mg ampicillin per liter at 37°C. Cell density was monitored using a Klett-Summerson colorimeter with a red number 66 filter. Polymer production was induced by the addition of IPTG to 0.4 mM at a cell density of 90 Klett units and cells were harvested by centrifugation 3 hours after induction. Cells were resuspended in 50 mM Tris-HCl buffer, pH 7.0 and lysed by a French press.

Purification. For purification of microbially produced protein-based polypeptides, a general methodology was developed whereby the polypeptides could be conveniently purified from the cell lysate based on their useful inverse temperature transition property [20]. First the bacterial cells are disrupted by sonication or French Press to release cell contents. The cell lysate is cooled to 4°C and centrifuged at high speed (10,000 x g, 60 min) to remove the cold insolubles. After the addition of an equal volume of TN buffer (100 mM Tris-HCl, pH 7.0, 0.5 M NaCl) to the cleared lysate, the solution is warmed to 37°C in a water bath to induce phase separation. The polymer is then recovered by centrifugation at 5,000 x g at 37°C for 15 minutes, resuspended and dissolved in water at 4°C. Typically, the process of cold and warm centrifugation is repeated two more times followed by dialysis against water and lyophilization. Qualitative analysis of the product is performed by SDS-PAGE followed by staining with 0.3 M $CuCl_2$ [21,22].

2.2 DETERMINATION OF T_t, THE TEMPERATURE OF THE INVERSE TEMPERATURE TRANSITION

2.2.1 *Temperature profiles of turbidity formation*

Temperature scans were carried out on an Aviv upgraded Cary-14 spectrophotometer. All of the polymer concentrations are 40 mg/ml. Scan rate was 0.5°C per min at a wavelength of 400 nm. Optical density arising from light scattering due to polymer aggregation was plotted against temperature. Scans were continued until optical density reached a maximum and began to decrease due to the coacervate phase separating and settling. Data were normalized to a maximum optical density of 100% turbidity. T_t was taken as the temperature value at 50% turbidity.

3. Results

A series of results are given, which convey the facility whereby following T_t becomes the means of achieving self-assembly.

3.1 SELF-ASSEMBLY BY THE ΔT_t -MECHANISM

The specific examples will be the self-assembly (loading) of drug delivery vehicles, the association of polymer chains of opposite sign, and the self association of polymer chains having both positive and negative charges in regular positions along the chain.

3.1.1 *The Self-Assembly of Charged Polymer with Oppositely Charged Drug*

Loading Profile for Cationic Naltrexone with Anionic Asp-containing Polymer. Genetically engineered D/5F, i.e., (GDGFP GVGVP GVGFP GFGFP GVGVP GVGFP)$_{12}$(GVGVP) exhibits a pKa of 5.6, shifted from a normal pKa of 3.9. When at a pH of 7.4 using 0.01 M phosphate as buffer and on raising the temperature toward 100°C, this polymer never exhibits an inverse temperature transition; it simply remains in solution. Operationally, the value of T_t for this anionic protein-based polymer is too high to be observed in water when fully ionized. When the cationic narcotic antagonist, Naltrexone, is added at a concentration of 2.65 mM (0.18 equivalents of drug to carboxylate), however, the transition temperature decreases to 37°C, and when further increased to 4.24 mM (0.29 equivalents) the value of T_t now occurs at 16°C, as seen in Figure 2A. The ion-pairing reduced the temperature of the inverse temperature transition and results in phase separation of polymer plus drug at room temperature. What has been achieved by increasing the pKa of the aspartic acid residue due to the presence of the Phe (F) residues is an increased affinity of drug for polymer. A drug delivery device

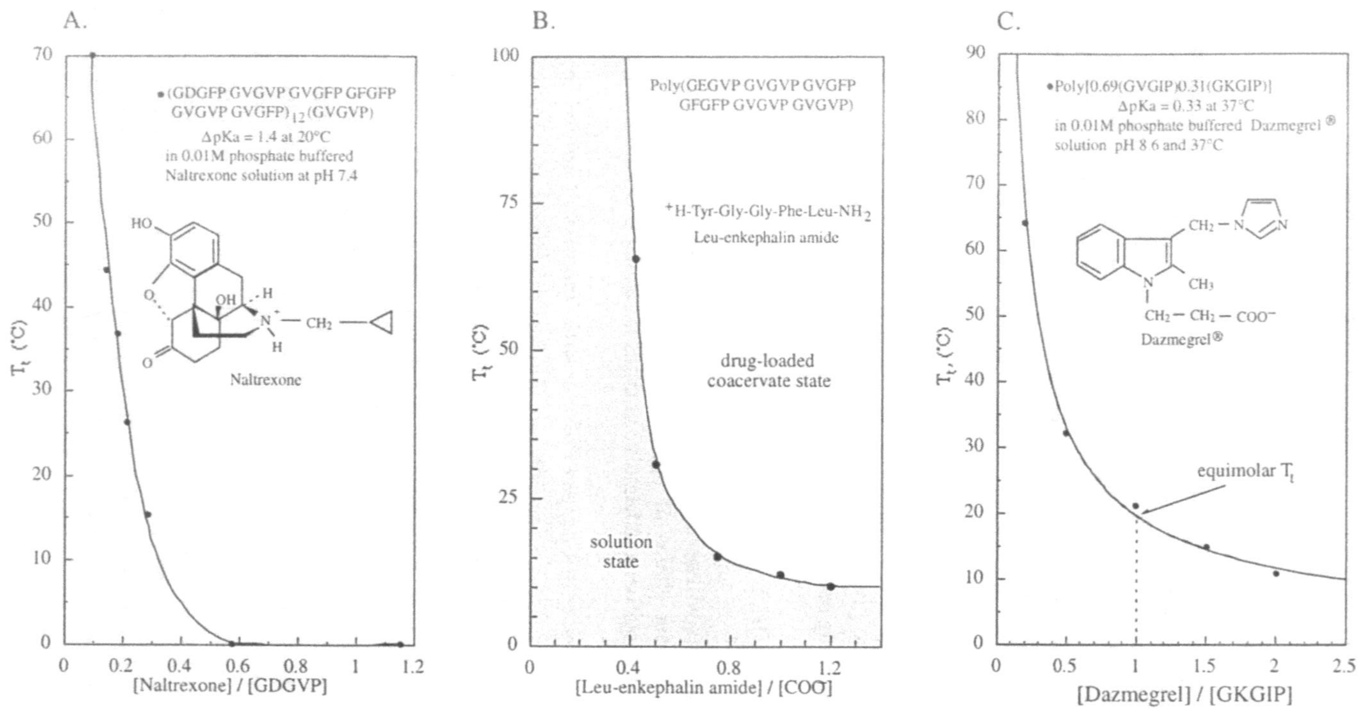

Figure 2: See text for discusstion. Parts B and C adapted with permission from Urry,D.W., Harris,C.M., Luan,C.X., Luan, C.H., Gowda,D.C., Parker, T.M., Peng,S.Q. and Xu, J., Chapter entitled "Transductional Protein-based Polymers as New Controlled Release Vehicles," Part VI: New Biomaterials for Drug Delivery, In Controlled Drug Delivery: The Next Generation, (Kinam Park, ed.) Am. Chem. Soc. Professional Reference Book, pp. 405–437, 1997.

self-assembles on combining a solution of the drug with a solution of polymer. The resulting vehicle releases a constant amount of drug from a constant surface area.

Loading Profile: Cationic Leu-enkephalin Amide with Anionic Glu-containing Polymer. The second protein-based polymer to be considered is chemically synthesized and anionic poly(GEGVP GVGVP GVGFP GFGFP GVGVP GVGVP), E/3F, of molecular weight greater than 50 kDa and a pKa of approximately 5 (as seen in Figure 1B). When at a pH of 7.4 using 0.01 M phosphate, it also exhibits a T_t that is too high to observe in water. When the positively charged opioid peptide Leu-enkephalin amide, $^{+}$H-Tyr-Gly-Gly-Phe-Leu-NH_2, is added at a concentration of 7.6 mM, the value of T_t appears at 31°C, and the T_t lowers further to 16°C on increasing the concentration of opioid peptide to 11.4 mM, as shown in Figure 2B. The result is again the self-assembly of a drug delivery vehicle that is capable of releasing a constant amount of Leu-enkephalin amide from a constant surface area. With the proper design of polymer a constant release of about 5 μmoles per day for up to 3 months has been achieved for a one cm^2 surface area.

Loading Profile: Anionic Drug Dazmegrel® with Cationic Lys-Containing Polymer. As seen in Figure 2C, the third polymer, chemically synthesized and cationic poly[0.69(GVGIP),0.31(GKGIP)], was loaded with the anionic drug, Dazmegrel®, an anionic inhibitor of thromboxane synthetase of relevance to pressure ulcer prevention. At 13.9 mM Dazmegrel®, T_t is 32°C, and at 41.7 mM Dazmegrel® T_t is lowered to 14°C. Thus, it also works to have a lysine-containing cationic polymer loaded with an anionic drug. Note, this co-polypentapeptide, with a random distribution of positively charged Lys (K) residues along its length, exhibits a lower affinity for the drug.

3.1.2 *Dependence of T_t of Lys-polytricosapeptide on Increasing Hydrophobicity*

The Lys-polytricosapeptide with increasing Phe Residues. Four polytricosapeptides are considered. All four have one Lys (K) per 30 residue repeat, and there is one polymer each with 0, 2, 3 and 4 Phe (F) residues per 30 mer, designated as K/0F, K/2F, K/3F and K/4F. The polymers of Figure 3A are in 0.01 M phosphate pH 7.4, where the Lys residue remains positively charged in all polymers. The value of T_t is 70 for K/0F. As one progresses to K/2F, to K/3F and to K/4F, the value of T_t decreases remarkably in systematic fashion. The T_t value for the K/5F polymer was too low to be determined. For the E/XF series under these conditions, all except the E/5F polymer have values of T_t that are too high to be determined.

3.1.3 *Self-Assembly by "Poising" of Polymers of Opposite Charge*

Poising. The pKa values for the lower molecular weight microbially-prepared Glu-containing polytricosamers are given in parentheses, following their short-hand designations, i.e., E/0F(4.6), E/2F(5.4), E/3F(5.8) and E/4F(6.4), E/5F(6.6). The pKa values, also given in parentheses, similarly decrease in a systematic manner for the Lys-polytricosamers, i.e., K/0F(10.0), K/2F(9.8), K/3F(9.5) and K/4F(8.9). The increasing ΔpKa values, when referenced to the pKa of the 0F polymers, are directly proportional to the lowered Gibbs free energy for ion-pairing. Increased ΔpKa is an example of poising, which can be used to effect the self-assembly of protein-based polymers.

Self-Assembly of K/0F with the Series E/0F, E/2F, E/3F, E/4F and E/5F. As seen in Figure 3B, there is a systematic lowering of T_t on increasing the Phe content of the associating polymer. The greater the ΔpKa, the lower the value of T_t. At 20°C, only the polymers with 3, 4 and 5 Phe residues per 30 mer would assemble; those with 0 and 2 Phe residues per 30 mer would remain in solution. Note that T_t is a linear function of the number of F residues.

Self-Assembly of E/0F with the Series K/0F, K/2F, K/3F, and K/4F. Seen in Figure 3C are an analogous set of curves to those of Figure 3B. Again the greater the poising, that is, the larger the ΔpKa, the greater is the driving force for assembly. The trigger for self-assembly is simply moving the value of T_t from above to below the operating temperature. Again note the linear relationship between T_t and number of F residues.

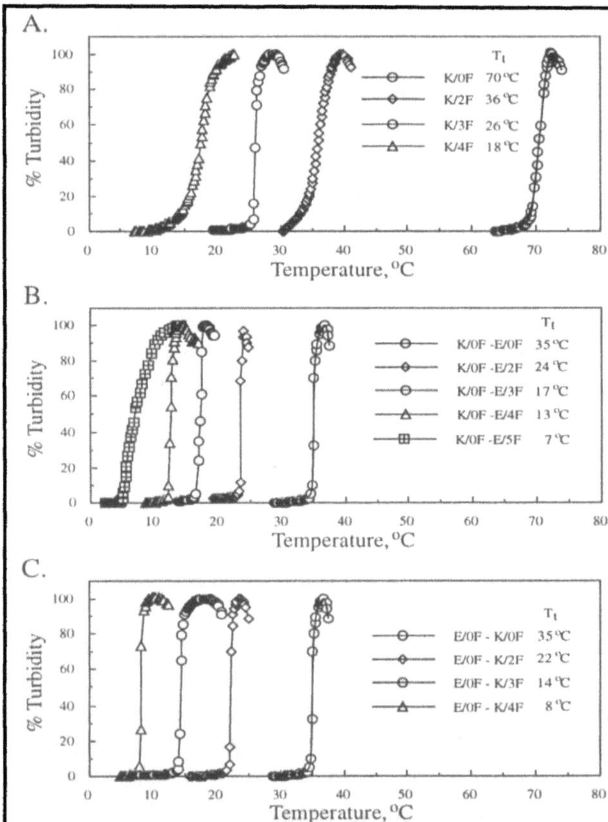

Figure 3: Temperature profiles for assembly, A as a function of pH, B showing association of oppositely charged polymers with a constant Lys(K)-containing polymer and varying hydrophobicity of the Glu(E)-containing polymer, and C showing association of oppositely charged polymers with a constant Glu(E)-containing polymer and varying hydrophobicity of the Lys(K)-containing polymer. See text for further discussion and polymer identifications.

3.1.4 *Variations of Self-Assembling Pairs of Oppositely Charged Polymers*

Here we consider pairs of polytricosapeptides by first noting their individual T_t values at pH 7.5 in 0.01 M phosphate and then demonstrating the lowering of T_t on combining solutions of soluble polymers. In this section, designed protein-based polymers are seen to remain in solution until a suitably designed counter polymer is introduced which triggers self-assembly.

The D/5F and K/5I Pair of Polytricosapeptides. In particular in Figure 4A is shown the genetically engineered polymer (GDGFP GVGVP GVGFP GFGFP GVGVP GVGFP)$_{12}$(GVGVP), i.e., D/5F. It remains soluble at all reachable temperatures in water buffered at pH 7.5 using 0.01 M phosphate. Also, the chemically synthesized polymer, poly(GKGIP GVGVP GVGIP GIGIP GVGVP GVGIP), i.e., K/5I, with a molecular weight of greater than 50 kDa, is soluble in water buffered at pH 7.5 using 0.01 M phosphate but exhibits its inverse temperature transition at 41°C. When 40 mg/ml solutions of each polymer are combined, the polymers self-assemble with the inverse temperature transition initiating at 14°C, as shown in Figure 4A. Thus, at 20°C both solutions are clear; the polymers are dispersed, but on combining the two solutions, the polymers self-assemble.

The E/5I and K/5I Pair of Polytricosapeptides. Next, using the same cationic polymer, K/5I, the effect of the ΔpKa due to differences in hydrophobicity is examined by considering an anionic polymer with the lesser perturbed Glu pKa; the Phe (F) residues are replaced by the less hydrophobic Ile (I) residues. In the case of polymer D/5F of Figure 4A, the ΔpKa is 1.4, whereas the ΔpKa for polymer E/5I of Figure 4B is only 0.5. Again, the T_t value for the carboxylate-containing polymer is greater than 80°C. Again individually, both polymers remain in solution below 35°C. When the two 40 mg/ml solutions are combined at 30°C, however, the polymers self-assemble. For the D/5F-K/5I pair of Figure 4A, the combined T_t value was 10°C lower, because the ΔpKa for D/5F is greater than that of E/5I. For the pair of polymers of Figure 4B, at 20°C there would be no self-assembly. Thus, the hydrophobic-induced pKa shift increases the affinity between polymers and drives self-assembly.

3.1.5 *Polytricosapeptides with both Positive and Negative Charges in Each 30 mer*

Finally, we consider two polytricosapeptides, (GVGVP GVGFP GEGFP GVGVP GVGVP GKGVP)$_n$ (GVGVP), designated as EK/2F, and (GVGVP GVGFP GEGFP GVGVP GVGFP GKGVP), designated as EK/3F. Each polymer contains both Glu (E) and Lys (K) periodically positioned in each repeating 30 mer. The first polymer contains two Phe (F) residues per 30 mer, and the second is more hydrophobic with three Phe (F) residues per 30 mer. As association of chains provides for ion-pairing, each polymer alone has an intrinsically low T_t value, as seen in Figure 4C. Interestingly, the T_t value for EK/3F (14°C) is the same as for the combined polymers, E/0F-K/3F (14°C) of Figure 3C, and closely the same as for the combined polymers, K/0F-E/3F (17°C) of Figure 3B. In these cases, the summed K, E and F residues per 30 mer is the same.

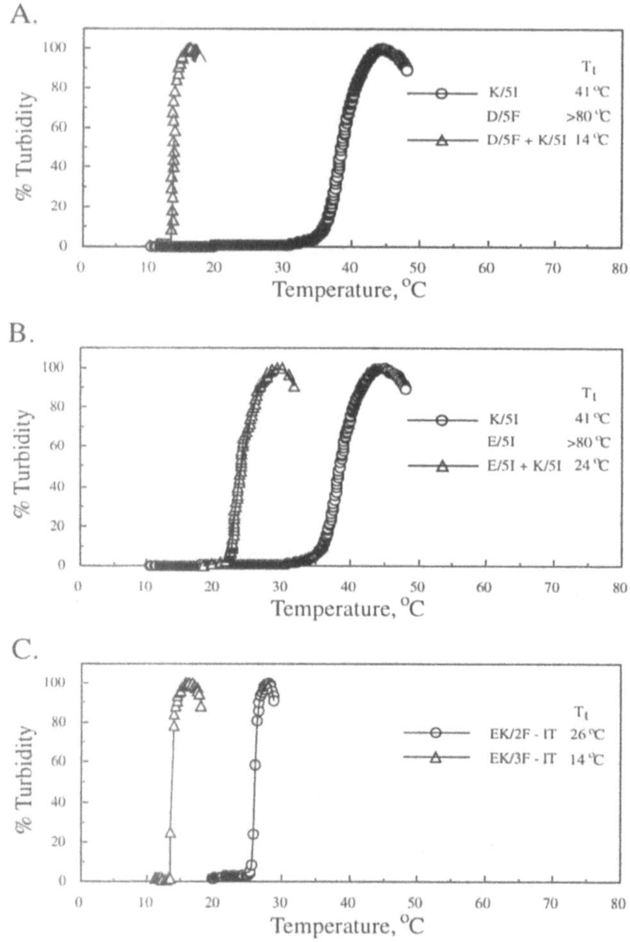

Figure 4: See Text for Discussion.

4. Discussion

Salts dissolve in water because the Gibbs free energy of a single ion surrounded by its full complement of hydration is lower than the free energy of the ion surrounded by oppositely charged ions in the crystal. Polymers with charges distributed along their length can be soluble for similar reasons. With polymers, however, the situation becomes complicated due to the properties of the other chemical groups constrained to co-exist along the polymer length. The result in amphiphilic polymers is an apolar-polar repulsive free energy of hydration, e.g., a competition between hydrophobic and charged groups for hydration in a polymer. Because of this, a charged side chain may not be able to obtain its full hydration shell. The result is an ionizable group with a hydrophobically shifted pKa that is poised for self-assembly by ion-pairing.

Under these circumstances charged groups come under stress to decrease their free energy, i.e., to become less charged by gaining or losing a proton (the experimentally observed pKa shifts) or to become more disposed to ion-pairing [23]. As we have seen, these forces can be employed to design (poise) polymers capable of self-assembly. Also, we have seen that the process

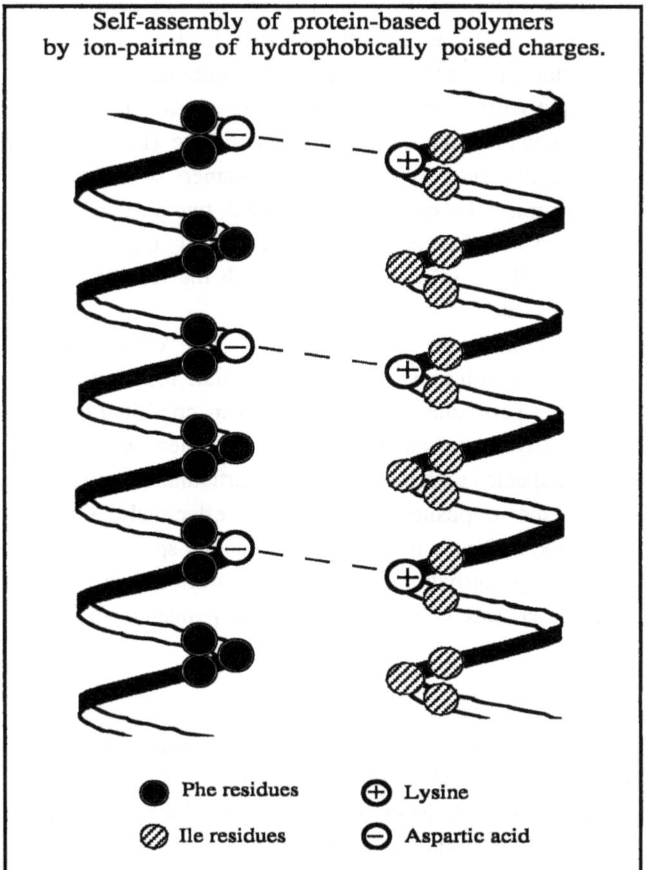

Figure 5: Schematic representation of the self-assembly of oppositely charged polymers. This assembly is driven by the presence of residues more hydrophobic than Val (V) which raise the free energy of the charged species (as evidenced by hydrophobic-induced pKa shifts) and favor ion-pairing

of self-assembly can be followed simply in terms of changes in T_t, that is, by means of the ΔT_t-mechanism.

A schematic representation of the polymer self-assembly achieved in Figure 4A is shown in Figure 5. In addition to the many reviewed structural studies that gave rise to the structure [24], there are a number of phenomenological reasons to represent such a regular structure. For example, when the Lys-containing polymer is comprised of the same pentamers, as K/5I in Figure 4A but with a random incorporation of the pentamers, combination with polymer E/5F of Figure 4A results in a smaller ΔT_t. Furthermore, the ΔpKa exhibited by a Glu (E) or Asp (D) residue, which drives the association with the polymer containing a pKa shifted Lys (K) residue, is much greater for the regular structure with proximal Phe (F) residues than when the polymer is comprised of randomly distributed pentamers of the same composition [25]. Finally, when assuming the structure of Figure 5, two polymers can have the same composition, but the one with five Phe residues structurally distal to the carboxyl of Asp exhibits less than one-half the pKa shift as that with the five Phe residues structurally proximal to the Asp residue.

In contrast to the perspective that "creation of structures" requires far-from-equilibrium conditions, we have seen that relatively small changes in the chemical energy of a particular chemical species in combination with the presence of a properly designed (hydrophobically "poised") polymer can drive supramolecular assembly of chemical species with polymer. The particular chemical species may itself be a polymer carrying the opposite charge, most effectively with a complementary structural periodicity, or it may be a smaller molecular species of the opposite charge.

Any perturbation that lowers T_t from above to below the working temperature can drive self-assembly. This is the ΔT_t-mechanism for the design of self-assembling structures.

5. Acknowledgments: The authors wish to acknowledge the financial support of the Office of Naval Research by means of grant and contracts N00014-89-J-1970, N00014-98-1-0656, N00014-98-C-0279 and the National Institutes of Health contracts 1 R43 DA09511-01 and N43-DK-4-2209.

6. References:

1. Urry, D.W., Trapane, T.L., and Prasad K.U. (1985) Phase-Structure Transitions of the Elastin Polypentapeptide-Water System Within the Framework of Composition-Temperature Studies, *Biopolymers* **24**, 2345-2356.
2. Urry D.W. (1993) Molecular Machines: How Motion and Other Functions of Living Organisms Can Result from Reversible Chemical Changes, Angew. Chem. (German) **105**, 859-883, 1993; Angew. Chem. Int. Ed. Engl., **32**, 819-841.

3. Urry, D.W., Gowda, D.C., Parker, T.M., Luan,C.H., Reid, M.C., Harris, C.M., Pattanaik,A., and Harris, R.D. (1992) Hydrophobicity Scale for Proteins Based on Inverse Temperature Transitions, *Biopolymers* **32**, 1243-1250.
4. Urry, D.W., Peng, S.Q., and Parker, T.M. (1993) Delineation of Electrostatic-and Hydrophobic-Induced pKa Shifts in Polypentapeptides: The Glutamic Acid Residue, *J. Am. Chem. Soc.* **115**, 7509-7510.
5. Urry, D.W., Peng, S.Q., Parker, T.M., Gowda, D.C., and Harris, R.D. (1993) Relative Significance of Electrostatic- and Hydrophobic-Induced pK_a Shifts in a Model Protein: The Aspartic Acid Residue, *Angew. Chem.* (German) **105**, 1523-1525, 1993; *Angew. Chem. Int. Ed. Engl.* **32**, 1440-1442.
6. Urry, D.W., Peng, S.Q., Gowda, D.C., Parker, T.M., and Harris, R.D. (1994) Comparison of Electrostatic- and Hydrophobic-induced pKa Shifts in Polypentapeptides: The Lysine Residue, *Chemical Physics Letters* **225**, 97-103.
7. Urry, D.W. (1992) "Free Energy Transduction in Polypeptides and Proteins Based on Inverse Temperature Transitions," *Prog. Biophys. Molec. Biol.* **57**, 23-57.
8. The nature of the water of hypdrophobic hydration is considered to be like the hydrogen bonded water forming pentagonal structures surrounding alkane gases as found by Stackelberg and Muller (1951) *Naturwissenscaften* **38**, 456, and at the surface of hydrophic side chains such as leucine (Leu, L) in the crystal structure of the protein, crambin, by M. Teeter (1984) *Proc. Natl. Acad. Sci.* **81**, 6014-6018.
9. Urry, D.W., Luan, C.H., Harris, R.D., and Prasad,K.U. (1990) Aqueous Interfacial Driving Forces in the Folding and Assembly of Protein (Elastin)-Based Polymers: Differential Scanning Calorimetry Studies, *Polym. Preprints, Div. Polym. Chem., Am. Chem. Soc.* **31**, 188-189.
10. Urry, D.W., Peng, S.Q., Xu, J. and McPherson, D.T. (1997) Characterization of Waters of Hydrophobic Hydration by Microwave Dielectric Relaxation, *JACS* **119**, 1161–1162.
11. Urry, D.W. (1997) Physical Chemistry of Biological Free Energy Transduction as Demonstrated by Elastic Protein-Based Polymers *J. Phys. Chem.* B, **101(51)**, 11007-11028.
12. Pattanaik, A., Gowda, D.C. and Urry, D.W., (1991) Phosphorylation and Dephosphorylation Modulation of an Inverse Temperature Transition, *Biochem. Biophys. Res. Comm.* **178**, 539-545.
13. Urry, D.W., Gowda, D.C., Peng, S.Q., and Parker, T.M. (1995) Non-linear Hydrophobic-induced pKa Shifts: Implications for Efficiency of Conversion to Chemical Energy, *Chem.* Phys. Letters, **239**, 67–74.
14. Urry, D.W., and Peng, S.Q. (1995) Non-linear Mechanical Force-induced pKa Shifts: Implications for Efficiency of Conversion to Chemical Energy, *J. Am. Chem. Soc.***117**, 8478–8479.
15. Urry, D.W., Long, M.M., Harris, R.D., and Prasad, K.U. (1986) Temperature Correlated Force and Structure Development in Elastomeric Polypeptides: The Ile^1 Analog of the Polypentapeptide of Elastin, *Biopolymers* **25**, 1939-1953.
16. Prasad, K.U., Iqbal, M.A., and Urry, D.W. (1985) Utilization of 1-Hydroxybenzotriazole in Mixed Anhydride Coupling Reactions, *Int. J. Pept. and Protein Res.* **25**, 408-413.
17. Luan, C.H., Parker, T.M., Gowda, D.C., and Urry, D.W. (1992) Hydrophobicity of Amino Acid Residues: Differential Scanning Calorimetry and Synthesis of the Aromatic Analogues of the Polypentapeptide of Elastin, *Biopolymers* **32**, 1251-1261.

18. Urry, D.W., Peng, S.Q., and Parker, T.M. (1992) Hydrophobicity-induced pK Shifts in Elastin Protein-Based Polymers, *Biopolymers* **32**, 373-379.

19. Maniatis, T., Fritsch, E.F., and Sambrook, J. (1982) Molecular Cloning: A Laboratory Manual, Cold Spring Harbor Laboratory Press, Cold Harbor, NY.

20. McPherson, D.T., Xu, J., and Urry, D.W. (1996) Product Purification by Reversible Phase Transition Following *E. coli* Expression of Genes Encoding up to 251 Repeats of the Elastomeric Pentapeptide GVGVP, *Protein Expression and Purification* **7**, 51–57.

21. Laemmli, U. K. (1970) Cleavage of Structural Proteins During the Assembly of the Head of Bacteriophage T4, *Nature* **227**, 680-685.

22. Lee, C., Levin, A., and Branton, D. (1987) Copper Staining: A Five-Minute Protein Stain for Sodium Dodecyl Sulfate-Polyacrylamide Gels, *Anal. Biochem.* **166**, 308-312.

23. Urry, D.W., and Luan, C.H. (1995) Proteins: Structure, Folding and Function, in Giorgio Lenaz (Vol. Ed.), *Bioelectrochemistry: Principles and Practice*, Birkhäuser Verlag AG, Basel, Switzerland, pp. 105–182.

24. Urry, D.W. (1991) Thermally Driven Self-assembly, Molecular Structuring and Entropic Mechanisms in Elastomeric Polypeptides, in P. Balaram and S. Ramaseshan, (eds.) *Mol. Conformation and Biol. Interactions*, Indian Acad. of Sci., Bangalore, India, pp. 555-583.

25. Urry, D.W., McPherson, D.T., Xu, J., Daniell, H., Guda, C., Gowda, D.C., Jing, N., and Parker, T.M. (1996) Protein-Based Polymeric Materials: Syntheses and Properties" in J.C. Salamone (ed.) *The Polymeric Materials Encyclopedia: Synthesis, Properties and Applications*, CRC Press, Boca Raton, pp. 7263–7279.

Self-assembling Peptide Systems in Biology and Biomedical Engineering

SHUGUANG ZHANG & MICHAEL ALTMAN

Center for Biomedical Engineering 56-341 & Department of Biology, Massachusetts Institute of Technology, 77 Massachusetts Avenue, Cambridge, MA 02139-4307 USA, shuguang@MIT.EDU

1. Abstract

Molecular self-assembly of peptide systems is emerging as a new approach in protein engineering, study of protein folding, and protein-protein interaction as well as nanotechnology, polymer science, nanomaterials science and engineering. Self-assembling peptide systems lie at the interface of all these fields. Many self-assembling peptide and protein systems have been developed, ranging from simple peptides to complex proteins. Molecular self-assembly systems represent a significant advance in the molecular engineering of simple molecular building blocks useful for a wide range of applications. This field is extremely broad and is growing at an accelerating pace. The key elements in molecular self-assembly are chemical complementarity and structural compatibility. Several types of self-assembling peptides have been molecular-designed thus far. Type I peptides undergo intermolecular self-assembly while type II peptides undergo self-assembly and disassembly, i.e. intermolecular and intramolecular self-assembly, under the influence of various conditions. Type III peptides undergo self-assembly on to surfaces. These self-assembling peptide systems are simple and versatile, while being easy to modify and produce. Such a nanosystem represents a significant advancement towards molecular engineering of protein fragments for a diversity of technological innovations.

2. Introduction

Molecular self-assembly, by definition, is the spontaneous organization of molecules under thermodynamic equilibrium conditions into structurally well-defined arrangements due to noncovalent interactions. These molecules undergo self-association forming hierarchical structures without external instruction. Molecular self-assembly is ubiquitous in nature and

A. Aggeli et al. (eds.), Self-assembling Peptide Systems in Biology, Medicine and Engineering, 343–360.

has recently emerged as a new strategy in chemical synthesis, polymer science and engineering [1-3].

Proteins and protein fragments have not been seriously considered to be useful materials for traditional engineering science, although their potential has been recognized. Several recent discoveries and rapid developments in biotechnology, however, have rekindled the field of biological materials engineering. In the last few years, considerable advances have been made in the use of peptides as building blocks to produce biological materials for a wide range of applications [4-11].

A new class of oligopeptide-based biological materials was serendipitously discovered from the self-assembly of ionic self-complementary oligopeptides [12]. The first peptide, EAK16-II, was found as a repetitive segment in a yeast protein, zuotin, that was initially characterized to preferentially bind to left-handed Z-DNA [13]. A number of peptide molecular self-assembly systems have been designed and developed based on these sequences. This systematic analysis provided insight into the chemical and structural principles of peptide self-assembly. These peptides are short, simple to design, extremely versatile and easy to synthesize. Three types of self-assembling peptides have been systematically studied thus far. It is believed additional different types will be discovered and developed in the coming years. This class of biological materials has considerable potential for a number of applications, including scaffolding for tissue repair and tissue engineering, drug delivery of molecular medicine, as well as biological surface engineering. Similar systems have also been described. In one of those systems, peptides undergo self-assembly to form a gel with regular β-sheet repeats of well-defined structure [6]. The self-assembly of peptide nanotubes that allow ions to pass through and to insert themselves into lipid bilayers have also been described [5, 14]. Furthermore, a number of fascinating biomimetic peptides and protein structures have been engineered, such as helical coil-coils, di-, tri- and tetra-helical bundles [4, 15-16].

3. Type I Self-assembling peptides

3.1. THE CHEMICAL AND STRUCTURAL PROPERTIES.

Type I peptides, also called "molecular Lego", form β-sheet structures in aqueous solution because they contain two distinct sides, one hydrophilic and the other hydrophobic. Just as Lego bricks that have pegs and holes can only be assembled in a particular orientation, these peptides have this property at the molecular level. Hydrophobic interactions can form only on one face of the peptide, much like the structure found in silk where many methyl groups of alanines overlap. The unique structural feature of these peptides is that they can form complementary ionic bonds with regular repeats on the hydrophilic face (Fig. 1).

The complementary ionic sides have been classified into several moduli, *i.e.* modulus I, II, III, IV, etc., and mixed moduli. This classification is based on the hydrophilic surface of the molecules, having alternating + and - charged amino acid residues, either alternating by 1, 2, 3, 4 and so on. For example, molecules of modulus I have - + - + - + - +, modulus II, - - + + - - + +, modulus, IV - - - - + + + +. These well-defined sequences allow them to undergo ordered self-assembly, resemblance of some situation found in well-studied polymer assemblies.

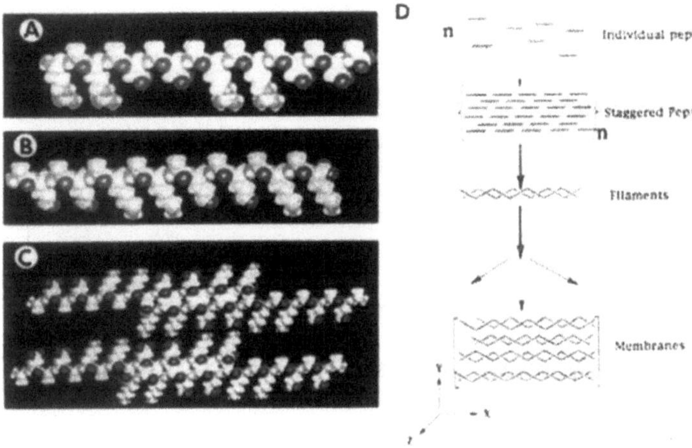

Fig. 1. Type I self-assembling peptides. Molecular models of the extended beta-strand structures of individual molecules are shown for RAD16 (A) and EAK16 (B). The distance between the charged side chains along the backbone is approximately 6.8 Å; the methyl groups of alanines are found on one side of the sheet and the charged residues are on the other side. Conventional beta-sheet hydrogen bond formation between the oxygen and hydrogen on the nitrogen of the peptide backbones are perpendicular to the page. (C) A proposed staggered assembly of molecular models for EAK16. The complementary ionic bonds and hydrophobic alanines are shown. Although an anti-parallel beta-sheet is illustrated, a parallel beta-sheet model or stacked peptide along the backbone is also possible. D) A proposed model of sequential events that could lead to assembly of macroscopic matrices. X, Y, and Z indicate three dimensions of the materials. Geometric shapes other than membrane can also be produced as suggested by the diverging thin arrows (see Figure 2).

Table 1. Type I self-assembling peptides that have been studied.

Name	Sequence (n-->c)	Ionic Modulus	Structure
	+ - + - + - + -		
RADA16-I	n-RADARADARADARADA-c	I	β
	+ - + - + - + -		

RGDA16-I	n-RADARGDARADARGDA-c	I	r.c.
	+ - + -		
RADA8-I	n-RADARADA-c	I	r.c.
	+ + - - + + - -		
RAD16-II	n-RARADADARARADADA-c	II	β
	+ + - -		
RAD8-II	n-RARADADA-c	II	r.c.
	- + - + - + - +		
EAKA16-I	n-AEAKAEAKAEAKAEAK-c	I	β
	- + - +		
EAKA8-I	n-AEAKAEAK-c	I	r.c.
	+ - + - + - + -		
RAEA16-I	n-RAEARAEARAEARAEA-c	I	β
	+ - + -		
RAEA8-I	n-RAEARAEA-c	I	r.c.
	+ - + - + - + -		
KADA16-I	n-KADAKADAKADAKADA-c	I	β
	+ - + -		
KADA8-I	n-KADAKADA-c	I	r.c.
	- - + + - - + +		
EAH16-II	n-AEAEAHAHAEAEAHAH-c	II	β

	− − + +		
EAH8-II	n-AEAEAHAH-c	II	r.c.
	− − + + − − + +		
EFK16-II	n-FEFEFKFKFEFEFKFK-c	II	β
	− + − +		
EFK8-II	n-FEFKFEFK-c	I	β
	− − + + − − + +		
ELK16-II	n-LELELKLKLELELKLK-c	II	β
	− − + +		
ELK8-II	n-LELELKLK-c	II	β
	− − + + − − + +		
EAK16-II	n-AEAEAKAKAEAEAKAK-c	II	β
	− − − − + +		
EAK12	n-AEAEAEAEAKAK-c	IV/II	α/β
	− − + +		
EAK8-II	n-AEAEAKAK-c	II	r.c.
	+ + + + − − − −		
KAE16-IV	n-KAKAKAKAEAEAEAEA-c	IV	β
	− − − − + + + +		
EAK16-IV	n-AEAEAEAEAKAKAKAK-c	IV	β

	+ - + - + -		
KLD12-I	n-KLDLKLDLKLDL-c	I	β
	+ + + + - - - -		
RAD16-IV	n-RARARARADADADADA-c	IV	β
	- - - - + + + +		
DAR16-IV	n-ADADADADARARARAR-c	IV	α/β
	- - - - + + + +		
DAR16-IV*	n-DADADADARARARARA-c	IV	α/β
	- - - - + + + +		
DAR32-IV	n-(ADADADADARARARAR)-c	IV	α/β

β, β-sheet; α, α-helix; r.c., random coil; N/A not applicable. *Both VE20 and RF20 are in β-sheet form when they are incubated in solution containing NaCl.

3.2. INFLUENCE OF SALTS.

Upon the addition of monovalent alkaline cations or the introduction of the peptide solutions into physiological media, these oligopeptides spontaneously assemble to form macroscopic structures that can be fabricated into various geometric shapes (Fig. 2) [17]. Scanning EM reveals that the matrices are made of interwoven filaments that are about 10-20 nm in diameter and pores about 50-100 nm in diameter [12, 18-19]. However, when alanines are changed to more hydrophobic residues, such as leucine and phenylalanine, reduced concentration of salt can also promote self-assembly. It was also found that different type of salt, such as LiCl, NaCl, KCl, CsCl, $MgCl_2$, and $CaCl_2$ resulted in different macroscopic textures of the peptide matrices.

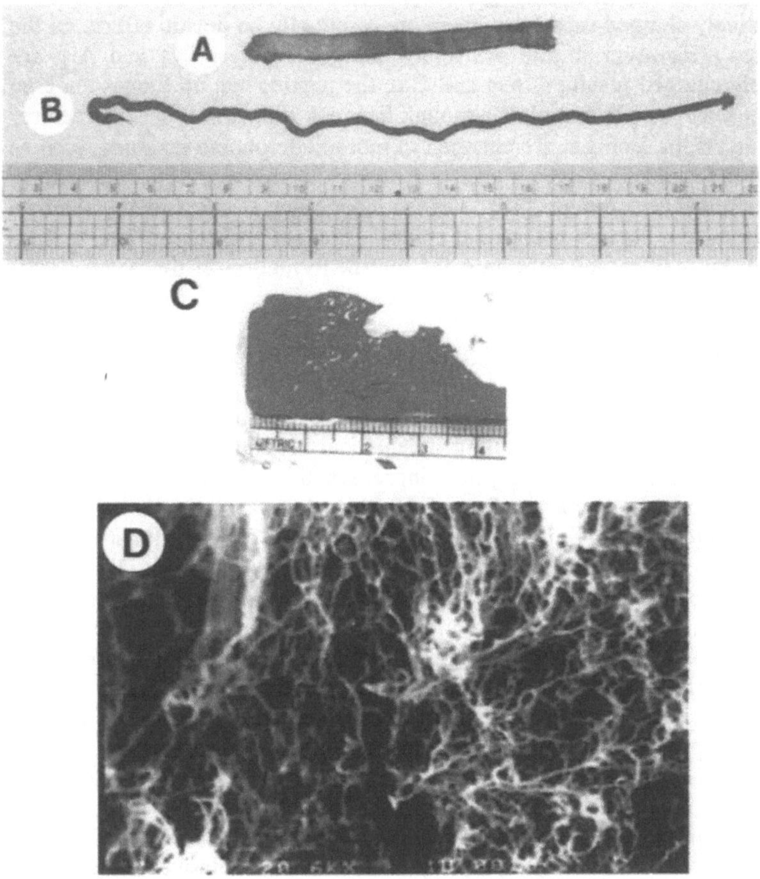

Fig. 2. Photographs of biological materials from the Type I self-assembling peptides. A) The peptide material is fabricated in tape form with 5 mm width, 0.2 mm thickness and 8 cm length. B) A peptide matrix is fabricated as a thread with a diameter of about 2 mm through a syringe. Threads with a length of greater than 20 cm have been produced from 1 ml peptide solutions. C) The peptide matrix fabricated as membrane, each scale is 1 mm. D) The SEM structure of RADA16-II. The material is self-assembled from individual interwoven fibers. The diameter of the fiber is about 10-20 nm and the enclosures are about 50-100 nm. Analyses using atomic force microscope (AFM) later confirmed the observations.

3.3. MOLECULAR STRUCTURE OF TYPE I SELF-ASSEMBLING PEPTIDES.

The molecular structure and proposed complementary ionic pairings of the Type I peptides between positively charged lysines and negatively charged glutamates in an overlap arrangement are modeled in Fig. 1. This structure represents an example of this class of self-assembling β-sheet peptides that undergo spontaneously association under physiological conditions. If the charged residues are substituted, *i. e.*, the positive charged lysines are replaced by positively charged arginines and the negatively charged glutamates are replaced by negatively charged aspartates, there are essentially no drastic effects on the self-assembly process. However, if the positively charged resides, Lys and Arg are replaced by negatively charged residues, Asp and Glu, the peptide can no longer undergo self-assembly to form macroscopic materials although they can still form β-sheet structures in the presence of salt. If the alanines are changed to more hydrophobic residues, such as Leu, Ile, Phe or Tyr, the molecules have a greater tendency to self-assemble and to form peptide matrices with enhanced strength [18]. On the other hand, if a few of the alanines are replaced by glycines, the peptide has a random coil structure and can no longer form matrices. Therefore, both sides are important for matrix formation in the Type I self-assembling peptides.

3.4. MECHANICAL PROPERTIES OF OLIGOPEPTIDE MATERIALS.

The mechanical properties for one peptide material, EFK8, has recently been examined [18]. This peptide material showed some interesting mechanical properties. Its fiber density increases as the concentration of the peptide increases but not the fiber diameter. Furthermore the mechanical strength of peptide materials is proportional to the concentration of the peptide with a Young's modulus about 2 kPa and 15 kPa at 0.3% and 1% in water, respectively, approximately in the order of collagen gels [18]. In the same manner, several peptides with Leu or Ile substitutions showed similar properties [unpublished results]. Interestingly, when polypeptides have only the negatively or positively charged residues with alternating Val, Leu, Phe or Tyr, they did not form visible materials [20-23]. Therefore, not only are the hydrophobic residues important, but both positive and negative charges on the same peptides are essential for peptide material formation. It should be pointed out that the peptide materials had no detectable swelling property when the material was delivered into a saline solution. This is likely due to the high water content of the material, in which >99% of the material is water. This is a very important factor if the materials are to be further developed as a scaffold for tissue engineering and tissue repair. This unique property of this material removes the chance of there being an unregulated expansion of the scaffold that could lead to adverse physiological effects on neighboring tissues.

3.5. PEPTIDE SCAFFOLD SUPPORT CELL ATTACHMENT, EXTENSIVE NEURITE OUTGROWTH AND ACTIVE SYNAPSES.

The peptide self-assembly phenomenon was a serendipitous observation during an experiment to test the cytotoxicity of these peptides in tissue cells. The cells exhibited a robust growth and no cytotoxicity was observed. However, a thin layer of material that was attached to the cells was observed under a phase contrast microscope. This thin layer of material occurred only in the duplicated cell culture dishes where the EAK16-II peptides were added. Whereas in other dishes where different peptides including EAK8-II (A single unit of the EAK repeat, see table 1) were used, this phenomenon was not observed. A number of mammalian cells have been tested and all have been found to be able to form stable attachments with the peptide materials (Table 2) [17]. Several peptide materials have been tested for their ability to support cell proliferation and differentiation. These results suggest that the peptide materials can not only support various types of cell attachments, but can also allow the attached cells to proliferate and differentiate. For example, as rat PC12 cells on peptide matrices were exposed to NGF, they underwent differentiation and exhibited extensive neurite outgrowth [19]. In addition, when primary mouse neuron cells were allowed to attach the peptide materials, the neuron cells projected lengthy axons that followed the surface contours of the self-assembled peptide materials and formed active synapses (Fig. 3), a key ingredient for neurite connection.

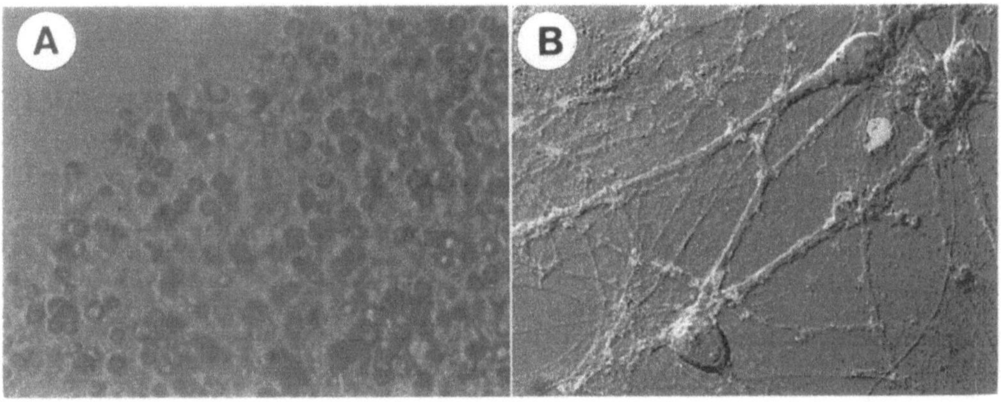

Fig. 3. Cell encapsulation and biological activity on the Type I peptide scaffold. A) Bovine primary chondrocyte cells in the peptide matrix. They eventually form a cartilage tissue after several weeks. B) Mouse hippocampal neuron cells form synapses on the peptide matrix. The discrete dots are the synaptic junctions.

The fundamental design principles of such self-assembling peptide systems can be readily extended to traditional organic polymers and polymer composites, where co-polymers can be designed and produced.

Table 2. Cell attachment to the Type I self-assembling peptide matrices.

Cell Type	Cell Line

Mouse Fibroblast	NIH-3T3
Chicken Embryo Fibroblast	CEF
Chinese Hamster Ovary	CHO
Human Cervical Carcinoma	Hela
Human Osteosarcoma	MG63
Human Hepatocellular Carcinoma	HepG2
Hamster Pancrease	HIT-T15
Human Embryonic Kidney	HEK293
Human neuroblastoma†	SH-SY5Y
Rat pheochromocytoma†	PC12
Bovine aortic endothelial cells*	
Mouse Cerebellum Granule Cells*†	
Mouse & rat Hippocampal Cells*†	
Human Foreskin Fibroblast*	
Human Epidermal Keratinocytes*	
Bovine chrondrocytes* (calf and adult cells)	

Various cell types attachment to the peptide matrices. Visual assessment of cell attachment was performed using phase contrast microscopy for over a period of two weeks. * refers to cells derived from primary sources. † refers to neuronal cells.

4. Type II self-assembling peptides

4.1. STRUCTURAL PROPERTIES OF TYPE II PEPTIDES.

This type of peptide was discovered while systematically analyzing a family of modulus IV peptides. Several Type II peptides are now being developed as "molecular switch" in which the peptides can drastically change their molecular structure in response to the environment (Fig.4). One of these peptides has 16 amino acids, DAR16-IV, which shows a β-sheet structure at ambient temperature with 5nm length but can undergo an abrupt structural transition at high temperatures to form a stable α-helical structure with 2.3 nm length [24-25]. Similar structural transformations can be induced by changes of pH. This suggests that secondary structures of some sequences, especially segments flanked by clusters of negative charges on the N-terminus and positive charges on the C-terminus, may undergo drastic conformational transformations under the appropriate conditions. These findings provide insight into protein-protein interactions during protein folding and the pathogenesis of some protein conformational diseases, including Kuru, scrapie, Huntington's, Parkinson's and Alzheimer's disease [26-28].

A

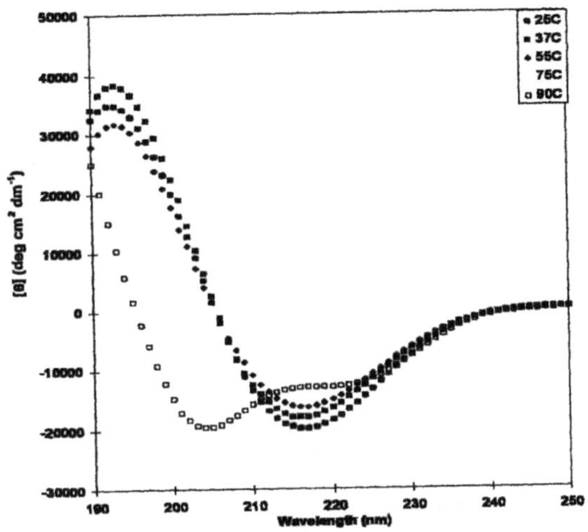

B

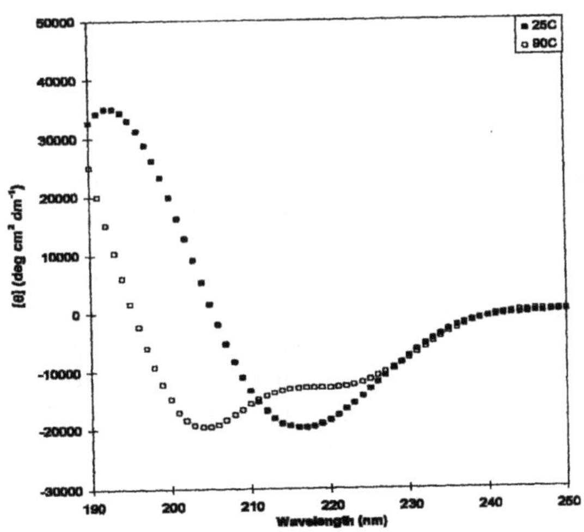

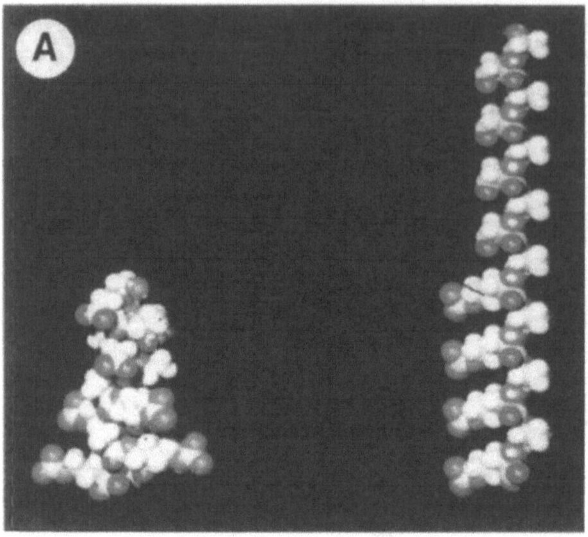

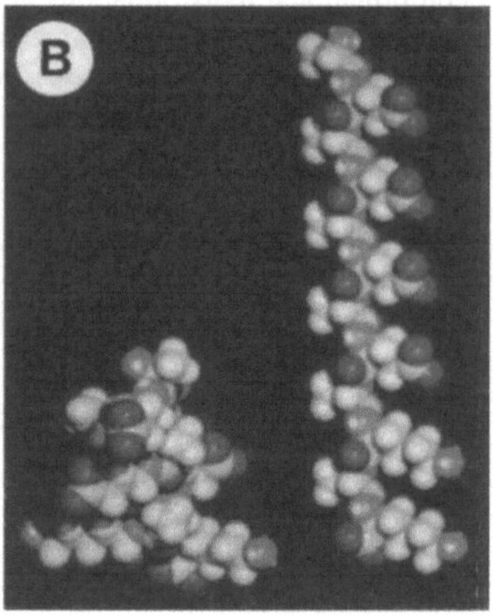

Fig. 4. Type II peptides. Temperature effect on DAR16-IV* structural transition. DAR16-IV* was incubated at various temperatures for 10 minutes and measured at 25°C. A) Structures of DAR16-IV* at from 25-90°C. At 25°C, it forms a stable beta-sheet. This beta-sheet structure is stable until 75°C. Here, the beta-sheet structure is abruptly converted to an alpha-helical structure with no detectable intermediate. The conversion in DAR16-IV* is much more abrupt than in EAK12-c, as only two distinct structural forms are observed. B) DAR16-IV exhibits two distinctive structures at two different temperatures, 25 and 90°C. C) Molecular models of DAR16-IV in beta-strand form and alpha-helical form. The length of the beta-strand is about 5 nm and the length of the alpha-helix is about 2.3 nm. D) Molecular models of EAK12 in beta-strand and alpha-helical form.

4.2. EXAMPLES OF DAR16-IV AND EAK12.

The peptides of DAR16-IV, (DADADADARARARARA), and EAK12 (AEAEAEAEAKAK) have a cluster of negatively charged glutamate residues close to N-terminus and a cluster of positively charged arginine residues near C-terminus. It is well known that many α-helices have a helical dipole moment with a partial negatively C-terminus toward a partial positively charged N-terminus [29]. Because of the unique sequence of DAR16-IV and EAK12, their side chain charges balance the helical dipole moment, therefore favoring helical structure formation. However, they also have alternating hydrophilic and hydrophobic residues as well ionic self-complementarity, which have been previously characterized to form stable β-sheet macrostructures. Thus the behavior of Type II peptides is likely to be more complex and dynamic than other stable β-sheet peptides. Additional molecules with such dipoles have been designed and studied, confirming the initial findings [25].

4.3. OTHER EXAMPLES.

Recently, a yeast alpha-agglutinin protein with the sequence of EYELENAKFFK has been found to undergo conformational changes from a beta-sheet structure to an atypical helical structure at high temperature and changes of pH (Zhao, personal communication). It would be interesting to see if more such examples could be found in other proteins. This structural dynamic behavior is quite surprising, but it forces us to reconsider the problem of protein folding and protein-protein interactions. It is believed that most secondary structures in proteins are stable once they are formed. However, proteins frequently undergo catalysis, transport, interaction with other substances and involved in a wide range of interactions. It is reasonable to speculate that protein secondary structures are not static, rather, they can adopt many conformations to accommodate their biological functions.

From a polymer and materials science and engineering point of view, these structural transformations can be viewed as possible molecular switches that can be regulated by changing temperature or pH. The length accompanying the structural change is approximately 2 fold so that one form could be viewed as On and the other Off, providing molecular switches for a new generation of nanoactuators.

Others have also reported similar findings that proteins and peptides can undergo self-assembly and disassembly or change their conformations depending on the environmental influence, such as its location, pH change and temperature or crystal lattice packing 30-32].

5. Type III self-assembling peptides

5.1. CHARACTERISTICS OF TYPE III PEPTIDES.

This type of peptides are like "molecular paint" and "molecular Velcro", undergoing self-assembly onto surface rather with themselves. They form monolayers on surfaces for specific cell pattern formation or to interact with other molecules. These oligopeptides have three distinct features (Fig. 5).

Fig. 5. Type III self-assembling peptides for biological surface engineering. Molecular models of the oligopeptide RADSC-14 with the sequence RADSRADSAAAAAC and of ethylene glycol thiolate (EG6SH). The N-terminal segment $(RADS)_2$ is the ligand for cell attachment, the five-alanine segment is a linker. The cysteine anchor is covalently bound to the gold atoms on the surface. Molecular models of the surface where both molecules form self-assembled monolayers with different heights. The extended lengths of RADSC-14 and EG6SH are approximately 5 and 4 nm, respectively.

The first feature is the terminal segment, containing ligands that incorporate a variety of functional groups for recognition by other molecules or cells. The second feature is the central linker where a variable spacer is not only used to allow freedom of interaction at a specified distance away from the surface but also to allow for flexibility or rigidity. The third feature is the surface anchor where a chemical group on the peptide can react with the surface to form a covalent bond. This simple system using Type III self-assembly peptides and other substances to engineer surfaces will be a useful tool in biomedical engineering and will provide new methods to study cell-cell communication and cell behavior (Fig. 6) [33].

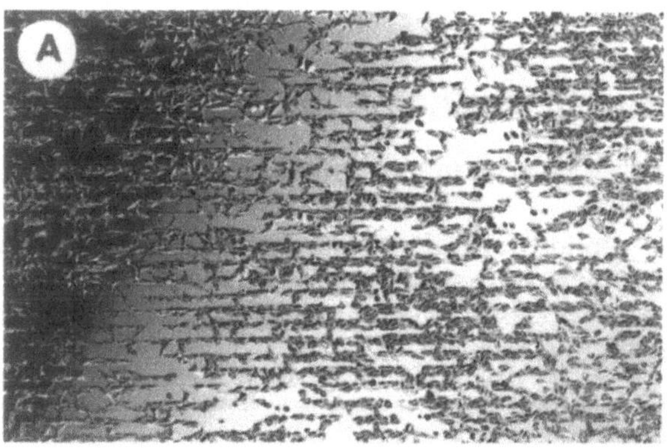

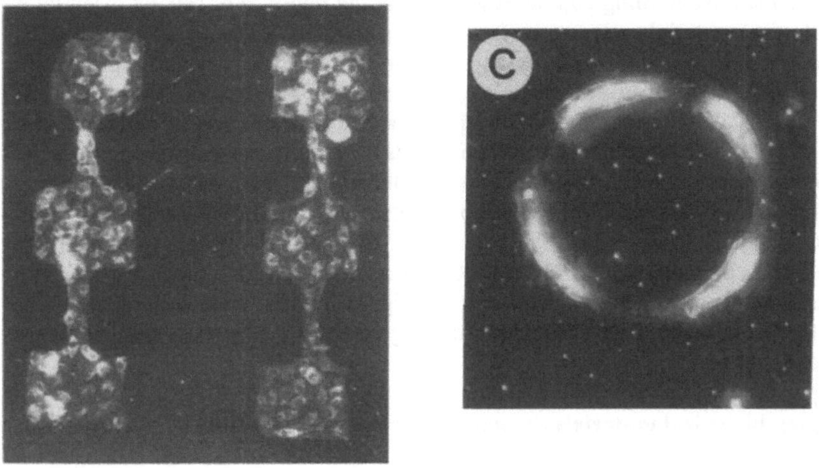

Fig. 6. Cells on designed surfaces. A) The images of linear cells were taken with a Normarski microscope. B) Bovine endothelial cells form square patterns connected with linear cell tracts. C) Four individually separated cells form a circle.

5.2. TECHNOLOGICAL IMPLICATIONS.

This simple system is an emerging technology that will have far reaching implications in biology. We are interested in developing this technology further for various studies. For example, we can design various patterns to address specific questions in cell biology on how communities of cells communicate through one or two cell connections (Fig. 6). By application of external stimuli, e.g. calcium, potassium, hormones, growth factors, cytotoxic substances or electric impulses — to one community of cells through micro-manipulation will allow the study of responses from the other community of cells through the messenger cells. This biological surface engineering technology using self-assembling peptides may also be useful in biomedical research and clinical applications as a new detection technique. For example, we can design a specific ligand as a "molecular hook" that can interact with specific molecules on cancer cell surfaces in high affinity so as to anchor cancer cells on the surfaces. This type of diagnostic device may be manufactured on a chip for rapid and sensitive detection. Furthermore, microcontact printing using self-assembling peptides is simply repetitious and, therefore, a robotic system can be potentially developed for printing specifically designed pattern on chips or other surfaces. Such a system may be generally useful for rapid and high throughput drug screens that influence cell behavior.

Others previously pioneered a similar kind of molecular surface self-assembly system through incorporating a segment of an organic linker for surface anchoring to absorb proteins on a surface [1, 34-35].

Type III self-assembling peptide research is at the interface of several disciplines including biological materials science, surface self-assembly engineering, and biology. It is significant for four reasons: 1) it is an example of the applications of molecular engineering of biological substrates to studying cell-material interaction. These types of studies will clarify how cells interact with surface substrates, and the molecular interactions of cells with their surface environments. 2) Since we can design specific surface patterns, we can control specific cell-cell interactions. This surface engineering using combination of self-assembling oligopeptides and microcontact printing will provide the opportunity to design many patterned surfaces. 3) Using this technique, we can study cell-cell communication in a well-controlled manner. This type of systematic study will be likely to open new avenues in the design of higher order architectures. 4) Microcontact printing using Type III self-assembling peptides is a process that is readily amenable to automation.

6. The emerging biological materials through molecular self-assembly

Developing new technologies often broadens the questions we can address and may thus deepen our understanding of seemingly intractable biological phenomena. For example, using various self-assembling peptide systems, we can dissect complex problems into smaller units to study them systematically. We believe that application of these simple and versatile peptide systems will make it possible to study complex and previously intractable biology phenomena, such as the study of cell-material interactions, the detailed mechanism of cell migration, cell mechanical compliance, cell-cell communication, and community cell behavior. Biological materials engineering through molecular design and the use of self-assembling biological building blocks are enabling technologies that will likely play an increasingly important role in future technology and will change our lives in the coming years.

Acknowledgment

We thank Drs. Todd Holmes, Lin Yan, Michael Lässle, Xing Su for advising and carrying out some experiments and Alexander Rich for encouragement. This work is supported in part by grants from the US Army Research Office, Hercules, Inc., and the Whitaker Foundation.

References

1. Whitesides, G.M., Mathias, J.P.& Seto, C.T. (1991) Molecular self-assembly and nanochemistry: a chemical strategy for the synthesis of nanostructures. *Science* 254:1312-1319.

2. Lehn JM 1993 Supramolecular chemistry. *Science* **260**, 1762-1763.

3. Huc, I. & Lehn, J-M. (1997) Virtual combinatorial libraries: dynamic generation of molecular and supramolecular diversity by self-assembly. *Proc Natl Acad Sci U S A* **94**, 2106-2110.

4. Hecht, M.H., Richardson, J.S., Richardson, D.C. (1990) De novo design, expression, and characterization of Felix: a four-helix bundle protein of native-like sequence. *Science* **249**, 884-891.

5. Ghadiri, M.R., Granja, J.R. & Buehler, L.K. (1994) Artificial transmembrane ion channels from self-assembling peptide nanotubes. *Nature* **369**, 301-304.

6. Aggeli, A. Bell, M., Boden, N., Keen, J.N., Knowles, P.F., McLeish, T.C., Pitkeathly, M., & Radford, S.E. (1997) Responsive gels formed by the spontaneous self-assembly of peptides into polymeric beta-sheet tapes. *Nature* **386**, 259-262.

7. Urry, D. W. (1997) Physical chemistry of biological free energy transduction as demonstrated by elastic protein-based polymers. *J Physical Chem. B* **101**, 11007-11028.

8. Mutter, M. & Tuchscherer, G. (1997) Non-native architectures in protein design and mimicry. *Cell Mol Life Sci* **53**, 851-863.

9. Petka, W.A., Harden, J.L., McGrath, K.P., Wirtz, D., Tirrell, D.A. (1998) Reversible hydrogels from self-assembling artificial proteins. *Science* **281**, 389-392.

10. Morse, D.E. (1999) Silicon biotechnology: harnessing biological silica production to construct new materials. *Trends Biotechnol* **17**,230-232.

11. Ball, P. (1994) Materials science: Polymers made to measure. *Nature* **367**, 323-324.

12. Zhang, S., Holmes, T., Lockshin, C. & Rich, A. (1993) Spontaneous assembly of a self-complementary oligopeptide to form a stable macroscopic membrane. *Proc Natl Acad Sci U S A* **90**, 3334-3338.

13. Zhang, S., Lockshin, C., Herbert, A., Winter, E. & A. Rich (1992) Zuotin, a putative Z-DNA binding protein in Saccharomyces cerevisiae. *EMBO J.* **11** 3787-3796.

14. Bieri, C., Ernst, O.P., Heyse, S., Hofmann, K.P., & Vogel, H. (1999) Micropatterned immobilization of a G protein-coupled receptor and direct detection of G-protein activation. *Nat Biotechnol* **17**, 1105-1108.

15. O'Shea, E.K., Rutkowski, R., & Kim, P.S. (1989) Evidence that the leucine zipper is a coiled coil. *Science* **243**, 538-542.

16. Baker, D., & DeGrado, W.F. (1999) Engineering and design. *Curr Opin Struct Biol* **9**, 485-486.

17. Zhang, S., Holmes, T.C., DiPersio, C.M., Hynes, R.O., Su, X., Rich, A. (1995) Self-complementary oligopeptide matrices support mammalian cell attachment. *Biomaterials* **16**, 1385-1393.

18. Leon, E.J., Verma, N., Zhang, S., Lauffenburger, D.A. & Kamm, R.D. (1998) Mechanical properties of a self-assembling oligopeptide matrix. *J. Biomaterials Science: Polymer Edition* **9**, 293-312.

19. Holmes, T.C. de Lacalle, S, Su, X., Liu, G. Rich, A. & Zhang, S. (2000) Extensive neurite outgrowth and active synapse formation on self-assembling peptide matrix Scaffolds. *Proc.Natl. Acad. Sci. USA* **97**, 6728-6733.

20. Brack, A. & Orgel, L.E. (1975) Beta structures of alternating polypeptides and their possible prebiotic significance. *Nature* **256**, 383-387.

21. Trudelle, Y. (1975) Conformational study of the sequential (Tyr-Glu)n copolymer in aqueous solution. *Polymer* **16**, 9-15.

22. St.Pierre, S., Ingwall, R.T., Varlander, M.S. & Goodman, M. (1978) Conformational studies of sequential poypeptides containing lysine and tyrosine. *Biopolymers* **17**, 1837.

23. Osterman, D. G. & Kaiser, E. T. (1985) Design and characterization of peptides with amphiphilic beta-strand structures. *J. Cell. Biochem.* **29**, 57-72.

24. Zhang, S. & Rich, A. (1997) Direct conversion of an oligopeptide from a β-sheet to an α-helix: a model for amyloid formation. *Proc Natl Acad Sci U S A* **94**, 23-28.

25. Altman, M., Lee, P., Rich, A. & Zhang, S. (2000) Conformational behavior of self-complementary peptides. *Protein Science* **9**, 1095-1106.

26. Gajdusek, C. (1996) Infectious amyloids: subacute spongiform encephalopathies as transmissible cerebral amyloidoses. in Field Virology 3rd Ed. (Ed. Fields, B.N. Knipe, D. M., Howley, P. M.) Lippincott-Raven publishers, Philadelphia.

27. Kelly, J.W. (1998) The alternative conformations of amyloidogenic proteins and their multi-step assembly pathways. *Curr Opin Struct Biol* **8**, 101-106.

28. Dobson, C. (1999) Protein misfolding, evolution and disease. *Trends in Biochem. Sci.* **24**, 329-332.

29. Hol, W.G., Halie, L.M., Sander, C. (1981) Dipoles of the alpha-helix and beta-sheet: their role in protein folding. *Nature* **294**, 532-536.

30. Minor, D.L. Jr, Kim, P.S. (1996) Context-dependent secondary structure formation of a designed protein sequence. *Nature* **380**, 730-734.

31. Tan, S. & Richmond, T. J. (1998) Crystal structure of the yeast MATalpha2/MCM1/DNA ternary complex. *Nature* **391**, 660-666.

32. Takahashi, Y., Ueno, A. & Mihara, H. (1999) Optimization of hydrophobic domains in peptides that undergo transformation from α-helix to β-fibril. *Bioorg Med Chem* **7**, 177-185.

33. Zhang, S,. Yan, L., Altman, M., Lassle, M., Nugent, H., Frankel, F., Lauffenburger, D.A,. Whitesides, G.M. & Rich, A. (1999) Biological surface engineering: a simple system for cell pattern formation. *Biomaterials* **20**, 1213-1220.

34. Massia, S.P., & Hubbell, J.A. (1991) Human endothelial cell interactions with surface-coupled adhesion peptides on a nonadhesive glass substrate and two polymeric biomaterials. *J Biomed Mater Res* **25**, 223-242.

35. Pakalns, T., Haverstick, K.L., Fields, G.B., McCarthy, J.B., Mooradian, D.L., Tirrell, M. (1999) Cellular recognition of synthetic peptide amphiphiles in self-assembled monolayer films. *Biomaterials* **20**, 2265-2279.

Index

MIX
Papier aus verantwortungsvollen Quellen
Paper from responsible sources
FSC® C105338

If you have any concerns about our products,
you can contact us on
ProductSafety@springernature.com

In case Publisher is established outside the EU,
the EU authorized representative is:
Springer Nature Customer Service Center GmbH
Europaplatz 3, 69115 Heidelberg, Germany

Printed by Libri Plureos GmbH
in Hamburg, Germany